Statistical Modeling

A Fresh Approach

DANIEL T. KAPLAN
MACALESTER COLLEGE

Cover Photo: The trunk of a scribbly gum eucalyptus tree on Fraser Island in Queensland, Australia. The scribbly gum moth lays its eggs between the old and new bark layers. The larvae burrow between the bark layers, leaving a winding tunnel that is revealed when the old bark falls away. [Photo credit: the author.]

To Maya

Contents

Preface

The purpose of this book is to provide an introduction to statistics that gives readers a sufficient mastery of statistical concepts, methods, and computations to apply them to authentic systems. By "authentic," I mean the sort of multivariable systems often encountered when working in the natural or social sciences, commerce, government, law, or any of the many contexts in which data are collected with an eye to understanding how things work or to making predictions about what will happen.

The world is complex and uncertain. We deal with the complexity and uncertainty with a variety of strategies including the scientific method and the discipline of statistics.

Statistics helps to deal with uncertainty, quantifying it so that you can assess how reliable — how likely to be repeatable — your findings are. The scientific method helps to deal with complexity: reduce systems to simpler components, define and measure quantities carefully, do experiments in which some conditions are held constant but others are varied systematically.

Beyond helping to quantify uncertainty and reliability, statistics provides another great insight of which most people are unaware. When dealing with systems involving multiple influences, it is possible and best to deal with those influences simultaneously. By appropriate data collection and analysis, the confusing tangle of influences can sometimes be straightened out (and it is possible to know when the attempt gives ambiguous and unreliable results). In other words, statistics goes hand-in-hand with the scientific method when it comes to dealing with complexity and understanding how systems work.

The statistical methods that can accomplish this are often considered advanced: multiple regression, analysis of covariance, logistic regression, etc. With appropriate software, any method is accessible in the sense of being able to produce a summary report on the computer. But a method is useful only when the user has a way to understand whether the method is appropriate for the situation, what the method is telling about the data, and what the method is not capable of revealing. Computer scientist Richard Hamming (1915-1998) said: "The purpose of computing is insight, not numbers." Without a solid understanding of the theory that underlies a method, the numbers generated by the computer may not give insight.

The methods of statistics can give tremendous insight, particularly the so-called advanced methods. For this reason, these methods need to be accessible both computationally and theoretically to the widest possible audience. Historically, the audience has been limited because few people have the algebraic skills needed to approach the methods in the way they are usually presented. But there are many paths to understanding and I have undertaken to find one that takes the greatest advantage of the actual skills that most people already have in abundance.

In trying to meet that challenge, I have made many unconventional choices. Theory becomes simpler when there is a unified framework for treating many aspects of statistics, so I have chosen to present just about everything in the context of models: descriptive statistics as well as inference as well as experimental design.

Another choice is the use of simple geometry to present theory. The underlying geometrical concepts are elementary: lengths, angles, projection, the Pythagorean theorem. Much of the statistical theory can be displayed in a two-dimensional sketch; once in a while three-dimensional visualization is needed. When I have presented the geometrical ideas in this manner to professionals in research groups or at statistics conferences and workshops, I find uniformly that people gain significant insight into their work. A common reaction is, "It can't be that easy." Yes, it can be and it should be.

Also unconventional, but hardly innovative, is the use of resampling and randomization to motivate the concepts of statistical inference. George Cobb [1] cogently describes the logic of statistical inference as the three-Rs: "Randomize, Repeat, Reject." In a decade of teaching statistics, I have found that students can understand this algorithmic logic much better than the derivations of algebraic formulas for means and standard deviations of sampling distributions.

As you might expect from the preceeding comments, algebraic notation and formulas are strongly de-emphasized in this book. I find that most people are not skilled in interpreting them and extracting meaning from them. In any event, formulas are no longer needed as a way to describe calculations since statistical work is now done on the computer.

And then there is software. Some people think that statistics should be taught without computers in order to help develop conceptual understanding. Others think that it is silly to ignore a technology that is universally used in practice and greatly expands our capabilities. Both points of view are right.

The main body of this book is presented in a way that makes little or no reference to software; the statistical concepts are paramount. But each chapter has a section on computational technique that shows how to get things done and aims to give the reader concrete skills in the analysis of data. An extensive set of exercises and classroom activities is published in workbook form, via the Internet. The computer is an effective teaching tool, so many of the exercises and activities make heavy use of it.

The software used is R, a modern, powerful and freely available system for statistical computations and graphics. The book assumes that you know nothing at all about scientific software and, accordingly, introduces R from basics. If you have experience with statistics, you probably already have a preferred software package. So long as that software will fit linear models with multiple explanatory variables and produce a more-or-less standard regression report, it can be used to follow this book. That said, I strongly encourage you to think about learning and using R. You can learn it easily by following the examples and can be doing productive statistics very quickly. Not only will you easily be able to fit models and get reports, but you can use R to explore ideas such as resampling and randomization. If you now use "educational" software, learning R will give you a professional-level tool for use in the future.

For many instructors, this book can support a nice *second* course in statistics — a follow-up to a conventional first introductory course. Increasingly, such a course is needed as more and more young people encounter basic statistical ideas in grade school and many of the topics of the conventional university course are absorbed into the high-school curriculum. At Macalester College, where I developed this book, mainstream students of biology, economics, political science, and so on use this book for their *first* statistics course. Accordingly, the book is written to be self-contained, making no assumption that readers have had any previous formal study in statistics.

Thanks and acknowledgments ...

I have been fortunate to have the assistance and support of many people. Some of the colleagues who have played important roles are David Bressoud, George Cobb, Dan Flath, Tom Halverson, Gary Krueger, Weiwen Miao, Phil Poronnik, Victor Addona, Karen Saxe, Michael Schneider, and Libby Shoop. Critical institutional support was given by Brian Rosenberg, Jan Serie, Dan Hornbach, Helen Warren, and Diane Michelfelder at Macalester and Mercedes Talley at the Keck Foundation.

I received encouragement from many in the statistics education community, including George Cobb, Joan Garfield, Dick De Veaux, Bob delMas, Julie Legler, Milo Schield, Paul Alper, Andy Zieffler, Sharon Lane-Getaz, Katie Makar, Michael Bulmer, and the participants in the monthly "Stat Chat" sessions. Helpful suggestions came from from Simon Blomberg, Dominic Hyde, Erik Larson, Julie Dolan, and Kendrick Brown. Michael Edwards helped with proofreading. Nick Trefethen and Dave Saville provided important insights about the geometry of fitting linear models.

Thanks also go to the hundred or so students at Macalester College who enrolled in the early, experimental sessions of Math 155 where many of the ideas in this book were first tried out. Among those students, I want to acknowledge particular help from Alan Eisinger, Caroline Ettinger, Bernd Verst, Wes Hart, Sami Saqer, and Michael Snavely.

Crucial early support for this project was provided by a grant from the Howard Hughes Medical Institute. An important Keck Foundation grant was crucial to the continuing refinement of the approach and the writing of this book.

Finally, my thanks and love to my wife, Maya, and daughters, Tamar, Liat, and Netta, who endured the many, many hours during which I was preoccupied by some or another statistics-related enthusiasm, challenge, or difficulty.

1

Models

ome models are useful. — George Box

truth. — Pablo Picasso

stical modeling, two words that themselves require

ing" ending) is a process of asking questions. "Sta-
ata — the statistical models you will construct will
fers also to a distinctively modern idea: that you can
ow and that doing so contributes to your understand-

person with a watch knows the time. A person with
." The statistical point of view is that it's better not
ches you can see how they disagree with each other.
ow precise the watches are. You don't know the time
precision tells you something about what you don't
certainty of the person with a single watch is merely
ence: the single watch provides no indication of what

Rutherford (1871-1937) famously said, "If your exper-
u ought to have done a better experiment." In other
good enough watch, you need only one: no statistics.
ics never hurts. A person with two watches that agree
the time, but has evidence that the watches are work-
sibly, the official world time is based on an average of
many atomic clocks. The individual clocks are fantastically precise; the point of averaging is to know when one or more of the clocks is drifting out of precision for some reason.

Why "statistical modeling" and not simply "statistics" or "data analysis?" Many people imagine that data speak for themselves and that the purpose of statistics is to extract the information that the data carry. Such people see data

analysis as an objective process in which the researcher should, ideally, have no influence. This can be true when very simple issues are involved, for instance how precise is the average of the atomic clocks used to set official time or what is the difference in time between two events. But many questions are much more complicated; they involve many variables and you don't necessarily know what is doing what to what.[2]

The conclusions you reach from data depend on the specific questions you ask. Like it or not, the researcher plays an active and creative role in constructing and interrogating data. This means that the process involves some subjectivity. But this is not the same as saying anything goes. Statistical methods allow you to make objective statements about how the data answer your questions. In particular, the methods help you to know if the data show anything at all.

The word "modeling" highlights that your goals, your beliefs, and your current state of knowledge all influence your analysis of data. The core of the scientific method is the formation of hypotheses that can be tested and perhaps refuted by experiment or observation. Similarly, in statistical modeling you examine your data to see whether they are consistent with the hypotheses that frame your understanding of the system under study.

Example 1.1: Grades A woman is applying to law school. The schools she applies to ask for her class rank, which is based on the average of her college course grades.

A simple statistical issue concerns the precision of the grade-point average. This isn't a question of whether the average was correctly computed or whether the grades were accurately recorded. Instead, imagine that you could send two essentially identical students to essentially identical schools. Their grade-point averages might well differ, reflecting perhaps the grading practices of their different instructors or slightly different choices of subjects or random events such as illness or mishaps or the scheduling of classes. One way to think about this is that the students' grades are to some extent random, contingent on factors that are unknown or perhaps irrelevant to the students' capabilities.

How do you measure the extent to which the grades are random? There is no practical way to create "identical" students and observe how their grades differ. But you can look at the variation in a single student's grades — from class to class — and use this as an indication of the size of the random influence in each grade. From this, you can calculate the likely range of the random influences on the overall grade-point average.

Statistical models let you go further in interpreting grades. It's a common belief that there are easy- and hard-grading teachers and that a grade reflects not just the student's work but the teacher's attitude and practices. Statistical modeling provides a way to use data on grades to see whether teachers grade differently and to correct for these differences between teachers. Doing this involves some subtlety, for example taking into account the possibility that strong students take different courses than weaker students.

Example 1.2: Nitrogen Fixing Plants Biologist Michael Anderson studies how plants fix nitrogen in the soil. All plants need nitrogen to grow. Since nitrogen is the primary component of air, there is plenty around. But it's hard for plants to get nitrogen from the air; they get it instead from the soil. Some plants, like alder and soybean, support nitrogen-fixing bacteria in nodules on the plant roots. The plant creates a hospitable environment for the bacteria; the bacteria, by fixing nitrogen in the soil, create a good environment for the plant. In a word, symbiosis.

Anderson is interested in how genetic variation in the bacteria influences the success with which they fix nitrogen. One can imagine using this information to breed plants and bacteria that are more effective at fixing nitrogen and thereby reducing the need for agricultural fertilizer.

Anderson has an promising early result. His extensive field studies indicate that different genotypes of bacteria fix nitrogen at different rates. Unfortunately, the situation is confusing since the different genotypes tend to populate different areas with different amounts of soil moisture, different soil temperatures, and so on. How can he untangle the relative influences of the genotype and the other environmental factors in order to decide whether the variation in genotype is genuinely important and worth further study?

Example 1.3: Sex Discrimination A large trucking firm is being audited by the government to see if the firm pays wages in a racially or sexually discriminatory way. The audit finds wage discrepancies between men and women for "office and clerical workers" but not for other job classifications such as technicians, supervisors, sales personnel, or "skilled craftworkers." It finds no discrepancies based on race.

A simple statistical question is whether the observed difference in average wages for men and women office and clerical workers is based on enough data to be reliable. In answering this question, it actually makes a difference what other groups the government auditors looked at when deciding to focus on sex discrimination in office and clerical workers.

Further complicating matters are the other factors that contribute to people's wages: the kind of job they have, their skill level, their experience. Statistical models can be used to quantify how these various factors contribute and to see whether they account for some or all of the wage discrepancy associated with sex. For instance, it turns out that men on average tend to have more job experience than women, and some or all of the men's higher average wages might be due to this.

Models can help you decide whether this potential explanation is plausible. For instance, if you see that both men's and women's wages increase with experience in the same way, you might be more inclined to believe that job experience is a legitimate factor rather than just a mask for discrimination.

1.1 Models and their Purposes

Many of the toys you played with as a child are models: dolls, balsa-wood airplanes with wind-up propellers, wooden blocks, model trains. But so are many serious objects of the adult world: architectural plans, bank statements, train schedules, the results of medical diagnostic tests, the signals transmitted by a telephone, the equations of physics, the genetic sequences used by biologists. There are too many to list.

What all models have in common is this:

A model is a representation for a particular purpose.

A model might be a physical object or it might be an idea, but it always stands for something else: it's a representation. Dolls stand for babies and other creatures, architectural plans stand for buildings and bridges, a white blood-cell count stands for the function of the immune system.

When you create a model, you have (or ought to have) a purpose in mind. Toys are created for the entertainment and (sometimes) edification of children. The various kinds of toys — dolls, blocks, model airplanes and trains — have a form that serves this purpose. Unlike the things they represent, the toy versions are small, safe, and inexpensive.

Models always leave things out and get some things — many things — wrong. Architectural plans are not houses; you can't live in them. But they are easy to transport, copy, and modify. That's the point. Telephone signals — unlike the physical sound waves that they represent — can be transported over long distances and even stored. A train schedule tells you something important but it obviously doesn't reproduce every aspect of the trains it describes; it doesn't carry passengers.

Statistical models revolve around data. But even so, they are first and foremost models. They are created for a purpose. The intended use of a model should shape the appropriate form of the model and determines the sorts of data that can properly be used to build the model.

There are three main uses for statistical models. They are closely related, but distinct enough to be worth enumerating.

Description. Sometimes you want to describe the range or typical values of a quantity. For example, what's a "normal" white blood cell count? Sometimes you want to describe the relationship between things. Example: What's the relationship between the price of gasoline and consumption by automobiles?

Classification or prediction. You often have information about some observable traits, qualities, or attributes of a system you observe and want to draw conclusions about other things that you can't directly observe. For instance, you know a patient's white blood-cell count and other laboratory measurements and want to diagnose the patient's illness.

Anticipating the consequences of interventions. Here, you intend to do something: you are not merely an observer but an active participant in the system. For example, people involved in setting or debating public policy have to deal with questions like these: To what extent will increasing the tax on gasoline reduce consumption? To what extent will paying teachers more increase student performance?

The appropriate form of a model depends on the purpose. For example, a model that diagnoses a patient as ill based on an observation of a high number of white blood cells can be sensible and useful. But that same model would give absurd predictions about intervention: Do you really think that lowering the white blood cell count by bleeding a patient will make the patient better?

To anticipate correctly the effects of an intervention you need to get the direction of cause and effect correct in your models. But for a model used for classification or prediction, it may be unnecessary to represent causation correctly. Instead, other issues, e.g. the reliability of data, can be the most important. One of the thorniest issues in statistical modeling — with tremendous consequences for science, medicine, government, and commerce — is how you can legitimately draw conclusions about interventions from models based on data collected without performing these interventions.

1.2 Observation and Knowledge

How do you know what you know? How did you find it out? How can you find out what you don't yet know? These are questions that philosophers have addressed for thousands of years. The views that they have expressed are complicated and contradictory.

From the earliest times in philosophy, there has been a difficult relationship between knowledge and observation. Sometimes philosophers see your knowledge as emerging from your observations of the world, sometimes they emphasize that the way you see the world is rooted in your innate knowledge: the things that are obvious to you.

This tension plays out on the pages of newspapers as they report the controversies of the day. Does the death penalty deter crime? Does increased screening for cancer reduce mortality? Will paying teachers more improve student outcomes?

Consider the simple, obvious argument for why severe punishment deters crime. Punishments are things that people don't like. People avoid what they don't like. If crime leads to punishment, then people will avoid committing crime.

Each statement in this argument seems perfectly reasonable, but none of them is particularly rooted in observations of actual and potential criminals. It's artificial — a learned skill — to base knowledge such as "people avoid punishment" on observation. It might be that this knowledge was formed by our own experiences, but usually the only explanation you can give is something like, "that's been my experience" or give one or two anecdotes.

When observations contradict opinions — opinions are what you think you know — people often stick with their opinions. Put yourself in the place of someone who believes that the death penalty really does deter crime. You are presented with accurate data showing that when a neighboring state eliminated the death penalty, crime did not increase. So do you change your views on the matter? Possibly, but possibly not. A skeptic can argue that it's not just punishment but also other factors that influence the crime rate, for instance the availability of jobs. Perhaps it was that a generally improving economic condition in the other state kept the crime rate steady even at a time when society is imposing lighter punishments.

It's difficult to use observation to inform knowledge because relationships are complicated and involve multiple factors. It isn't at all obvious how people can discover or demonstrate causal relationships through observation. Suppose one school district pays teachers well and another pays them poorly. You observe that the first district has better student outcomes than the second. Can you legitimately conclude that teacher pay accounts for the difference? Perhaps something else is at work: greater overall family wealth in the first district (which is what enabled them to pay teachers more), better facilities, smaller classes, and so on.

Historian Robert Hughes concisely summarized the difficulty of trying to use observation to discover causal relationships. In describing the extensive use of hanging in 18th and 19th century England, he wrote, "One cannot say whether public hanging did terrify people away from crime. Nor can anyone do so, until we can count crimes that were never committed." [3, p.35] To know whether hanging did deter crime, you would need to observe a **counterfactual**, something that didn't actually happen: the crimes in a world without hanging. You can't observe counterfactuals. So you need somehow to generate observations that give you data on what happens for different levels of the causal variable.

A modern idea is the **controlled experiment**. In its simplest ideal form, a controlled experiment involves changing one thing — teacher pay, for example — while *holding everything else constant*: family wealth, facilities, etc.

The experimental approach to gaining knowledge has had great success for example in medicine and science. For many people, experiment is the essence of science. But experiments are hard to perform and sometimes not possible at all. How do you "hold everything else constant?" Partly for this reason, you rarely see reports of experiments when you read the newspaper, unless the article happens to be about a scientific discovery.

Scientists pride themselves on recording their observations carefully and systematically in lab notebooks. Laboratories are filled with high-precision instrumentation. The quest for precision culminates perhaps in the physicist's fundamental quantities: the speed of light is reported to be $299,792,500 \pm 1000$ meters per second, the mass of the electron reported as $9.10938215 \pm 0.00000045 \times 10^{-31}$ kg. Each of these is precise to about 50 parts in a billion. Contrast this extreme precision with the humble speed measurements from a policeman's radar gun (perhaps a couple of miles or kilometers per hour — one part in 50) or the weight indicated on a bathroom scale (give or take a kilogram or a couple of pounds — about one part in 100 for an adult).

All such observations and measures are the stuff of **data**, the records of observations. Observations do not become data by virtue of high precision or expensive instrumentation or the use of metric rather than traditional units. For many purposes, data of low precision is used. An ecologist's count of the number of mating pairs of birds in a territory is limited by the ability to find nests. A national census of a country's population, conducted by the government can be precise to only a couple of percent. The physicist counting neutrinos in huge observatories buried under mountains to shield them from extraneous events waits for months for her results. These results are precise to only one part in two.

The precision that is needed in data depends on the purpose for which the data will be used. The important question for the person using the data is whether the precision, whatever it be, is adequate for the purpose at hand. To answer this question, you need to know how to measure precision and how to compare this to a standard reflecting the needs of your task. The scientist with expensive instrumentation and the framer of social policy both need to deal with data in similar ways to understand and interpret the precision of their results.

It's common for people to believe that conclusions drawn from data apply to certain areas — science, economics, medicine — but aren't terribly useful in other areas. In teaching, for example, almost all decisions are based on "experience" rather than observation. Indeed, there is often strong resistance to making formal observations of student progress as interfering with the teaching process.

This book is based on the idea that techniques for drawing valid conclusions from observations — data — are valuable for two groups of people. The first group is scientists and others who routinely need to use statistical methods to analyze experimental and other data.

The second group is everybody else. All of us need to draw conclusions from our experiences, even if we're not in a laboratory. It's better to learn how to do this in valid ways, and to understand the limitations of these ways, than to rely on an informal, unstated process of opinion formation. It may turn out that in any particular area of interest there are no useful data. In such situations, you won't be able to use the techniques. But at least you will know what you're missing. You may be inspired to figure out how to supply it or to recognize it when it does come along, and you'll be aware of when others are misusing data.

As you will see, the manner in which the data are collected plays a central role in what sorts of conclusions can be legitimately made; data do not always speak for themselves. You will also see that strongly supported statements about causation are difficult to make. Often, all you can do is point to an "association" or a "correlation," a weaker form of statement.

Statistics is sometimes loosely described as the "science of data." This description is apt, particularly when it covers both the collection and analysis of data, but it does not mean much until you understand what data are. That's the subject of the next chapter.

1.3 The Main Points of this Book

Statistics is about variation. Describing and interpreting variation is a major goal of statistics.

You can create empirical, mathematical descriptions not only of a single trait or variable but also of the relationships between two or more traits. Empirical means based on measurements, data, observations.

Models let you split variation into components: "explained" versus "unexplained." How to measure the size of these components and how to compare them to one another is a central aspect of statistical methodology. Indeed, this provides a definition of statistics:

> *Statistics is the explanation of variation in the context of what remains unexplained.*

By collecting data in ways that require care but are quite feasible, you can estimate how reliable your descriptions are, e.g., whether it's plausible that you should see similar relationships if you collected new data. This notion of reliability is very narrow and there are some issues that depend critically on the context in which the data were collected and the correctness of assumptions that you make about how the world works.

Relationships between pairs of traits can be studied in isolation only in special circumstances. In general, to get valid results it is necessary to study entire systems of traits simultaneously. Failure to do so can easily lead to conclusions that are grossly misleading.

Descriptions of relationships are often **subjective** — they depend on choices that you, the modeler, make. These choices are generally rooted in your own beliefs about how the world works, or the theories accepted as plausible within some community of inquiry.

If data are collected properly, you can get an indication of whether the data are consistent or inconsistent with your subjective beliefs or — and this is important — whether you don't have enough data to tell either way.

Models can also be used to check out the sensitivity of your conclusions to different beliefs. People who disagree in their views of how the world works often may not be able to reconcile their differences based on data, but they will be able to decide objectively whether their own or the other party's beliefs are reasonable given the data.

Notwithstanding everything said above about the strong link between your prior, subjective beliefs and the conclusions you draw from data, by collecting data in a certain context — experiments — you can dramatically simplify the interpretation of the results. It's actually possible to remove the dependence on identified subjective beliefs by intervening in the system under study experimentally.

This book takes a different approach than most statistics texts. Many people want statistics to be presented as a kind of automatic, algorithmic way to process data. People look for mathematical certainty in their conclusions. After all, there are right-or-wrong answers to the mathematical calculations that people (or computers) perform in statistics. Why shouldn't there be right-or-wrong answers to the conclusions that people draw about the world?

The answer is that there can be, but only when you are dealing with narrow circumstances that may not apply to the situations you want to study. An insistence on certainty and provable correctness often results in irrelevancy.

The point of view taken in this book is that it is better to be useful than to be provably certain. The objective is to introduce methods and ideas that can help you deal with drawing conclusions about the real world from data. The methods and ideas are meant to guide your reasoning; even if the conclusions you draw are not guaranteed by proof to be correct, they can still be more useful than the alternative, which is the conclusions that you draw without data, or the conclusions you draw from simplistic methods that don't honor the complexity of the real system.

1.4 Introduction to Computation with R

Modern statistics is done on the computer. There was a time, 60 years ago and before, when computation could only be done by hand or using balky mechanical calculators. The methods of applied statistics developed during this time reflected what could be done using such calculators, not necessarily what was best for illuminating the system under study. These methods took on a life of their own — they became the operational definition of statistics. They continue to be taught today, using electronic calculators or personal computers or even just using paper and pencil. For the old statistical methods, computers are merely a labor saving device.

But not for modern statistics. The statistical methods at the core of this book cannot be applied in a authentic and realistic way without powerful computers. Thirty years ago, many of the methods could not be done at all unless you had access to the resources of a government agency or a large university. But with the revolutionary advances in computer hardware and numerical algorithms over the last half-century, modern statistical calculations can be performed on an ordinary home computer or laptop. (Even a cell phone has phenomenal computational power, often besting the mainframes of thirty years ago.) Hardware and software today pose no limitation; they are readily available.

Each chapter of this book includes a section on computational technique. Many readers will be completely new to the use of computers for scientific and statistial work, so the first chapters cover the foundations, techniques that are useful for many different aspects of computation. Working through the early chapters is essential for developing the skills that will be used later in actual statistical work. It will take a few hours, but this investment will pay off handsomely.

Chances are, you use a computer almost every day: for email, word-processing, managing your music or your photograph collection, perhaps even using a spreadsheet program for accounting. The software you use for such activities makes it easy to get started. Possibly you have never even looked at an instruction manual or used the "help" features on your computer.

When you use a word processor or email, the bulk of what you enter into the computer — the content of your documents and email — is without meaning to the computer. This is not at all to say that it is meaningless. Your documents and letters are intended for human readers; most of the work you do is directed so that the recipients can understand them. But the computer doesn't need to understand what you write in order to format it, print it, or transmit it over the Internet. Indeed, the computer would be equally effective at handling random text generated by typing monkeys.

When doing scientific and statistical computing, things are different. What you enter into the computer is instructions to the computer to perform calculations and re-arrangements of data. Those instructions have to be comprehensible to the computer. If they make no sense or if they are inconsistent or ill formed, the computer won't be able to carry out your instructions. Worse, if the instructions make sense in some formal way but don't convey your actual intentions, the computer will perform some operation but the result will mislead you.

The difficulty with using software for mathematics and statistics is in making sure that your instructions make sense and do what you want them to do. This difficulty is not a matter of bad software design; it's intrinsic to the problem of communicating your intentions to the computer. The same difficulty would arise in word processing if the computer had to make sense of your writing, rejecting it when a claim is unconvincing or when a sentence is ambiguous. Statistical computing pioneer John Chambers refers to the "Prime Directive" of software[4]: "to program in such a way that computations can be understood and trusted."

Much of the design of packages for scientific and statistical work is oriented around the difficulty of communicating intentions. A popular approach is based on the computer mouse: the program provides a list of possible operations — like the keys on a calculator — and lets the user choose which operation to apply to some selected data. This style of user interface is employed, for example, in spreadsheet software, letting users add up columns of numbers, make graphs, etc. The reason this style is popular is that it can make things extremely easy ... so long as the operation that you want has been included in the software. But things get very hard if you need to construct your own operation and it can be difficult to understand or trust the operations performed by others.

Another style of scientific computation — the one used in this book — is based on language. Rather than selecting an option with a mouse, you construct a **command** that conveys both the operation that you want and the data to which you want to apply that operation. There are dramatic advantages to this language-based style of computation:

- It lets you **connect** computations to one another, so that the output of one operation can become the input to another.
- It lets you **repeat** the operation on new or modified data, allowing you to

automate tedious tasks and, importantly, to verify the correctness of your computations on data where you already know the answer.

- It lets you **accumulate** the results of previous operations, treating those results as new data.
- It lets you **document** concisely what was done so that you can demonstrate that what you said you did is what you actually did. In this way, you or others can repeat the analysis later if necessary to confirm your results.
- It lets you **modify** the computation in a controlled way to correct it or to vary some aspect of it while holding other aspects exactly the same as in the original.

In order to use the language-based approach, you will need to learn a few principles of the language itself: some vocabulary, some syntax, some grammar. This is much, much easier for the computer language than for a natural language like English or Chinese; it will take you only a couple of hours before you are fluent enough to do useful computations. In addition to letting you perform scientific computations in ways that use the computer and your own time and effort effectively, the principles that you will learn are broadly applicable to many computer systems and can provide significant insight even to how to use mouse-based interfaces.

1.4.1 The R Environment

The software package used in this book is called R. The R package provides an environment for doing statistical and scientific computation at a professional level. It was designed for statistics work, but suitable for other forms of scientific calculations and the creation of high-quality scientific graphics.[5]

There are several other major software packages widely used in statistics. Among the leaders are SPSS, SAS, and STATA. Each of them provides the computational power needed for statistical modeling. Each has its own advantages and its own devoted group of users.

One reason for the choice of R is that it offers a command-based computing environment. That makes it much easier to write about computing and also reveals better the structure of the computing process.[6] Also nice is that R is available for free and works on the major types of computers, e.g., Windows, Macintosh, and Unix/Linux. You can get information about how to install R on your computer at `www.r-project.org`.

In making your own choice, the most important thing is this: *choose something*! Readers who are familiar with SPSS, SAS, or STATA can use the information in each chapter's computational technique section to help them identify the facilities to look for in those packages.

Another package that's often used in statistical work is the spreadsheet program Excel. This package has its own advantages. It's effective for entering data and has nice facilities for formatting tables. The visual layout of the data seems to be intuitive to many people. Many businesses use Excel and it's widely taught in high schools. Unfortunately, it's very difficult to use for statistical analyses of any sophistication. Indeed, even some very elementary tasks such as making a histogram are difficult to do in Excel and the results are usually unsatisfactory from a graphical point of view. Worse, Excel is very hard to use reliably. There are lots of opportunities to make mistakes that will go undetected; Excel encourages bad programming practices that make software unreliable.

1.4.2 Setting R Up

You can skip this section if you want to jump ahead to read about how R works.

Better, though, if you try out the commands as they are shown, using the R software on your own computer.

To do this, you will need to start the R software. If R is not already installed on your computer (this is likely if you are using your own computer and have never used R before), the first step is to install it.

If you are used to installing software, you will find R follows the usual pattern.

1. Use a web browser to go to `www.r-project.org`. Select the Download/CRAN link on the left of the page. This will bring you to a list of download sites.

2. Choose one in your own region of the world. This will bring up a page with a choice of Linux, Mac OS X, or Windows. Choose whichever is appropriate for your computer.

3. For Windows: choose the link for the "base" distribution, and then download the link that looks like "R-2.9.0-win32.exe." The name may be slighly different, depending on what new versions have been released. Run this program, accepting the defaults.

   For Macintosh, follow the link that looks like "R-2.9.0.dmg" and follow the instructions. Again, the name may be slightly different, depending on what new versions have been released.

   If you are using Linux, you probably don't need any instructions on how to install software.

Once R is installed, you start it like most programs, by clicking on an icon. On a Windows computer, this will be under the "start" button, on a Macintosh this will be in the applications folder.

When you start R, a new window appears on your screen. It will look something like Figure 1.1. You're ready to go!

Actually, there is one more thing that you can do that will make things easier later on, when you start to analyze the data sets that go along with this book. Download one file from a web site and put it in some convenient place on your computer, for instance your "desktop" or a folder that you make for this book. The file is at
`www.macalester.edu/~kaplan/ISM/ISM.Rdata`
This file contains various data sets that you will be using.

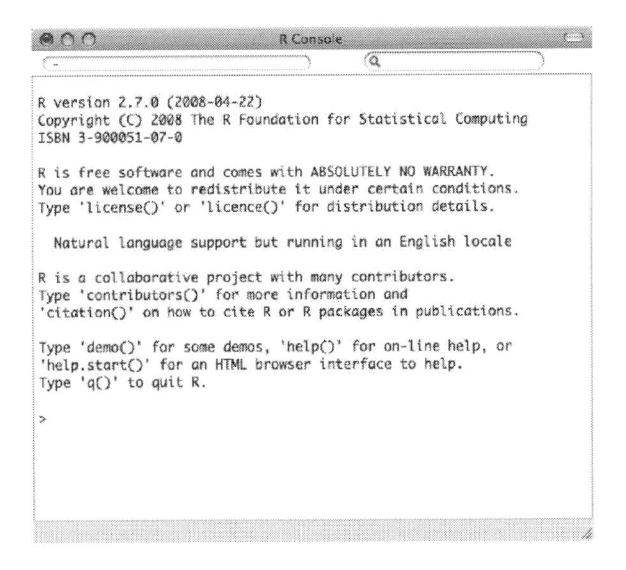

Figure 1.1: The R command window.

Once you have downloaded the file, called `ISM.Rdata`, double-clicking on it in the ordinary way will start R or load the data into an already started session of R.

1.4.3 Invoking an Operation

People often think of computers as *doing* things: sending email, playing music, storing files. Your job in using a computer is to tell the computer *what* to do. There are many different words used to refer to the "what": a procedure, a task, a function, a routine, and so on. I'll use the word **computation**. Admittedly, this is a bit circular, but it is easy to remember: computers perform computations.

Complex computations are built up from simpler computations. This may seem obvious, but it is a powerful idea. An **algorithm** is just a description of a computation in terms of other computations that you already know how to perform. To help distinguish between the computation as a whole and the simpler parts, it is helpful to introduce a new word: an **operator** performs a computation.

It's helpful to think of the computation carried out by an operator as involving four parts:

1. The name of the operator
2. The input arguments
3. The output value
4. Side effects

A typical operation takes one or more **input arguments** and uses the information in these to produce an **output value**. Along the way, the computer might take some action: display a graph, store a file, make a sound, etc. These actions are called **side effects**.

To tell the computer to perform an computation — call this **invoking an operation** or giving a **command** — you need to provide the name and the input arguments in a specific format. The computer then returns the output value. For example, the command `sqrt(25)` invokes the square root operator (named `sqrt`) on the argument 25. The output from the computation will, of course, be 5.

The syntax or invoking an operation consists of the operator's name, followed by round parentheses. The input arguments go inside the parentheses.

The software program that you use to invoke operators is called an **interpreter**. (The interpreter is the program you are running when you start R.) You enter your commands as a dialog between you and the interpreter. To start, the interpreter prints a prompt, after which you type your command:

PROMPT ⟶ > `sqrt(25)` ⟵ COMMAND

When you press "Enter," the interpreter reads your command and performs the computation. For commands such as this one, the interpreter will print the output value from the computation:

> `sqrt(25)`

OUTPUT MARKER ⟶ [1] 5 ⟵ OUTPUT VALUE

NEXT PROMPT ⟶ >

The dialog continues as the interpreter prints another prompt and waits for your further command.

To save space, I'll usually show just the give-and-take from one round of the dialog:

```
> sqrt(25)
[1] 5
```

(Go ahead! Type `sqrt(25)` after the prompt in the R interpreter, press "enter," and see what happens.)

Often, operations involve more than one argument. The various arguments are separated by commas. For example, here is an operation named `seq` that produces a sequence of numbers:

```
> seq(3,10)
 [1]  3  4  5  6  7  8  9 10
```

The first argument tells where to start the sequence, the second tells where to end it.

The order of the arguments is important. Here is the sequence produced when 10 is the first argument and 3 the second:

```
> seq(10,3)
[1] 10  9  8  7  6  5  4  3
```

For some operators, particularly those that have many input arguments, some of the arguments can be referred to by name rather than position. This is particularly useful when the named argument has a sensible default value. For example, the seq operator can be instructed how big a jump to take between successive items in the sequence. This is accomplished using an argument named by:

```
> seq(3,10,by=2)
[1] 3 5 7 9
```

Depending on the circumstances, all four parts of a operation need not be present. For example, the date operation returns the current time and date; no input arguments are needed.

```
> date()
[1] "Wed Apr 16 06:18:06 2008"
```

Note that even though there are no arguments, the parentheses are still used. Think of the pair of parentheses as meaning, "Do this."

Naming and Storing Values

Often the value returned by an operation will be used later on. Values can be stored for later use with the **assignment operator**. This has a different syntax that reminds the user that a value is being stored. Here's an example of a simple assignment:

```
> x = 16
```

This command has stored the value 16 under the name x. The syntax is always the same: an equal sign (=) with a name on the left and a value on the right.

Such stored values are called **objects**. Making an assignment to an object defines the object. Once an object has been defined, it can be referred to and used in later computations.

Notice that an assignment operation does not return a value or display a value. Its sole purpose is to have the side effects of defining the object and thereby storing a value under the object's name.

To refer to the value stored in the object, just use the object's name itself. For instance:

```
> x
[1] 16
```

Doing a computation on the value stored in an object is much the same:

```
> sqrt(x)
[1] 4
```

You can create as many objects as you like and give them names that remind you of their purpose. Some examples: wilma, ages, temp, dog.houses, foo3. There are some rules for object names:

- Use only letters and numbers and the two punctuation marks "dot" (`.`) and "underscore" (`_`).
- Do NOT use spaces anywhere in the name.
- A number or underscore cannot be the first character in the name.
- Capital letters are treated as distinct from lower-case letters. The objects named `wilma` and `Wilma` are different.

For the sake of readability, keep object names short. But if you really must have an object named something like `agesOfChildrenFromTheClinicalTrial`, feel free.

Objects can store all sorts of things, for example a sequence of numbers:

```
> x = seq(1,7)
```

When you assign a new value to an existing object, as just done to `x`, the former value of that object is erased from the computer memory. The former value of `x` was 16, but after the above assignment command it is

```
> x
 [1]  1  2  3  4  5  6  7
```

The value of an object is changed only *via* the assignment operator. Using an object in a computation does not change the value. For example, suppose you invoke the square-root operator on `x`:

```
> sqrt(x)
 [1] 1.00 1.41 1.73 2.00 2.24 2.45 2.65
```

The square roots have been returned as a value, but this doesn't change the value of `x`:

```
> x
 [1]  1  2  3  4  5  6  7
```

If you want to change the value of `x`, you need to use the assignment operator:

```
> x = sqrt(x)
> x
 [1] 1.00 1.41 1.73 2.00 2.24 2.45 2.65
```

Connecting Computations

The brilliant thing about organizing operators in terms of input arguments and output values is that the output of one operator can be used as an input to another. This lets complicated computations be built out of simpler ones.

For example, suppose you have a list of 10000 voters in a precinct and you want to select a random sample of 20 of them for a survey. The `seq` operator can

Aside. 1.1 Assignment vs Algebra

An assignment command like `x=sqrt(x)` can be confusing to people who are used to algebraic notation. In algebra, the equal sign describes a relationship between the left and right sides. So, $x = \sqrt{x}$ tells us about how the quantity x and the quantity $\sqrt{x}$ are related. Students are usually trained to "solve" such relationships, going through a series of algebraic steps to find values for x that are consistent with the mathematical statement. (For $x = \sqrt{x}$, the solutions are $x = 0$ and $x = 1$.) In contrast, the assignment command `x=sqrt(x)` is a way of replacing the previous values stored in `x` with new values that are the square root of the old ones.

be used to generate a set of 10000 choices. The `sample` operator can be used to select some of these choices at random.

One way to connect the computations is by using objects to store the intermediate outputs.

```
> choices = seq(1,10000)
> sample( choices, 20 )
 [1] 5970 8476 9340 8266 6909
 [6] 3692 8979 1640 4266 5580
[11] 1208 6141 4973 5575 8498
[16] 1001  923 3246 4194 2126
```

You can also pass the output of an operator *directly* as an argument to another operator. Here's another way to accomplish exactly the same thing as the above.

```
> sample( seq(1,10000), 20 )
```

Numbers and Arithmetic

The language has a concise notation for arithmetic that looks very much like the traditional one:

```
> 7+2
[1] 9
> 3*4
[1] 12
> 5/2
[1] 2.5
> 3-8
[1] -5
> -3
[1] -3
> 5^2
[1] 25
```

Arithmetic operators, like any other operators, can be connected to form more complicated computations. For instance,

```
> 8+4/2
[1] 10
```

To a human reader, the command 8+4/2 might seem ambiguous. Is it intended to be (8+4)/2 or 8+(4/2)? The computer uses unambiguous rules to interpret the expression, but it's a good idea for you to use parethesis so that you can make sure that what you intend is what the computer carries out:

```
> (8+4)/2
[1] 6
```

Traditional mathematical notation uses superscripts and radicals to indicate exponentials and roots, e.g., 3^2 or $\sqrt{3}$ or $\sqrt[3]{8}$. This special typography doesn't work well with an ordinary keyboard, so R and most other computer languages uses a different notation:

```
> 3^2
[1] 9
> sqrt(3)
[1] 1.73
> 8^(1/3)
[1] 2
```

There is a large set of mathematical functions: exponentials, logs, trigonometric and inverse trigonometric functions: Some examples:

Traditional	Computer
e^2	exp(2)
$\log_e(100)$	log(100)
$\log_{10}(100)$	log10(100)
$\log_2(100)$	log2(100)
$\cos(\frac{\pi}{2})$	cos(pi/2)
$\sin(\frac{\pi}{2})$	sin(pi/2)
$\tan(\frac{\pi}{2})$	tan(pi/2)
$\cos^{-1}(-1)$	acos(-1)

Numbers can be written in **scientific notation**. For example, the "universal gravitational constant" that describes the gravitational attraction between masses is 6.67428×10^{-11} (with units meters-cubed per kilogram per second squared). In the computer notation, this would be written G=6.67428e-11. The Avogradro constant, which gives the number of atoms in a mole, is $6.02214179 \times 10^{23}$ per mole, or 6.02214178e23.

The computer language does not directly support the recording of units. This is unfortunate, since in the real world numbers often have units and the units matter. For example, in 1999 the Mars Climate Orbiter crashed into Mars because the design engineers specified the engine's thrust in units of pounds, while the guidance engineers thought the units were newtons.

Computer arithmetic is accurate and reliable, but it often involves very slight rounding of numbers. Ordinarily, this is not noticeable. However, it can become apparent in some calculations that produce results that are zero. For example, mathematically $\sin(\pi) = 0$, however the computer does not duplicate this mathematical relationship exactly:

```
> sin(pi)
[1] 1.22e-16
```

Whether a number like this is properly interpreted as "close to zero," depends on the context and, for quantities that have units, on the units themselves. For instance, the unit "parsec" is used in astronomy in reporting distances between stars. The closest star to the sun is Proxima, at a distance of 1.3 parsecs. A distance of 1.22×10^{-16} parsecs is tiny in astronomy but translates to about 2.5 meters — not so small on the human scale.

In statistics, many calculations relate to probabilities which are always in the range 0 to 1. On this scale, `1.22e-16` is very close to zero.

There are two "special" numbers. `Inf` stands for ∞, as in

```
> 1/0
[1] Inf
```

`NaN` stands for "not a number," and is the result when a numerical operation isn't defined, for instance

```
> 0/0
[1] NaN
> sqrt(-9)
[1] NaN
```

Aside. 1.2 Complex Numbers

Mathematically oriented readers will wonder why R should have any trouble with a computation like $\sqrt{-9}$; the result is the imaginary number $3i$. R works with complex numbers, but you have to tell the system that this is what you want to do. To calculate $\sqrt{-9}$, use `sqrt(-9+0i)`.

Types of Objects

Most of the examples used so far have dealt with numbers. But computers work with other kinds of information as well: text, photographs, sounds, sets of data, and so on. The word **type** is used to refer to the kind of information.

Modern computer languages support a great variety of types. There are four types that will be most important here:

numeric The numbers of the sort already encountered.

character Text data.

logical Answers to yes/no questions.

data frames Collections of data more or less in the form of a spreadsheet table.

It's important to know about the types of data because operators expect their input arguments to be of specific types. When you use the wrong type of input, the computer might not be able process your command.

Character Data

You indicate character data to the computer by enclosing the text in double quotation marks. For example:

```
> filename = "swimmers.csv"
```

There is something a bit subtle going on in the above command, so look at it carefully. The purpose of the command is to create an object, named `filename`, that stores a little bit of text data. Notice that the name of the object is not put in quotes, but the text characters are.

Whenever you refer to an object name, make sure that you don't use quotes, for example:

```
> filename
[1] "swimmers.csv"
```

If you make a command with the object name in quotes, it won't be treated as referring to an object. Instead, it will merely mean the text itself:

```
> "filename"
[1] "filename"
```

Similarly, if you omit the quotation marks from around text, the computer will treat it as if it were an object name and will look for the object of that name. For instance, the following command directs the computer to look up the value contained in an object named `swimmers.csv` and insert that value into the object `filename`.

```
> filename = swimmers.csv
Error: object "swimmers.csv" not found
```

As it happens, there was no object named `swimmers.csv` because it had not been defined by any previous assignment command. So, the computer generated an error.

For the most part, you will not need to use very many operators on text data; you just need to remember to include text, such as file names, in quotation marks. Sometimes, you will want to convert non-text items to text in order to display them in graphs. There is a special operator, `as.character` for doing this:

```
> as.character(3)
[1] "3"
```

The quotes in the output show that it is character type rather then numeric type. This isn't terribly important to the human reader, but the computer regards "3" as a different thing than 3. For instance, you can do arithmetic on numbers but not on characters.

```
> 3 + 2
[1] 5
> as.character(3) + 2
Error in as.character(3) + 2 :
 non-numeric argument to binary operator
```

Data Frames

A data frame is a collection of values arranged as a table. For example, here is part of a data frame that records an experiment on the uptake of carbon dioxide by the grass species *Echinochloa crus-galli*.

```
Plant         Type  Treatment conc uptake
  Mc1 Mississippi    chilled  350    18.9
  Mc3 Mississippi    chilled  250    17.9
  Mn3 Mississippi nonchilled   95    11.3
  Mn3 Mississippi nonchilled  350    27.9
  Qn1       Quebec nonchilled  500    35.3
  Qc2       Quebec    chilled  500    38.6
  Qc3       Quebec    chilled  175    21.0
  Qn3       Quebec nonchilled  500    42.9
  Qn3       Quebec nonchilled  175    32.4
... and so on for 84 lines altogether
```

A data frame is a kind of tabular organization of data. In this example, it records several variables: the geographic origin of the plant (Mississippi or Quebec), and whether the plant had been chilled overnight before the uptake measurement was made, the ambient atmospheric CO_2 level and the uptake of CO_2 by the plant.

Each of the components of the data frame could be stored by an object of character type or of numerical type, for instance, `Treatment` is character and `conc` is numerical. The data frame brings the various components together in one place, facilitating processing and analysis of the data.

The information for a data frame is often stored in a spreadsheet file and read into R for analysis. Until you learn how to read in such files, you can use some of the built-in data frames intended for example purposes. If you want to follow along the examples with the CO_2 data frame, use this command to create an object named `CO2` that contains the data frame:

```
> data(CO2)
```

Columns of Data Frames

Perhaps the most common operation on a data frame is to refer to the values in a single column. This can be done using a special syntax involving the $ sign. To refer to the `conc` column in the `CO2` data frame, you would use the command `CO2$conc`. To refer to the `Treatment` column, use `CO2$Treatment`. Think of this style of reference as analogous to naming a person with a first name and a last name: the name of the data frame object comes first and the variable name second, separated by the $, something like Einstein$Albert.

Each component is just like an ordinary object and can be used in any way you would use an object:

```
> length(CO2$conc)
[1] 84
> mean(CO2$conc)
[1] 435
> max(CO2$uptake)
[1] 45.5
> table( CO2$Treatment)
nonchilled    chilled
        42         42
```

Logical Data and Logical Operators

Many computations involve selections of subsets of data that meet some criterion. For example, in studying the health of newborn babies, you might want to focus only on those below a certain birthweight or perhaps those babies whose mother smoked during pregnancy. The question of whether the case satisfies the criterion boils down to a yes-or-no answer.

Logic is the study of valid inference; it is intimately tied up with the idea of truth versus falsehood. In computer languages, **logical data** refers to a type of data that can represent whether something is true or false. To illustrate, consider the simple sequence 1 to 10

```
> x = seq(1,7)
```

Now a simple question about the values in `x`: Are any of them less than π? Here's how you can ask that question:

```
> x < pi
 [1] FALSE FALSE FALSE  TRUE  TRUE  TRUE  TRUE
```

A computer command like `x<pi` is not the same as an algebraic statement like $x < \pi$. The algebraic statement describes the relationship between x and π, namely that x is less than π. The computer command asks a question: Is the value of `x` less than the value of `pi`? Asking this question invokes a computation; the returned value is the answer to the question, either `TRUE` or `FALSE`.

Here are some of the operators for asking such questions:

`x<y`	Is x less than y?
`x<=y`	Is x less than or equal to y?
`x>y`	Is x greater than y?
`x>=y`	Is x greater than or equal to y?
`x==y`	Is x equal to y?
`x!=y`	Is x unequal to y?

Notice the double equal signs in `x==y`. A single equal sign would be the assignment operator.

Mostly, these comparison operators apply to numbers. The `==` and `!=` operators also apply to character strings. To illustrate, I'll define two objects `v` and `w`:

```
> v = "hello"
> w = seq(1,15,by=3)
> w
 1  4  7 10 13
> w < 12
[1]  TRUE  TRUE  TRUE  TRUE FALSE
> w > 7
[1] FALSE FALSE FALSE  TRUE  TRUE
> w != 4
[1]  TRUE FALSE  TRUE  TRUE  TRUE
> v == "goodbye"
[1] FALSE
> v != "hi"
[1] TRUE
> v == "hello"
[1] TRUE
> v == "Hello"
[1] FALSE
```

The `FALSE` in the last line results from taking into account the difference between upper-case and lower-case letters.

Sometimes you need to combine more than one logical result. For example, to ask, "Is `w` between 7 and 12?" involves combining two separate questions: "Is `w` greater than 7 AND is it less than 12?" In the computer language, this question would be stated `w>7&w<12`.

```
> w > 7 & w < 12
[1] FALSE FALSE FALSE  TRUE FALSE
```

There are also logical operators for "or" and "not". For instance, you might ask whether `w` is less than 5 OR greater then 9:

```
> w < 5 | w > 9
[1]  TRUE  TRUE FALSE  TRUE  TRUE
```

The "not" operator just flips `TRUE` and `FALSE`:

```
> !(w < 5 | w > 9)
[1] FALSE FALSE  TRUE FALSE FALSE
```

As this example illustrates, you can group logical operations in just the same way as arithmetic operations.

Missing Data

When recording data from an experiment or an observational study, it sometimes happens that a particular measurement can't be made or is lost or is otherwise unavailable. In R, such **missing data** can be recorded with the special code `NA`. As you might expect, arithmetic and other operations on missing data can't be sensibly performed: giving NA as an input produces NA as an output.

```
> 7 == NA
[1] NA
> NA == 'Hello'
[1] NA
> NA == NA
[1] NA
```

In order to test whether there is missing data, a special operator `is.na` can be used:

```
> is.na(NA)
[1] TRUE
```

Collections

R can work with collections of numbers and character strings. Some operators work on each item in the collection, while others combine the items together in some way. To illustrate, I'll define three small collections, `x`, `y`, and `fruits`:

```
> x = seq(1,7)
> x
 [1]  1  2  3  4  5  6  7

> y = c(7,8,9)
> y
[1] 7 8 9

> fruits = c("apple","berry","cherry")
> fruits
[1] "apple"  "berry"  "cherry"
```

The `c` operator used in defining `y` and `fruits` is useful for creating a small collection "by hand." Often, however, collections will be created by reading in data from a file or using some other operator. I'll introduce these as needed for specific tasks.

Arithmetic and comparison operators often work item-by-item on the collection. For example:

```
> x + 100
 [1] 101 102 103 104 105 106 107
> sqrt(y)
[1] 2.645751 2.828427 3.000000
> fruits == "cherry"
[1] FALSE FALSE  TRUE
```

If the operator involves two collections, they have to be the same size, or R will reuse the smaller collection to match the size of the larger one:

```
> x == y
 [1] FALSE FALSE FALSE FALSE FALSE FALSE  TRUE
Warning message:
In x == y : longer object length is not a
            multiple of shorter object length
```

The warning message is displayed when some aspect of the computation is deemed suspect or odd. Pay attention to such messages since they may signal that the computation the interpreter carried out is not the one you intended.

It's usually obvious what sorts of operators will combine the items of the collection rather than working on them item by item. Here are some examples:

```
> x
[1] 1  2  3  4  5  6  7
> y
[1] 7  8  9
> mean(x)
[1] 4
> median(y)
[1] 8
> min(x)
[1] 1
> max(x)
[1] 7
> sum(x)
[1] 28
> any( fruits == "cherry")
[1] TRUE
> all( fruits == "cherry")
[1] FALSE
```

The `length` operator tells how many items there are in the collection:

```
> length(x)
[1] 7
> length(y)
```

```
[1] 3
> length(fruits)
[1] 3
```

When there are too many items in a collection to display conveniently in one line, the R interpreter will break up the display over multiple lines.

```
> seq(3,19)
 [1]  3  4  5  6  7  8  9 10 11
[10] 12 13 14 15 16 17 18 19
```

At the start of each line, the number in brackets tells the index of the item that starts that line. In the above, for instance, the item 3 is displayed following a `[1]` because 3 is the first item in the collection. Similarly, the `[10]` indicates that the item 12 is the tenth item in the collection. The brackets are just for display purposes; they are not part of the collection itself.

Defining your own operators

Occasionally, you may need to define your own operators. This is convenient if you need to repeat an operation many times or if you need to define a mathematical function.

It's important to keep in mind the difference between an operator and a command. A command is an instruction to perform a particular computation on a particular input argument or set of input arguments. The input arguments always have to be values, though of course you can refer to the value by giving the name of an object that has already been assigned a value.

In contrast, in defining an operator, you can treat the arguments abstractly; just a name without a value having been assigned. To illustrate, here is a command that creates the mathematical function $f(x) = 3x^2 + 2$ and stores it in an operator named `f`:

```
> f = function(x) { 3*x^2 + 2 }
```

One you have defined the function, you can invoke it in the standard way. For example:

```
> f(3)
[1] 29
> f(10)
[1] 302
```

There are some novel features to the syntax used to define a new operator. First, the arguments to `function` aren't treated as values but as pure names. Second, the contents of the curly braces { and } — the function contents — are the commands that will be evaluated when the function is invoked.

It doesn't matter what names you use in the function contents so long as they match the names used in the arguments to `function`. For example, here is another operator, called `g`, that will perform exactly the same computation as `f` when invoked:

```
> g = function(marge) { 3*marge^2 + 2 }
```

When you invoke an operator, the interpreter carries out several steps. Consider the invocation

```
> g(7)
```

In carrying out this command, the interpreter will:

1. Temporarily define or redefine an object `marge` that has the value 7.
2. Execute the function contents.
3. Return the value of these contents as the return value of the command.
4. Discard the definition or redefinition in (1).

Operators can have more than one argument. For instance, here is an operator `hypotenuse` that computes the length of the hypotenuse of a right triangle given the lengths of the legs

```
> hypotenuse = function(a,b) { sqrt( a^2 + b^2 ) }
```

When programmers create new operators that they expect to use on many different occasions, they put the commands to define the operators into a text file called a **source file**. This file can be read into R using a special operator, called `source` that causes the commands to be executed, thereby defining the new operators.

Saving and Documenting Your Work

Up to now, the commands used as examples have been simple one-line statements. As you work further, you will build more elaborate computations by combining simpler ones. It will become important to be able to document what you have done, providing a record so that others can confirm your results and so that you and others can modify your work as needed.

One way in which a record is created of your interaction with the computer is the dialog in the interpreter console itself. In some ways this is analogous to a document created by a word processor: for example, you can copy the contents and paste it into another document.

But the idea of dialog-as-document is flawed. For example, in a word-processor, when you correct a mistake the old version is erased. But in the R dialog, to fix a mistake you give a new command — the old, mistaken command is still there in the dialog.

You should keep in mind that there are several different components, some of which are more appropriate than others for your documentation.

Your Commands The commands that you execute are what defines the computation being performed. These commands themselves are a valuable form of documentation.

The objects you create These objects, and the values that are stored in them, reflect the **state** of the computation. If you want to pick up on your work where you left off, you can save these objects. This is called "saving the **workspace**."

Side effects This refers to the output printed by the interpreter and plots. Sometimes you will want to include this in your documentation, but usually just select elements.

1.4.4 Using Customizations to R

One of the features that makes R so powerful is that new commands can easily be added to the system. This makes it possible, for instance, to customize the software to make routine tasks easier. Such customizations have been written specifically for the people following this book. To use them — and you will need them in later chapters — you should download a file from the web:
`www.macalester.edu/~kaplan/ISM/ISM.Rdata`
This file contains various data sets that you be using.

Once you have downloaded the file, called `ISM.Rdata`, double-click on it in the ordinary way to start R or to load the data into an already started session of R.

Chapter 2

Data: Cases, Variables, Samples

The tendency of the casual mind is to pick out or stumble upon a sample which supports or defies its prejudices, and then to make it the representative of a whole class. — Walter Lippmann (1889-1974)

The word "data" is plural. This is appropriate. Statistics is, at its heart, about variability: how things differ from one another.

Figure 2.1 shows a small collection of sea shells ollected during an idle quarter hour sitting at one spot on the beach on Malololailai island in Fiji. All the

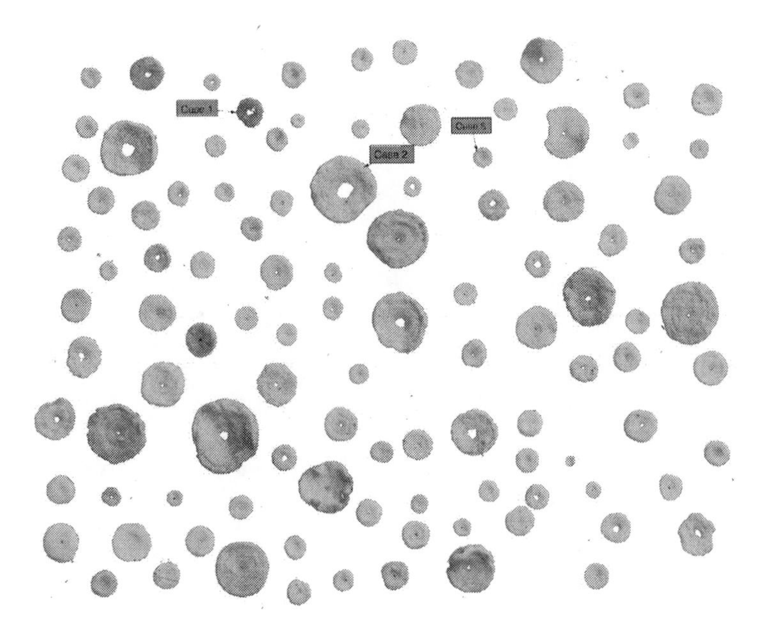

Figure 2.1: A collection of 103 sea shells. (The photo is printed to 3/4 scale.)

shells in the collection are similar; the collector picked up only the small disk-shaped shells with a hole in the center. But the shells also differ from one another: in overall size and weight, in color, in smoothness, in the size of the hole, etc.

Any data set is something like the shell collection. It consists of **cases**: the objects in the collection. Each case has one or more attributes or qualities, called **variables**. This word "variable" emphasizes that it is differences or variation that is often of primary interest.

Usually, there are many possible variables. The researcher chooses those that are of interest, often drawing on detailed knowledge of the system that is under study. The researcher measures or observes the value of each variable for each case. The result is a **table**, also known as a **data frame**: a sort of spreadsheet. Within the data frame, each row refers to one case, each column to one variable.

A data frame for the sea shell collection might look like this:

Case	diameter	weight	color	hole
1	4.3 mm	0.010 mg	dark	medium
2	12.0 mm	0.050 mg	light	very large
3	3.8 mm	0.005 mg	light	none

and so on — there are 103 cases altogether

Each individual shell has a case number that identifies it; three of these are shown in the figure. These case numbers are arbitrary; the position of each case in the table — first, second, tenth, and so on — is of no significance. The point of assigning an identifying case number is just to make it easier to refer to individual cases later on.

If the data frame had been a collection of people rather than shells, the person's name or another identifying label could be used to identify the case.

There are many sorts of data that are recorded in different formats. For example, photographic images are stored as arrays of pixels without any reference to cases and variables. But the data frame format is very general and can be used to arrange all sorts of data, even if it is not the format always used in practice. Even data that seems at first not to be suitable to arrange as a data frame is often stored this way, for example the title/artist/genre/album organization found in music MP3 players or the geographic features (place locations and names, boundaries, rivers, etc.) found in geographic information systems (GIS).

2.1 Kinds of Variables

Most people tend to think of data as numeric, but variables can also be descriptions, as the sea shell collection illustrates. The two basic types of data are:

Quantitative: Naturally represented by a number, for instance diameter, weight, temperature, age, and so on.

Categorical: A description that can be put simply into words or categories, for instance male versus female or red vs green vs yellow, and so on. The value for each case is selected from a fixed set of possibilities. This set of possibilities are the **levels** of the categorical variable.

Categorical variables show up in many different guises. For example, a data frame holding information about courses at a college might have a variable subject with levels biology, chemistry, dance, economics, and so on. The variable semester could have levels Fall2008, Spring2009, Fall2009, and so on. Even the instructor is a categorical variable. The instructor's name or ID number could be used as the level of the variable.

Quantitative variables are numerical but they often have units attached to them. In the shell data frame, the variable diameter has been recorded in millimeters, while weight is given in milligrams. The usual practice is to treat quantitative variables as a pure number, without units being given explicitly. The information about the units — for instance that diameter is specified in millimeters — is kept in a separate place called a **code book**.

The code book contains a short description of each variable. For instance, the code book for the shell data might look like this:

Code book for shells from Malololailai island, collected on January 12, 2008.

diameter: the diameter of the disk in millimeters.

weight: the shell weight in milligrams.

color: a subjective description of the color. Levels: light, medium, and dark.

hole: the size of the inner hole. Levels: none, small, large, very large.

On the computer, a data frame is usually stored as a spreadsheet file, while the corresponding code book can be stored separately as a text file.

Sometimes quantitative information is represented by categories as in the hole variable for the sea shells. For instance, a data frame holding a variable income might naturally be stored as a number in, say, dollars per year. Alternatively, the variable might be treated categorically with levels of, say, "less than $10,000 per year," "between $10,000 and $20,000," and so on. Almost always, the genuinely quantitative form is to be preferred. It conveys more information even if it not correct to the last digit.

The distinction between quantitative and categorical data is essential, but there are other distinctions that can be helpful even if they are less important.

Some categorical variables have levels that have a *natural* order. For example, a categorical variable for temperature might have levels such as "cold," "warm," "hot," "very hot." Variables like this are called **ordinal**. Opinion surveys often ask for a choice from an ordered set such as this: strongly disagree, disagree, no opinion, agree, strongly agree.

For the most part, this book will treat ordinal variables like any other form of categorical variable. But it's worthwhile to pay attention to the natural ordering of an ordinal variable. Indeed, sometimes it can be appropriate to treat an ordinal variable as if it were quantitative.

2.2 Data Frames and the Unit of Analysis

When collecting and organizing data, it's important to be clear about what is a case. For the sea shells, this is pretty obvious; each individual shell is an individual case. But in many situations, it's not so clear.

A key idea is the **unit of analysis**. Suppose, for instance, that you want to study the link between teacher pay, class size, and the performance of school children. There are all sorts of possibilities for how to analyze the data you collect. You might decide to compare different schools, looking at the average class size, average teacher pay, and average student performance in each school. Here, the unit of analysis is the school.

Or perhaps rather than averaging over all the classes in one school, you want to compare different classes, looking at the performance of the students in each class separately. The unit of analysis here is the class.

You might even decide to look at the performance of individual students, investigating how their performance is linked to the individual student's family income or the education of the student's parents. Perhaps even include the salary of student's teacher, the size of the student's class, and so on. The unit of analysis here is the individual student.

What's the difference between a unit of analysis and a case? A case is a row in a data frame. In many studies, you need to bring together different data frames, each of which may have a different notion of case. Returning to the teacher's pay example, one can easily imagine at least three different data frames being involved, with each frame storing data at a different level:

1. A frame with each class being a case and variables such as the size of the class, the school in which the class is taught, etc.

2. A frame with each teacher being a case and variables such as the teacher's salary, years of experience, advanced training, etc.

3. A frame with each student being a case and variables such as the student's test scores, the class that the student is in, and the student's family income and parent education.

Once you choose the unit of analysis, you combine information from the different data frames to carry out the data analysis, generating a single data frame in which cases are your chosen unit of analysis. The choice of the unit of analysis can be determined by many things, such as the availability of data. As a general rule, it's best to make the unit of analysis as small as possible. But there can be obstacles to doing this. You might find, for instance, that for privacy reasons (or less legitimate reasons of secrecy) the school district is unwilling to release the data at the individual student level, or even to to release data on individual classes.

In the past, limitations in data analysis techniques and computational power provided a reason to use a coarse unit of analysis. Only a small amount of data could be handled effectively, so the unit of analysis was made large. For example, rather than using tens of thousand of individual students as the unit of analysis, a few dozen schools might be used instead. Nowadays these reasons are obsolete. The methods that will be covered in this book allow for a very fine unit of analysis. Standard personal computers have plenty of power to perform the required calculations.

2.3 Populations and Samples

A data frame is a collection, but a collection of what? Two important statistical terms are "population" and "sample." A **population** is the set of all the possible objects or units which might have been included in the collection. The root of the word "population" refers to people, and often one works with data frames in which the cases are indeed individual people. The statistical concept is broader; one might have a population of sea shells, a population of houses, a population of events such as earthquakes or coin flips.

A **sample** is a selection of cases from the population. The **sample size** is the number of cases in the sample. For the shells in Figure 2.1, the sample size is $n = 103$.

A **census** is a sample that contains the entire population. The most familiar sort of census is the kind to count the people living in a country. The United States and the United Kingdom have a census every ten years. Countries such as Canada, Australia, and New Zealand hold a census every five year.

Almost always, the sample is just a small fraction of the population. There are good reasons for this. It can be expensive or damaging to take a sample: Imagine a biologist who tried to use all the laboratory rats in the world for his or her work! Still, when you draw a sample, it is generally because you are interested in finding out something about the population rather than just the sample at hand. That is, you want the sample to be genuinely representative of the population. (In some fields, the ability to draw conclusions from a sample that can be generalized is referred to as **external validity** or **transferability**.)

The process by which the sample is taken is important because it controls what sorts of conclusions can legitimately be drawn from the sample. One of the most important ideas of statistics is that a sample will be representative of the population if the sample is collected at random. In a **simple random sample**, each member of the population is equally likely to be included in the sample.

Ironically, taking a random sample, even from a single spot on the beach, requires organization and planning. The sea shells were collected haphazardly, but this is not a genuinely random sample. The bigger shells are much easier to see and pick up than the very small ones, so there is reason to think that the small shells are under-represented: the collection doesn't have as big a proportion of them as in the population. To make the sample genuinely random, you need to have access in some way to the entire population so that you can pick any member with equal probability. For instance, if you want a sample of students at a particularly university, you can get a list of all the students from the university registrar and use a computer to pick randomly from the list. Such a list of the entire set of possible cases is called a **sampling frame**.

In a sense, the sampling frame is the *definition* of the population for the purpose of drawing conclusions from the sample. For instance, a researcher studying cancer treatments might take the sampling frame to be the list of all the patients who visit a particular clinic in a specified month. A random sample from that sampling frame can reasonably be assumed to represent that particular population, but not necessarily the population of all cancer patients.

You should always be careful to define your sampling frame precisely. If you decide to sample university students by picking randomly from those who enter the front door of the library, you will get a sample that might not be typical for *all* university students. There's nothing wrong with using the library students for your sample, but you need to be aware that your sample will be representative of just the library students, not necessarily all students.

When sampling at random, use formal random processes. For example, if you are sampling students who walk into the library, you can flip a coin to decide whether to include that student in your sample. When your sampling frame is in the form of a list, it's wise to use a computer random number generator to select the cases to include in the sample.

A **convenience sample** is one where the sampling frame is defined mainly in a way that makes it easy for the researcher. For example, during lectures I often sample from the set of students in my class. These students — the ones who take statistics courses from me — are not necessarily representative of all university students. It might be fine to take a convenience sample in a quick, informal, preliminary study. But don't make the mistake of assuming that the convenience sample is representative of the population. Even if you believe it yourself, how will you convince the people who are skeptical about your results?

When cases are selected in an informal way, it's possible for the researcher to introduce a non-representativeness or **sampling!bias**. For example, in deciding which students to interview who walk into the library, you might consciously or subconsciously select those who seem most approachable or who don't seem to be in a hurry.

There are many possible sources of sampling bias. In surveys, sampling bias can come from non-response or self-selection. Perhaps some of the students who you selected randomly from the people entering the library have declined to participate in your survey. This **non-response** can make your sample non-representative. Or, perhaps some people who you didn't pick at random have walked up to you to see what you are up to and want to be surveyed themselves. Such **self-selected** people are often different from people who you would pick at random.

In a famous example of self-selection bias, the newspaper columnist Ann Landers asked her readers, "If you had it to do over again, would you have children?" Over 70% of the respondents who wrote in said "No." This result is utterly different from what was found in surveys on the same question done with a random sampling methodology: more than 90% said "Yes." Presumably the people who bothered to write were people who had had a particularly bad experience as parents whereas the randomly selected parents are representative of the whole population. Or, as Ann Landers wrote, "[T]he hurt, angry, and disenchanted tend to write more readily than the contented ...." (See [7].)

Non-response is often a factor in political polls, where people don't like to express views that they think will be unpopular.

It's hard to take a genuinely random sample. But if you don't, you have no guarantee that your sample is representative. Do what you can to define your sampling frame precisely and to make your selections as randomly as possible from that frame. By using formal selection procedures (e.g., coin flips, computer random number generators) you have the opportunity to convince skeptics who might otherwise wonder what hidden biases were introduced by informal selections. If you believe that your imperfect selection may have introduced biases — for example the suspected under-representation of small shells in my collection — be up-front and honest about it. In surveys, you should keep track of the non-response rate and include that in whatever report you make of your study.

Example 2.1: Struggling for a Random Sample Good researchers take great effort to secure a random sample. One evening I received a phone call at home from the state health department. They were conducting a survey of access to health care, in particular how often people have illnesses for which they don't get treatment. The person on the other end of the phone told me that they were dialing numbers randomly, checked to make sure that I live in the area of interest, and asked how many adults over age 18 live in the household. "Two," I replied, "Me and my wife." The researcher asked me to hold a minute while she generated a random number. Then the researcher asked to speak to my wife. "She isn't home right now, but I'll be happy to help you," I offered. No deal.

The sampling frame was adults over age 18 who live in a particular area. Once the researcher had made a random selection, as she did after asking how many adults are in my household, she wasn't going to accept any substitutes. It took three follow-up phone calls over a few days — at least that's how many I answered, who knows how many I wasn't home for — before the researcher was able to contact my wife. The researcher declined to interview me in order to avoid self-selection bias and worked hard to contact my wife — the randomly selected member of our household — in order to avoid non-response bias.

2.4 Longitudinal and Cross-Sectional Samples

Data are often collected to study the links between different traits. For example, the data in the following table are a small part of a larger data set of the speeds of runners in a ten-mile race held in Washington, D.C. in 2004. The variable net gives the time from the start line to the finish line, in seconds. Such data might be used to study the link between age and speed, for example to find out at what age people run the fastest and how much they slow down as they age beyond that.

state	net	age	sex
DC	6382	23	F
VA	5080	26	F
DC	4742	27	M
Kenya	2962	27	M
DC	6291	29	F
MD	6405	32	M
VA	6608	34	F
DC	5921	37	M
MD	5549	41	F
MD	5486	46	F
VA	8374	53	F
PA	6026	53	M
VA	5526	60	M
VA	5585	61	M
VA	5931	65	M

... and so on.

This sample is a **cross section**, a snapshot of the population that includes people of different ages. Each person is included only once.

Another type of sample is **longitudinal**, where the cases are tracked over time, each person being included more than once in the data frame. A longitudinal data set for the runners might look like this:

state	net	age	sex	year
DC	6382	23	F	2004
DC	6516	24	F	2005
DC	6493	25	F	2006
DC	6526	26	F	2007
DC	6571	27	F	2008
MD	6405	32	M	2004
MD	6819	34	M	2006
MD	6753	35	M	2007

... and so on.

If your concern is to understand how individual change as they age, it's best to collect data that show such change in individuals. Using cross-sectional data to study a longitudinal problem is risky. Suppose, as seems likely, that younger runners who are slow tend to drop out of racing as they age, so the older runners who do participate are those who tend to be faster. This could bias your estimate of how running speed changes with age.

2.5 Computational Technique

2.5.1 Reading and Writing Data

Data used in statistical modeling are usually organized into tables, often created using spreadsheet software. Most people presume that the same software used to create a table of data should be used to display and analyze it. This is part of the reason for the popularity of spreadsheet programs such as Excel.

For statistical modeling, it's helpful to take another approach that strictly separates the processes of data collection and of data analysis: use one program to create data files and another program to analyze the data stored in those files. By doing this, one guarantees that the original data are not modified accidentally in the process of analyzing them. This also makes it possible to perform many different analyses of the data; modelers often create and compare many different models of the same data.

The spreadsheet programs that can create data files use a variety of different formats. Many of these formats are proprietary and include various features that make it difficult for any other software to read the file. A good, simple, general purpose format supported by spreadsheet software and by statistical software is called the **comma separated value** or **CSV** format.

The next sections describe how to read data from a CSV file into R and how to use a spreadsheet program to create new data.

Reading CSV Files into R

For the data sets associated with this book and its exercises, an easy way to import data into R is with the `ISMdata` operator that's included in the extensions to R found in the `ISM.Rdata` workspace file. (See Section 1.4.4 on page 30. You must load the workspace file before you can use `ISMdata`.)

`ISMdata` lets you refer to a file by a short name in quotes without worrying about where it is located. It knows where to locate the data files used with this book, whether they be on your computer or on the web. For example, the data set `hdd-minneapolis.csv` is stored on the Internet. You can read it into an objected named `hdd` with this statement:

```
> hdd = ISMdata("hdd-minneapolis.csv")
Not in library.  Trying to find it on the web ...
File was read from the web.
```

If you do not have an Internet connection, then `ISMdata` will be able to locate only files on your computer.

It is possible to use `ISMdata` to read in your own data that you have created and stored in CSV files. To do this, you need to tell `ISMdata` where the data is located. The easiest way to do this is to use a mouse-based file navigator. To do this, invoke `ISMdata` with no input argument:

```
> swim = ISMdata()
```

Since there was no character string file name given as an argument, `ISMdata` brings up a file navigator to let you select the file interactively:

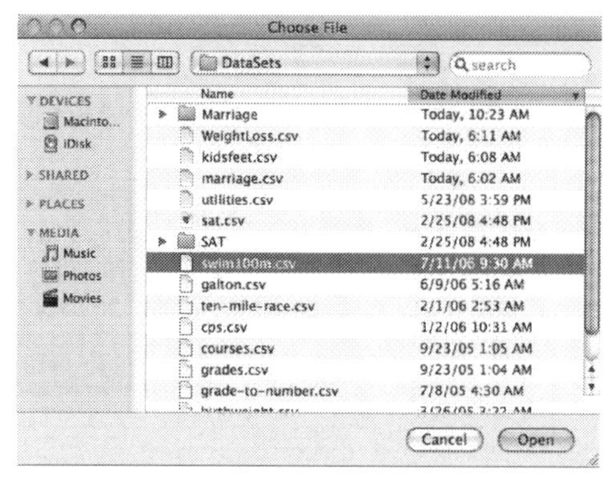

Selecting a file interactively.

Selecting and opening the desired file will cause it to be read into R as a data frame. Make sure to choose the correct CSV file. If you choose some other sort of file by accident, `ISMdata` will struggle to read it in, displaying various junk on your screen.

The statement above will cause the resulting data frame to be called `swim`. But be careful. If you accidentally choose a different file (e.g., `kidsfeet.csv`) the data from that file will be read in and stored under the name `swim`, even though they have nothing to do with swimming.

[Optional] Although `ISMdata` will work well for most purposes, some users might prefer the flexibility offered by the built-in R operators for reading in files. These include `read.csv`, `scan`, and a variety of operators for importing files from other packages, such as the `read.spss` operator in the "foreign" library package.

For obvious reasons, the most of the examples used in this book are based on data that has already been collected and stored in CSV files. In your own investigations, you will generally need to create your own data sets. Instructions and suggestions for creating CSV files and codebooks are given in the exercises.

2.5.2 Simple Operations with Data Frames

To illustrate, here are data on world-record swimming times:

```
> swim = ISMdata("swim100m.csv")
```

This data set, stored in the object named `swim`, has three variables:

```
> names(swim)
[1] "year" "time" "sex"
```

Year refers to the calendar year in which the record was set; time is the record time itself, in seconds; sex records whether the record is for men or women.

```
> head(swim)
  year time sex
1 1905 65.8   M
2 1908 65.6   M
3 1910 62.8   M
... and so on.
```

Numerical Operations

There are two basic kinds of variables: quantitative and categorical. R treats these variables differently, as it should since operations that make perfect sense for a quantitative variable (such as the `sum` or the `mean` or `median`) make little sense for categorical variables. For instance, it makes sense to compute

```
> max(swim$year)
[1] 2004
```

But it does not make sense to compute the mean `sex`:

```
> max(swim$sex)
Error in Summary.factor
  max not meaningful for factors
```

The word **factor** is how R refers to categorical variables. Once you know that, the second line of the error message makes more sense.

Adding a New Variable

Sometimes you will compute a new quantity from the variables and you want to treat this as a new variable. You can do this by assignment to the data frame, using the $ and giving a new name for the variable. As a trivial example, here is how to convert the swim time from seconds to minutes:

```
> swim$minutes = swim$time / 60
```

One this has been done, the new variable appears just like the old ones:

```
> names(swim)
[1] "year"    "time"    "sex"     "minutes"
```

You could also, if you want, redefine an existing variable, for instance:

```
> swim$time = swim$time / 60
```

Such assignment operations do not change the original file from which the data were read, only the data frame in the current session of R.

Extracting Subsets of Data

Selecting a subset of cases is done with the `subset` operator. For instance, here's how to create a data frame with just the women's records:

```
> women = subset( swim, sex=='F' )
```

The `subset` operator takes two arguments: the first is a data frame from which to extract the subset. The second is a logical (`TRUE/FALSE`) criterion for each case, saying whether to include it.

Notice that in this example, the name `sex` was used, rather than the full `swim$sex`. The `subset` operator allows the shorthand since the first argument sets a context for evaluating any names in the second argument. Other operators also allow this sort of shorthand.

The `subset` operator creates a new data frame which you can assign to a name; it does not modify the original data frame. You can have as many data frames as you want in an R session, so there is little reason to modify the original. But if you want to, you can do it by re-assignment:

```
> swim = subset( swim, sex=='F' )
```

After this command, the male records are no longer in `swim`. If you want them back, you have to re-read the original data file.

2.5.3 Sampling

Suppose you have a sampling frame with 1000 cases, arranged as a spreadsheet with one row for each case. You want to select a random set of 20 of these. The `shuffle` operator lets you pick random members of a set.

A common use for `shuffle` is to pick random cases from a data frame, just as you might do when sampling randomly from a sampling frame.

```
> shuffle(swim, 5)
   year  time sex
55 1976 55.65   F
39 1924 72.20   F
12 1944 55.90   M
18 1964 52.90   M
46 1936 64.60   F
```

The results returned by `shuffle` will never contain the same case more than once, just as if you were dealing cards from a shuffled deck. In contrast, `resample` replaces each case after it is dealt so that it can appear more than once in the result. This is called **sampling with replacement**. This will be useful later on when studying the statistical properties of the sampling process. For example, `resample` allows you to generate a sample of any size, even one that is bigger than the data set from which the sample is being drawn.

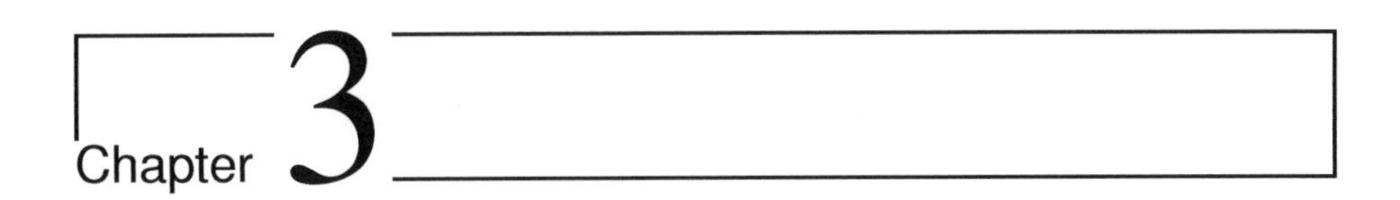

Describing Variation

Variation itself is nature's only irreducible essence. Variation is the hard reality, not a set of imperfect measures for a central tendency. Means and medians are the abstractions. — Stephen Jay Gould

A statistical model partitions variation into parts. People describe the partitioning in different ways depending on their purposes and the conventions of the field in which they work: explained variation versus unexplained variation; described variation versus undescribed; predicted variation versus unpredicted; signal versus noise; common versus individual. This chapter describes ways to quantify variation in a single variable. Once you can quantify variation, you can describe how models divide it up.

To start, consider a familiar situation: the variation in human heights. Everyone is familiar with height and how heights vary from person to person, so variation in height provides a nice example to compare intuition with the formal descriptions of statistics. Perhaps for this reason, height was an important topic for early statisticians. In the 1880s, Francis Galton, one of the pioneers of statistics, collected data on the heights of about 900 adult children and their parents in London. Figure 3.1 shows part of his notebook.

Galton was interested in studying the relationship between a full-grown child's height and his or her mother's and father's height. One way to study this relationship is to build a model that accounts for the child's height — the **response variable** — by one or more **explanatory variables** such as mother's height or father's height or the child's sex. In later chapters you will construct such models. For now, though, the objective is to describe and quantify how the response variable varies across children without taking into consideration any explanatory variables.

Galton's height measurements are from about 200 more-or-less normal families in the city of London. In itself, this is an interesting choice. One might think that the place to start would be with exceptionally short or tall people, looking at what factors are associated with these extemes.

FAMILY HEIGHTS. from RFF
(add 60 inches to every entry in the Table)

	Father	Mother	Sons in order of height	Daughters in order of height.
1	18·5	7·0	13·2	9·2, 9·0, 9·0
2	15·5	6·5	13·5, 12·5	5·5, 5·5
3	15·0	about 4·0	11·0	8·0
4	15·0	4·0	10·5, 8·5	7·0, 4·5, 3·0
5	15·0	−1·5	12·0, 9·0, 8·0	6·5, 2·5, 2·5

Figure 3.1: Part of Francis Galton's notebook recording the heights of parents and their adult children. [8]

Adults range in height from a couple of feet to about nine feet. (See Figure 3.2.) One way to describe variation is by the **range of extremes**: an interval that includes every case from the smallest to the largest.

What's nice about describing variation through the extremes is that the range includes every case. But there are disadvantages. First, you usually don't always have a **census**, measurements on the entire population. Instead of a census, you typically have only a **sample**, a subset of the population. Usually only a small proportion of the population is included in a sample. For example, of the millions of people in London, Galton's sample included only 900 people. From such a sample, Galton would have had no reason to believe the either the tallest or shortest person in London, let alone the world, happened to be included.

A second disadvantage of using the extremes is that it can give a picture that is untypical. The vast majority of adults are between $4\frac{1}{2}$ feet and 7 feet tall. Indeed, an even narrower range — say 5 to $6\frac{1}{2}$ feet — would give you a very good idea of typical variation in heights. Giving a comprehensive range — 2 to 9 feet — would be misleading in important ways, even if it were literally correct.

A third disadvantage of using the extremes is that even a single case can have a strong influence on your description. There are about six billion people on earth. The discovery of even a single, exceptional 12-foot person would cause you substantially to alter your description of heights even though the population — minus the one new case — remains unchanged.

It's natural for people to think about variation in terms of records and extremes. People are used to drawing conclusions from stories that are newsworthy and exceptional; indeed, such anecdotes are mostly what you read and hear about. In statistics, however, the focus is usually on the unexceptional cases, the typical cases. With such a focus, it's important to think about **typical variation** rather than extreme variation.

Figure 3.2: Some extremes of height. Angus McAskill (1825-1863) and Charles Sherwood Stratton (1838-1883). McAskill was 7 feet 9 inches tall. Stratton, also known as Tom Thumb, was 2 feet 6 inches in height.

3.1 Coverage Intervals

One way to describe typical variation is to specify a fraction of the cases that are regarded as typical and then give the **coverage interval** or range that includes that fraction of the cases.

Imagine arranging all of the people in Galton's sample of 900 into order, the way a school-teacher might, from shortest to tallest. Now walk down the line, starting at the shortest, counting heads. At some point you will reach the person at position 225. This position is special because one quarter of the 900 people in line are shorter and three quarters are taller. The height of the person at position 225 — 64.0 inches — is the 25th **percentile** of the height variable in the sample.

Continue down the line until you reach the person at position 675. The height of this person — 69.7 inches — is the 75th percentile of height in the sample.

The range from 25th to 75th percentile is the **50-percent coverage interval**: 64.0 to 69.7 inches in Galton's data. Within this interval is 50% of the cases.

For most purposes, a 50% coverage interval excludes too much; half the cases are outside of the interval.

Scientific practice has established a convention, the **95-percent coverage interval** that is more inclusive than the 50% interval but not so tied to the extremes as the 100% interval. The 95% coverage interval includes all but 5% of the cases, excluding the shortest 2.5% and the tallest 2.5%.

To calculate the 95% coverage interval, find the 2.5 percentile and the 97.5

Position in list k	Height (inches)	# of previous cases $k-1$	Percentile
1	59.0	0	0
2	61.5	1	10
3	62.0	2	20
4	63.0	3	30
5	64.7	4	40
6	65.5	5	50
7	68.0	6	60
8	69.0	7	70
9	72.0	8	80
10	72.0	9	90
11	72.0	10	100

Table 3.1: Finding sample percentiles by sorting and counting.

percentile. In Galton's height data, these are 60 inches and 73 inches, respectively.

There is nothing magical about using the coverage fractions 50% or 95%. They are just conventions that make it easier to communicate clearly. But they are important and widely used conventions.

To illustrate in more detail the process of finding coverage intervals, let's look at Galton's data. Looking through all 900 cases would be tedious, so the example involves just a few cases, the heights of 11 randomly selected people:

```
72.0 62.0 68.0 65.5 63.0 64.7 72.0 69.0 61.5 72.0 59.0
```

Table 3.1 puts these 11 cases into sorted order and gives for each position in the sorted list the corresponding percentile. For any given height in the table, you can look up the percentile of that height, effectively the fraction of cases that came previously in the list. In translating the position in the sorted list to a percentile, convention puts the smallest case at the 0th percentile and the largest case at the 100th percentile. For the kth position in the sorted list of n cases, the percentile is taken to be $(k-1)/(n-1)$.

With $n = 11$ cases in the sample, the sorted cases themselves stand for the 0th, 10th, 20th, ..., 90th, and 100th percentiles. If you want to calculate a percentile that falls in between these values, you (or the software) interpolate between the samples. For instance, the 75th percentile would, for $n = 11$, be taken as half-way between the values of the 70th and 80th percentile cases.

Coverage intervals are found from the tabulated percentiles. The 50% coverage interval runs from the 25th to the 75th percentile. Those exact percentiles happen not be be in the table, but you can estimate them. Take the 25th percentile to be half way between the 20th and 30th percentiles: 62.5 inches. Similarly, take the 75th percentile to be half way between the 70th and 80th percentiles: 70.5 inches.

Thus, the 50% coverage interval for this small subset of $N = 11$ cases is 62.5

to 70.5 inches. For the complete set of $N = 900$ cases, the interval is 64 to 69 inches — not exactly the same, but not too much different. In general you will find that the larger the sample, the closer the estimated values will be to what you would have found from a census of the entire population.

This small $N = 11$ subset of Galton's data illustrates a potential difficulty of a 95% coverage interval: The values of the 2.5th and 97.5th percentiles in a small data set depend heavily on the extreme cases, since interpolation is needed to find the percentiles. In larger data sets, this is not so much of a problem.

A reasonable person might object that the 0th percentile of the sample is probably not the 0th percentile of the population; a small sample almost certainly does not contain the shortest person in the population. There is no good way to know whether the population extremes will be close to the sample extremes and therefore you cannot demonstrate that the estimates of the extremes based on the sample are valid for the population. The ability to draw demonstrably valid inferences from sample to population is one of the reasons to use a 50% or 95% coverage interval rather than the range of extremes.

Different fields have varying conventions for dividing groups into parts. In the various literatures, one will read about **quintiles** (division into 5 equally sized groups, common in giving economic data), **stanines** (division into 9 unevenly sized groups, common in education testing), and so on.

In general-purpose statistics, it's conventional to divide into four groups: **quartiles**. The dividing point between the first and second quartiles is the 25th percentile. For this reason, the 25th percentile is often called the "first quartile." Similarly, the 75th percentile is called the "third quartile."

The most famous percentile is the 50th: the **median**, which is the value that half of the cases are above and half below. The median gives a good representation of a typical value. In this sense, it is much like the **mean**: the average of all the values.

Neither the median nor the mean describes the variation. For example, knowing that the median of Galton's sample of heights is 66 inches does not give you any indication of what is a typical range of heights. In the next section, however, you'll see how the mean can be used to set up an important way of measuring variation: the typical distance of cases from the mean.

3.2 The Variance and Standard Deviation

The 95% and 50% coverage intervals are important descriptions of variation. For constructing models, however, another format for describing variation is more widely used: the variance and standard deviation.

To set the background, imagine that you have been asked to describe a variable in terms of a *single* number that typifies the group. Think of this as a very simple model, one that treats all the cases as exactly the same. For human heights, for instance, a reasonable model is that people are about 5 feet 8 inches tall (68 inches).

Aside. 3.1 What's Normal?

You've just been told that a friend has hypernatremia. Sounds serious. So you look it up on the web and find out that it has to do with the level of sodium in the blood:

> For adults and older children, the normal range [of sodium] is 137 to 145 millimoles per liter. For infants, the normal range is somewhat lower, up to one or two millimoles per liter from the adult range. As soon as a person has more than 145 millimoles per liter of blood serum, then he has hypernatremia, which is too great a concentration of sodium in his blood serum. If a person's serum sodium level falls below 137 millimoles per liter, then they have hyponatremia, which is too low a concentration of sodium in his blood serum. [From http://www.ndif.org/faqs.]

You wonder, what do they mean by "normal range?" Do they mean that your body stops functioning properly once sodium gets above 145 millimoles per liter? How much above? Is 145.1 too large for the body to work properly? Is 144.9 fine?

Or do they mean a coverage interval? But what kind of coverage interval: 50%, 80%, 95%, 99%, or somethings else? And in what population? Healthy people? People who go for blood tests? Hospitalized people?

If 137 to 145 were a 95% coverage interval for healthy people, then 19 of 20 healthy people would fall in the 137 to 145 range. Of course this would mean that 1 out of 20 healthy people would be out of the normal range. Depending on how prevalent sickness is, it might even mean that most of the people outside of the normal range are actually healthy.

People frequently confuse "normal" in the sense of "inside a 95% coverage interval" with "normal" in the sense of "functions properly." It doesn't help that publications often don't make clear what they mean by normal. In looking at the literature, the definition of hypernatremia as being above 145 millimoles of sodium per liter appears in many places. Evidently, a sodium level above 145 is very uncommon in healthy people who drink normal amounts of water, but it's hard to find out from the literature just how uncommon it is.

If it seems strange to model every case as the same, you're right. Typically you will be interested in relating one variable to others. To model height, depending on your purposes, you might consider explanatory variables such as age and sex as pediatricians do when monitoring a child's growth. Galton, who was interested in what is now called genetics, modeled child's height (when grown to adulthood) using sex and the parents' heights as explanatory variables. One can also imagine using economic status, nutrition, childhood illnesses, participation in sports, etc. as explanatory variables. The next chapter will introduce such models. At this point, though, the descriptions involve only a single variable — there are no other variables that might distinguish one case from another. So the only possible model is that all cases are the same.

Any given individual case will likely deviate from the single model value. The value of an individual can be written in a way that emphasizes the common

Figure 3.3: The heights of nine brothers were recorded on a tintype photograph in the 1880s. The version here has been annotated to show the mean height and the individual deviations from the mean.

model value shared by all the cases and the deviation from that value of each individual:

$$\text{individual case} = \text{model value} + \text{deviation of that case.}$$

Writing things in this way partitions the description of the individual into two pieces. One piece reflects the information contained in the model. The other piece — the deviation of each individual — reflects how the individual is different from the model value.

As the single model value of the variable, it's reasonable to choose the mean. For Galton's height data, the mean is 66.7 inches. Another equally good choice would be the *median*, 66.5 inches. (Chapter 6 deals with the issue of how to choose the "best" model value, and how to define "best.")

Once you have chosen the model value, you can find how much each case deviates from the model. For instance, Figure 3.3 shows a group of nine brothers. Each brother's height differs somewhat from the model value; that difference is the deviation for that person.

In the more interesting models of later chapters, the model values will differ from case to case and so part of the variation among individual cases will be captured by the model. But here the model value is the same for all cases, so the deviations encompass all of the variation.

The word "deviation" has a negative connotation: a departure from an accepted norm or behavior. Indeed, in the mid-1800s, early in the history of statistics, it was widely believed that "normal" was best quantified as a single model value. The deviation from the model value was seen as a kind of mistake or imperfection. Another, related word from the early days of statistics is "error." Nowadays, though, people have a better understanding that a range of behaviors is normal; normal is not just a single value or behavior.

The word **residual** provides a more neutral term than "deviation" or "error" to describe how each individual differs from the model value. It refers to what is left over when the model value is taken away from the individual case. The word "deviation" survives in statistics in a technical terms such as **standard deviation**, and **deviance**. Similarly, "error" shows up in some technical terms such as **standard error**.

Height (inches)	Model Value (inches)	Residual (inches)	Square-Residual (inches^2)
72.0	66.25	5.75	33.06
62.0	66.25	-4.25	17.06
68.0	66.25	1.75	3.05
65.5	66.25	-0.75	0.56
63.0	66.25	-3.25	10.56
64.7	66.25	-1.55	2.40
72.0	66.25	5.75	33.06
69.0	66.25	2.75	7.56
61.5	66.25	-4.75	22.56
72.0	66.25	5.75	33.06
59.0	66.25	-7.25	52.56
			Sum of Squares: 216.5

Table 3.2: Calculation of the sum of squares of the residuals for the subset of $N = 11$ cases from Galton's data used in Table 3.1. (For this small subset, the mean height is 66.25 inches, somewhat different from the mean of 66.7 inches in the complete set.)

Any of the measures from Section 3.1 can be used to describe the variation in the residuals. The range of extremes is from -10.7 inches to 12.3 inches. These numbers describe how much shorter the shortest person is from the model value and how much taller the tallest. Alternatively, you could use the 50% interval (-2.7 to 2.8 inches) or the 95% interval (-6.7 to 6.5 inches). Each of these is a valid description.

In practice, instead of the coverage intervals, a very simple, powerful, and perhaps unexpected measure is used: the **mean square** of the residuals. To produce this measure, add up the square of the residuals for the individual cases. This gives the **sum of squares** of the residuals as shown in Table 3.2. Such sums of squares will show up over and over again in statistical modeling.

It may seem strange to square the residuals. Squaring the residuals changes their units. For the height variable, the residuals are naturally in inches (or centimeters or meters or any other unit of length). The sum of squares, however is in units of square-inches, a bizarre unit for relating information about height. A good reason to compute the square is to emphasize the interest in *how far* each individual is from the mean. If you just added up the raw residuals, negative residuals (for those cases below the mean) would tend to cancel out positive residuals (for those cases above the mean). By squaring, all the negative residuals become positive square-residuals. It would also be reasonable do this by taking an absolute value rather than squaring, but the operation of squaring captures in a special way important properties of the potential randomness of the residuals.

Finding the sum of squares is an intermediate step to calculating an important measure of variation: the **mean square**. The mean square is intended to report a typical square residual for each case. The obvious way to get this is to divide the sum of squares by the number of cases, N. This is not exactly what is done by statisticians. For reasons that you will see later, they divide by $N - 1$ instead. (When N is big, there is no practical difference between using N and $N - 1$.) For the subset of the Galton data in Table 3.2, the mean square residual is 21.65 square-inches.

Later in this book, mean squares will applied to all sorts of models in various ways. But for the very simple situation here — the every-case-the-same model — the mean square has a special name that is universally used: the **variance**.

The variance provides a compact description of how far cases are, on average, from the mean. As such, it is a simple measure of variation.

It is undeniable that the unfamiliar units of the variance — squares of the natural units — make it hard to interpret. There is a simple cure, however: take the square root of the variance. This is called, infamously to many students of statistics, the **standard deviation**. For the Galton data in Table 3.2, the standard deviation is 4.65 inches (the square-root of 21.65 square-inches).

The term "standard deviation" might be better understood if "standard" were replaced by a more contemporary equivalent word: "typical." The standard deviation is a measure of how far cases typically deviate from the mean of the group.

Historically, part of the appeal of using the standard deviation to describe variation comes from the ease of calculating it using arithmetic. Percentiles require more tedious sorting operations. Nowadays, computers make it easy to calculate 50% or 95% intervals directly from data. Still, standard deviations remain an important way of describing variation both because of historical convention and, more important for this book, because of its connection to concepts of modeling and randomness encountered in later chapters.

3.3 Displaying Variation

One of the best ways to depict variation is graphically. This takes advantage of the human capability to capture complicated patterns at a glance and to discern detail. There are several standard formats for presenting variation; each has its own advantages and disadvantages.

A simple but effective display plots out each case as a point on a number line. This **rug plot** is effectively a graphical listing of the values of the variable, case-by-case.

From the rug plot of Galton's heights in Figure 3.4, you can see that the shortest person is about 56 inches tall, the tallest about 79 inches. you can also see that there is a greater density in the middle of the range. Unfortunately, the rug plot also suppresses some important features of the data. There is one tick at each height value for which there is a measurement, but that one tick might correspond to many individuals. Indeed, in Galton's data, heights were rounded off to a quarter of an inch, so many different individuals have exactly the same recorded height. This is why there are regular gaps between the tick marks.

Another simple display, the **histogram**, avoids this overlap by using two

Figure 3.4: A rug plot of Galton's height data. Each case is one tick mark.

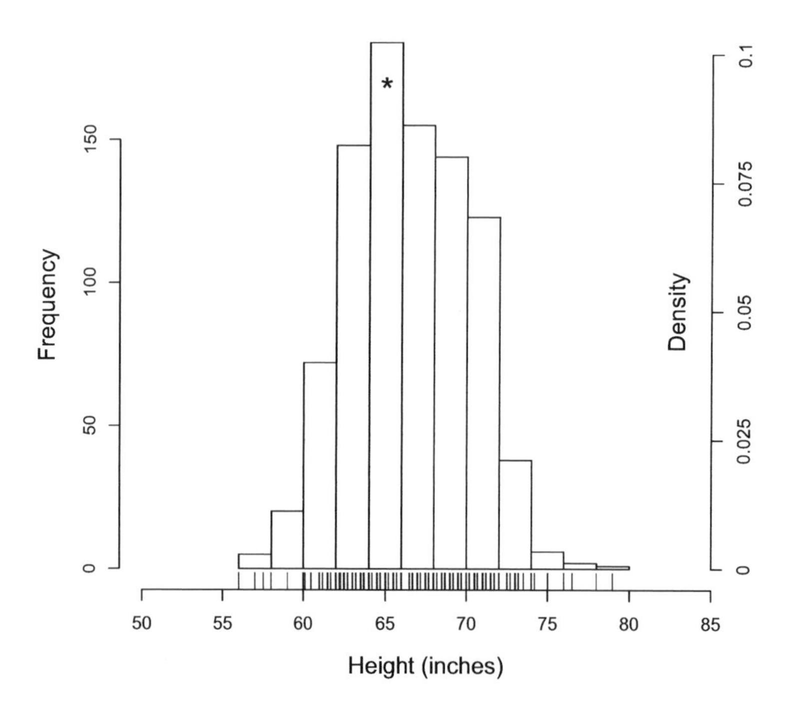

Figure 3.5: A histogram of Galton's height data. A rug plot is underneath. There are two axes: one ("Frequency") arranged so that the height of the bars reflects the number of cases that fall into the respective bins, the other ("Density" or "Relative Frequency") so that the area of the bars reflects the proportion of cases that fall into the bins.

different axes. As in the rug plot, one of the axes shows the variable. The other axis displays the number of individual cases within set ranges, or "bins," of the variable.

There are two widely used ways to mark the vertical axis in a histogram. The simplest is called **frequency** (or **absolute frequency**) and is arranged so that each bar's height gives the direct count of individuals that fall into the corresponding bin. For example, consider the histogram shown in Figure 3.5 and, in particular, the bar marked with a * that covers the bin from 64 to 66 inches. The bar's height on the frequency axis is 184, meaning that there are 184 individuals between 64 and 66 inches. Adding up the heights of all the bars gives the total number of cases being plotted in the histogram: 898 cases in this example.

The other format for the vertical axis of a histogram is called **density** (or **relative frequency**). In the density format, each bar's *area* shows the proportion of cases that fall into the bin. The density format makes it easy to compare histograms from different size samples.

It takes a bit of thought to understand the units of measuring density. In general, the word "density" refers to an amount per unit length, or area, or volume. For example, the mass-volume density of water is the mass of water per unit of volume, typically given as kilograms per liter. The sort of density used for histograms is the fraction-of-cases per unit of the binning variable.

To see how this plays out in Figure 3.5, consider again the bin for heights between 64 and 66 inches, marked in the figure with a *. As shown by the frequency axis, there are 184 individual cases that fall into that bin. Since there are 898 cases altogether, the fraction of cases in the bin is $\frac{184}{898} = 0.205$. These cases are spread out over a range of 2 inches (64 to 66) in the variable. So, the density will be 0.205 per 2 inches, or 0.1025 per inch.

The units of a fraction density will always be the reciprocal of the units of the variable that is being studied. Since Galton's height data was measured in inches, the units of density are inches^{-1}.

When calibrating a histogram in terms of density, it is the area of the bar, not the bar height, that gives the fraction of cases spanned by the bar. Adding together the areas of all the bar will always give an area of 1, since all of the cases fall into one bar or another.

3.3.1 Shapes of Distributions

The histogram of Galton's height data displays a bell-shaped pattern that is typical of many variables: the individuals are distributed with most near the center and fewer and fewer near the edges. The pattern is so common that a mathematical idealization of it is called the **normal distribution**.

The boundary points between bins in a histogram are somewhat arbitrary. It is not unheard of for writers to modify the divisions in order to make one bar higher; the careful reader should be alert to such shenanigans. In addition, the many divisions between bars make the plot somewhat busy graphically.

A modern alternative to the histogram is the density plot, an example of which is shown in Figure 3.6. This is very much like the histogram, but dispenses with the discrete bins. The resulting curve follows the same overall shape as the histogram, but is smoother. This is less distracting to the eye. The vertical axis of the density plot is always calibrated in density format and so the area under the curve is always 1.

Still another useful graphical format is a **box plot**, which shows the minimum, first quartile, median, third quartile, and maximum. Figure 3.7 shows a box plot of the Galton height data. Note that the variable itself is being plotted

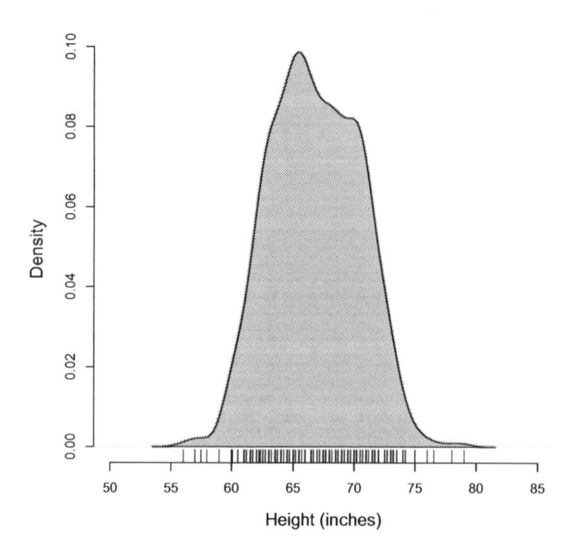

Figure 3.6: A density plot of Galton's height data.

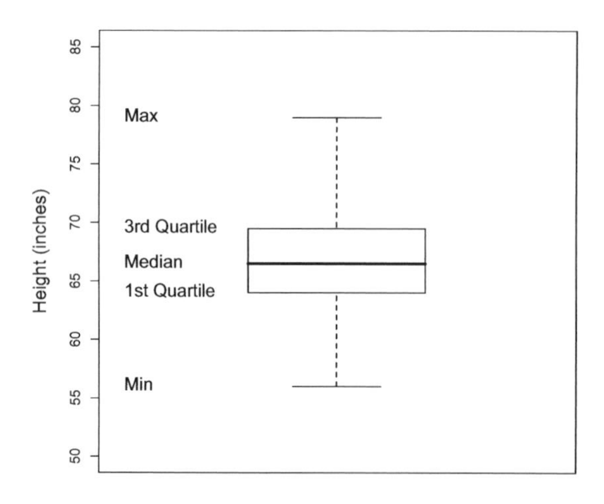

Figure 3.7: A box plot of Galton's height data.

on the vertical axis, contrasting with the other graphical depictions where the variable has been plotted on the horizontal axis.

The stylized display of the box plot is most useful when comparing the distribution of a variable across two or more different groups. For example, Figure 3.8 shows a box plot of the height data, breaking out separately the males and

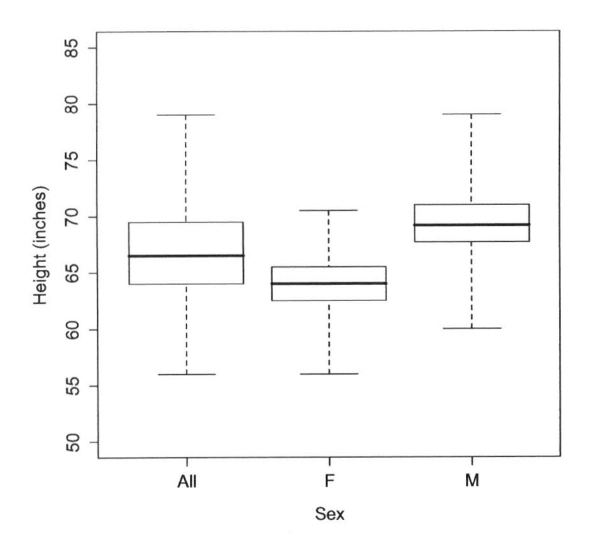

Figure 3.8: A box plot of height versus sex.

females. This plot is showing two variables; height is plotted as the response variable, with sex as an explanatory variable. The box plot shows that the variation among the males or among the females — as measured say by the inter-quartile interval or the min-to-max range — is less than the variation of the data when the person's sex is ignored. That is, some of the variation in the height variable is accounted for by the explanatory variable sex. Such partititioning of variation between the response and explanatory variables will be a major theme of modeling.

3.4 Normal and Non-normal Distributions

The symmetrical bell-shaped distribution seen, for example, in the Galton height data is so common that it is called "normal." It's important to remember that despite the name "normal," many variables display a very different pattern.

For example, consider a very mundane variable: the monthly natural gas utility bill for the author's house. The distribution (plotted in Figure 3.11) shows the main concentration of bills near about $20 per month. But this is by no means typical. For many months — winter months — the bill is much higher. The overall shape is called **right skew** because the tail on the right side of the peak is much longer than the tail on the left side.

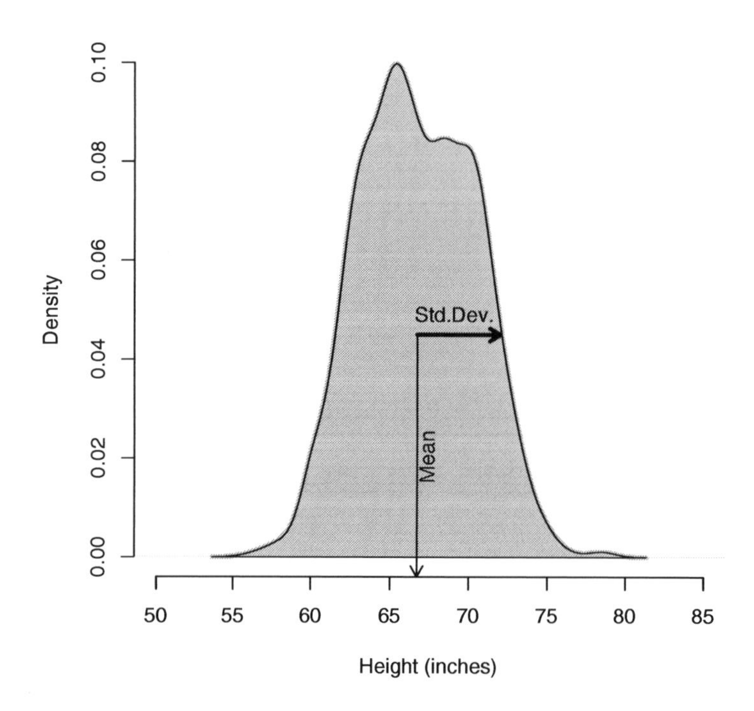

Figure 3.9: Eye-balling the mean and standard deviation from a density plot. The mean value is at the center of a bell-shaped distribution. The standard deviation is roughly the half-width at half-height.

There is nothing abnormal or strange about these data, even if the distribution does not have the so-called "normal" shape. In most months, a typical value for the gas bill is \$25 to \$50, but there are a few winter months where much, much more gas is consumed for heating.

Measurements such as the mean or standard deviation are most meaningful when the underlying data are normally distributed, or close to it. For strongly skewed data, or for data containing **outliers**, the median can offer a better indication than the mean of a typical value. For such data, rather than using the standard deviation to quantify spread, it may be preferable to use measures such as the **interquartile interval**, or **IQR** for short, which gives the length of the 50% coverage interval.

3.5 Categorical Variables

All of the measures of variation encountered so far are for quantitative variables; the kind of variable for which it's meaningful to average or sort.

Categorical variables are a different story. One thing to do is to display the

Aside. 3.2 Back of the Envelope: Center and Spread

One way to interpret the mean and standard deviation of a variable is in terms of the **center** and **spread** of the distribution. This is most straightforward when the distribution of the variable is symmetric and roughly bell-shaped. In this case, the value of the mean falls at the center of the distribution, as shown in Figure 3.9.

The standard deviation measures the spread of the distribution. Arithmetically, the standard deviation is set by the square root of the variance which is, in turn, the mean square of the deviations from the mean. Graphically, there is a simple and useful approximation for a bell-shaped distribution: the standard deviation is roughly the half-width at half-height of the distribution as shown in Figure 3.9.

Both the mean and the standard deviation have units: the same as the units of the variable itself. When estimating the numerical values of the mean and standard deviation from the graph, use the scale of the variable shown on the horizontal axis.

Such "eye-ball" estimates are rough. The half-width-at-half-height method is merely a rule of thumb. For the Galton height data plotted in Figure 3.9, the rule overestimates the standard deviation: the actual standard deviation is about 3.6, not 5 as the figure suggests. There is no point in eye-balling an estimate when you can do the calculation. But it often happens that instead of having access to the actual data, you have only a graphical summary.

There is also a graphical interpretation of coverage intervals in terms of density plots. The 95% interval supports 95% of the area under the distribution as shown in Figure 3.10. Similarly, the 50%-interval supports half the area under the distribution.

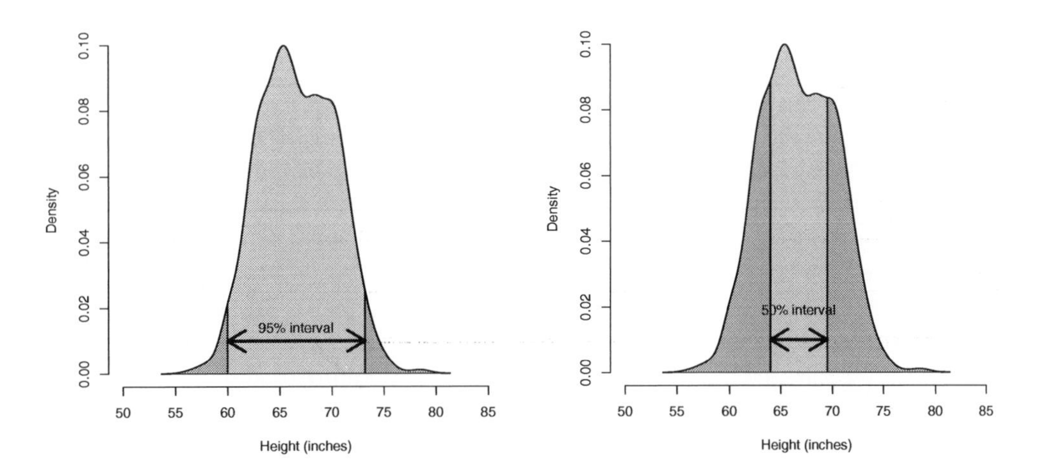

Figure 3.10: Coverage intervals shown on density plots. Left: The 95% coverage interval supports the central 95% of the area under the density curve; the dark-shaded area at the tails is only 5% of the total area. Right: The 50% coverage interval supports the central 50% of the area, with 25% of the area on each side.

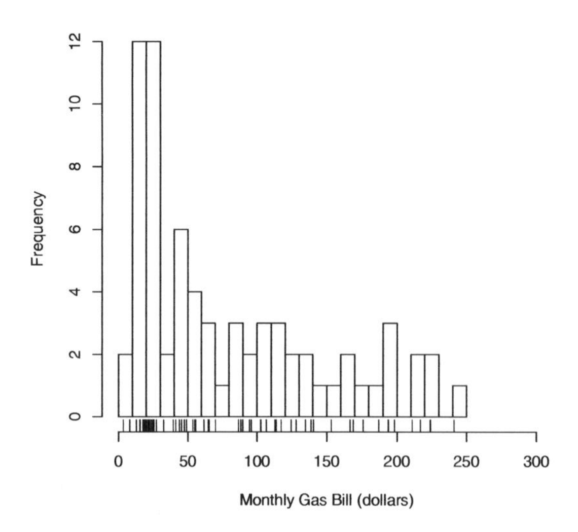

Figure 3.11: Histogram of the monthly gas bill from the utility data. The distribution is skewed to the right.

variation using tables and bar charts. For example, Galton's height data has a categorical variable sex with levels F and M. Either a table or a chart are effective ways to show how many cases there are for each level.

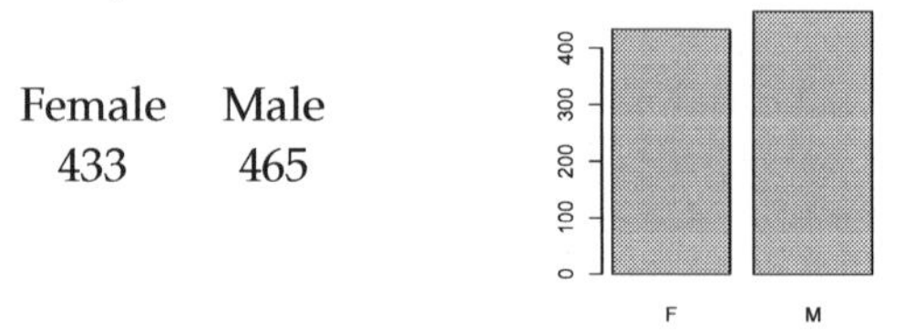

Female	Male
433	465

Describing categorical variation quantitatively is a more difficult matter. Recall that for quantitative variables, one can define a "typical" value, e.g., the mean or the median. It makes sense to measure residuals as how far an individual value is from the typical value, and to quantify variation as the average size of a residual. This led to the variance and the standard deviation as measures of variation.

For a categorical variable like sex, concepts of mean or median or distance or size don't apply. Neither F nor M is typical and in-between doesn't really make sense. Even if you decided that, say, M is typical — there are somewhat more of them in Galton's data — how can you say what the "distance" is between F and M?

Statistics textbooks hardly ever give a quantitative measure of variation in categorical variables. That's understandable for the reasons described above. But it's also a shame because many of the methods used in statistical modeling rely on quantifying variation in order to evaluate the quality of models.

There are quantitative notions of variation in categorical variables. You won't have use for them directly, but they are used in the software that fits models of categorical variables.

Solely for the purposes of illustration, here is one measure of variation in categorical variables with two levels, say F and M as in `sex` in Galton's data. Two-level variables are important in lots of settings, for example diagnostic models of whether or not a patient has cancer or models that predict whether a college applicant will be accepted.

Imagine that altogether you have N cases, with k being level F and $N - k$ being level M. (In Galton's data, N is 898, while $k = 433$ and $N - k = 465$.)

A simple measure of variation in two-level categorical variables is called, somewhat awkwardly, the **unalikeability**. This is

$$\text{unalikeability} = 2\frac{k}{N}\frac{N-k}{N}.$$

Or, more simply, if you write the proportion of level F as p_F, and therefore the proportion of level M as $1 - p_F$,

$$\text{unalikeability} = 2p_F(1 - p_F).$$

For instance, in the variable `sex`, $p_F = \frac{433}{898} = 0.482$. The unalikeability is therefore $2 \times 0.482 \times .518 = 0.4993$.

Some things to notice about unalikeability: If all of the cases are the same (e.g., all are F), then the unalikeability is zero — no variation at all. The highest possible level of unalikeability occurs when there are equal numbers in each level; this gives an unalikeability of 0.5.

Where does the unalikeability come from? One way to think about unalikeability is as a kind of numerical trick. Pretend that the level F is 1 and the level M is 0 — turn the categorical variable into a quantitative variable. With this transformation, `sex` looks like `0 1 1 1 0 0 1 1 0 1 0` and so on. Now you can quantify variation in the ordinary way, using the variance. It turns out that the unalikeability is exactly twice the variance.

Here's another way to think about unalikeability: although you can't really say *how far* F is from M, you can certainly see the difference. Pick two of the cases at random from your sample. The unalikeability is the probability that the two cases will be different: one an M and the other an F.

There is a much more profound interpretation of unalikeability that is introduced in Chapter 18, which covers modeling two-level categorical variables. For now, just keep in mind that you can measure variation for any kind of variable and use that measure to calculate how much of the variation is being captured by your models.

3.6 Computational Technique

To illustrate computer techniques for describing variability, consider the data that Galton collected on the heights of adult children and their parents. The file `galton.csv` stores these data in a modern, case/variable format.

```
> galton = ISMdata("galton.csv")
```

3.6.1 Simple Statistical Calculations

Simple numerical descriptions are easy to compute. Here are the mean, median, standard deviation and variance of the children's heights (in inches).

```
> mean( galton$height )
[1] 66.76069
> median( galton$height )
[1] 66.5
> sd( galton$height )
[1] 3.582918
> var( galton$height )
[1] 12.83730
```

Notice that the variance (`var`) is just the square of the standard deviation (`sd`). In principle, it's unnecessary to have both operators. Having both is merely a convenience.

A percentile tells where a given value falls in a distribution. For example, a height of 63 inches is on the short side in Galton's data:

```
> pdata( 63, galton$height )
[1] 0.1915367
```

Only about 19% of the cases have a height less than or equal to 63 inches. The `pdata` operator takes one or more values as a first argument and finds where they fall in the distribution of values in the second argument.

A quantile refers to the same sort of calculation, but inverted. Instead of giving a value in the same units as the distribution, you give a probability: a number between 0 and 1. The `qdata` operator then calculates the value whose percentile would be that value:

```
> qdata( .20, galton$height )
 20%
63.5
```

Remember that the probability is given as a number between 0 and 1, so use 0.50 to indicate that you want the value which falls at the 50th percentile.

- The 25th and 75th percentile in a single command — in other words, the 50 percent coverage interval:

```
> qdata(c(0.25, 0.75), galton$height )
 25%  75%
64.0 69.7
```

- The 2.5th and 97.5th percentile — in other words, the 95 percent coverage interval:

```
> qdata(c(0.025, 0.975), galton$height )
 2.5% 97.5%
   60    73
```

The interquartile range is the width of the 50 percent coverage interval:

```
> IQR(galton$height)
[1] 5.7
```

Some other useful operators are `min`, `max`, and `range`.

For convenience, the `summary` operator gives a quick description of a quantitative variable:

```
> summary(galton$height)
   Min. 1st Qu.  Median    Mean 3rd Qu.    Max.
  56.00   64.00   66.50   66.76   69.70   79.00
```

In exercises in later chapters, you will need to compute the sum of squares of quantitative variables and of residuals. This is done by connecting simple computations: squaring then summing.

```
> sum( galton$height^2 )
[1] 4013892
```

In all of these examples, the operator has been applied directly to a variable in a data frame. Of course, any of these operators can be applied to any set of data. For example, here is the standard deviation of the numbers $1, 2, 3, \cdots, 100$ created by `seq`:

```
> mean( seq(1,100))
[1] 50.5
```

3.6.2 Residuals and Sums of Squares

The residual from the mean can be computed like this:

```
> resids = galton$height - mean(galton$height)
```

Remember that each case has its own residual. There are 898 cases in the Galton data, so there are 898 residuals.

```
> resids
[1]     6.43931   2.43931   2.23931   2.23931
[5]     6.73931   5.73931  -1.26069  -1.26069
        ... and so on ...
[893]   1.93931   1.73931   0.93931  -2.76069
[897]  -3.26069  -3.76069
```

The sum of squares is

```
> sum(resids^2)
[1] 11515.06
```

3.6.3 Simple Statistical Graphics

There are several basic types of statistical graphics to display the distribution of a variable: histograms, density plots, and boxplots. These are easily mastered by example.

Technical note: These graphics are implemented by the "lattice" package in R. This is loaded automatically with the `ISM.Rdata` workspace, so you don't have to worry about it. But if you don't use the `ISM.Rdata` workspace, you will need to load lattice manually with the command `library(lattice)`.

Histograms

Constructing a histogram involves dividing the range of a variable up into bins and counting how many cases fall into each bin. This is done in an almost entirely automatic way:

```
> histogram( galton$height )
```

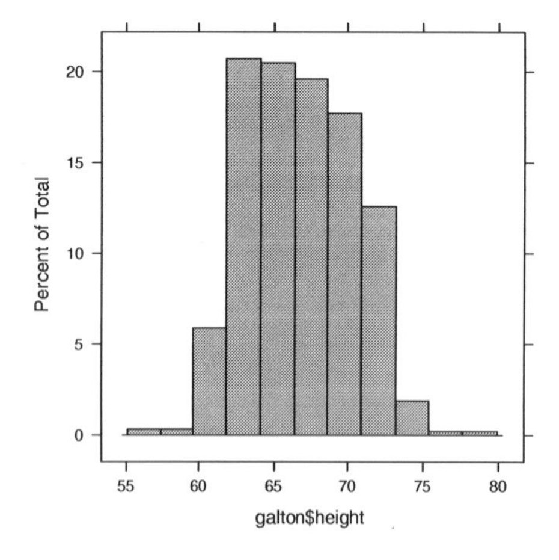

When constructing a histogram, R makes an automatic but sensible choice of the number of bins. If you like, you can control this yourself. For instance:

```
> histogram( galton$height, breaks=25 )
```

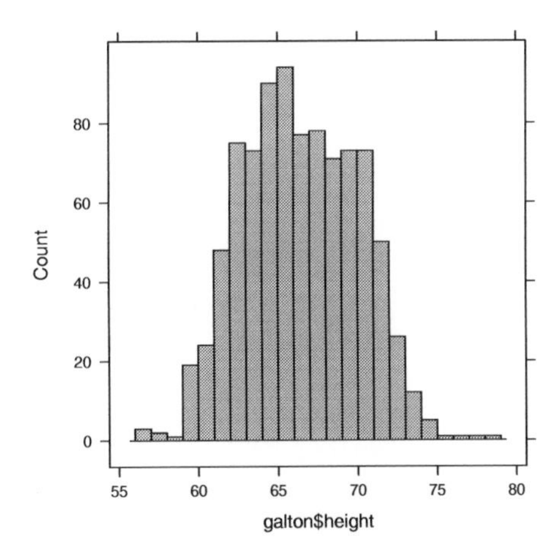

The horizontal axis of the histogram is always in the units of the variable. For the histograms above, the horizontal axis is in "inches" because that is the unit of the `galton$height` variable.

The vertical axis is conventionally drawn in one of three ways: controlled by an optional argument named `type`.

Absolute Frequency or Counts A simple count of the number of cases that falls into each bin. This mode is set with `type="count"` as in

```
> histogram( galton$height, type="count")
```

Relative Frequency The vertical axis is scaled so that the height of the bar give the proportion of cases that fall into the bin. This is the default.

Density The vertical axis *area* of the bar gives the relative proportion of cases that fall into the bin. Set `type="density"` as in `histogram(galton$height, type="density")`.

In a density plot, areas can be interpreted as probabilities and the area under the entire histogram is equal to 1.

Other useful optional arguments set the labels for the axes and the graph as a whole and color the bars. For example,

```
> histogram(galton$height, type="density",
    xlab="Height (inches)",
    main="Distribution of Heights",
    col="blue")
```

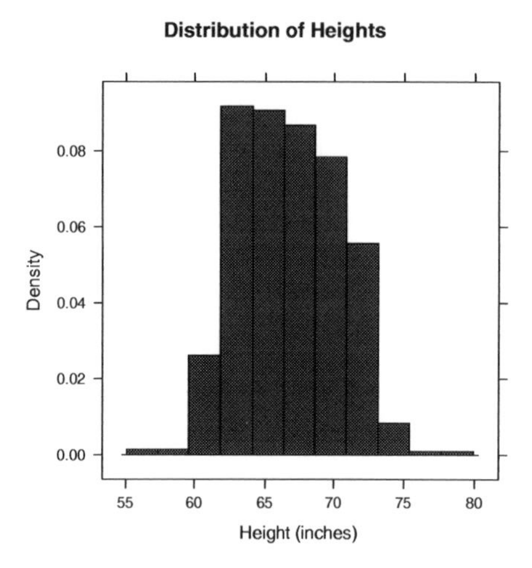

The above command is so long that it has been broken into several lines for display purposes. R ignores the line breaks, holding off on executing the command until it sees the final closing parentheses. Notice the use of quotation marks to delimit the labels and names like `"blue"`.

Density Plots

A **density plot** avoids the need to create bins and plots out the distribution as a continuous curve. Making a density plot involves two operators. The **density** operator performs the basic computation which is then displayed using either the `plot` or the `lines` operator. For example:

```
> densityplot( galton$height )
```

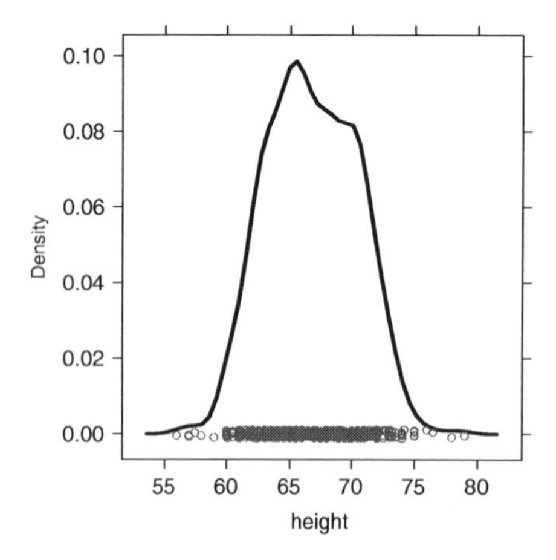

If you want to suppress the rug-like plotting of points at the bottom of the graph, use `densityplot(galton$height,plot.points=FALSE)`.

Box-and-Whisker Plots

Box-and-whisker plots are made with the `bwplot` command:

```
> bwplot( galton$height )
```

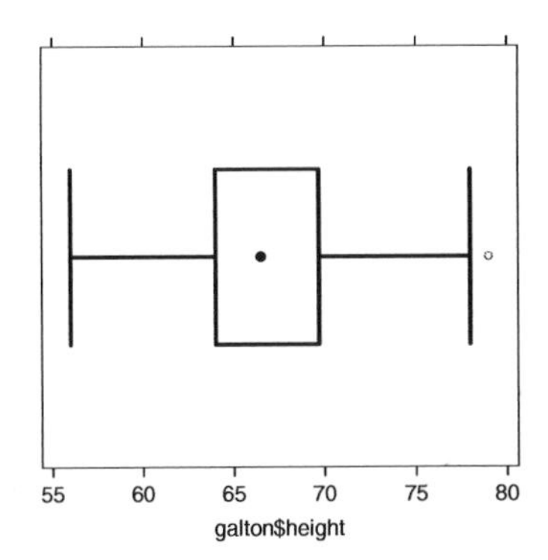

The median is represented by the heavy dot in the middle. Outliers, if any, are marked by dots outside the whiskers.

The real power of the box-and-whisker plot is for comparing distributions. This will be raised again more systematically in later chapters, but just to illustrate, here is how to compare the heights of males and females:

```
> bwplot( height ~ sex, data=galton )
```

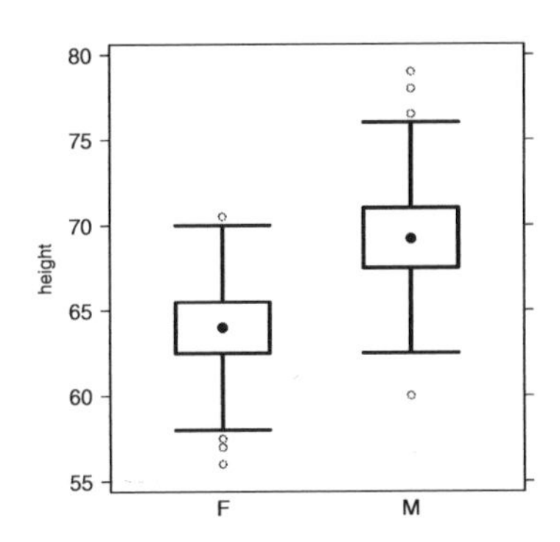

3.6.4 Displays of Categorical Variables

For categorical variables, it makes no sense to compute descriptive statistics such as the mean, standard deviation, or variance. Instead, look at the number of cases at each level of the variable.

```
> table( galton$sex )
  F   M
433 465
```

By processing such a table with the `prop.table` operator, you can calculate the proportion of cases at each level.

```
> prop.table( table( galton$sex ))
     F      M
0.4822 0.5178
```

The barchart operator will produce graphics from tables.

```
> barchart( table(galton$sex) )
```

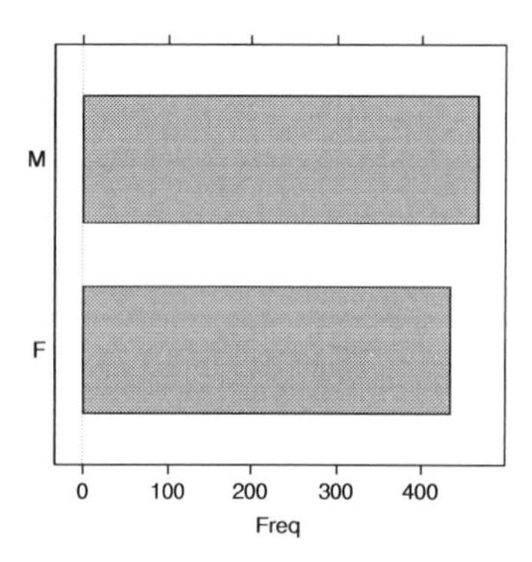

3.6.5 Outliers

The outlier operator uses the same rule of thumb as in bwplot to identify which cases are outliers. It returns a logical variable, TRUE or FALSE for each case.

```
> outlier(galton$height)
  [1] FALSE FALSE FALSE FALSE FALSE
  [6] FALSE FALSE FALSE FALSE FALSE
 ... and so on.
```

The direct output of outlier is rarely what you want. Typically you will want to use outlier to help in counting how many outliers there are or to look at the outlier cases themselves, or to extract cases that are not outliers:

```
> table( outlier( galton$height ) )
FALSE  TRUE
  897     1
> subset( galton, outlier(galton$height) )
    family father mother sex height nkids
289     72     70     65   M     79     7
> cleaned = subset( galton, !outlier(galton$height) )
```

There was just one outlier (according to the rule of thumb). The object named cleaned contains those cases that were not outliers with respect to height. Other cases might be outliers with respect to mother or father — the outlier program looks at only one variable. (The ! is the logical operator meaning "not,", so !outlier(galton$height) refers to the cases that are not outliers with respect to height.) That it's easy to remove outliers does not mean that you should do so without careful thought.

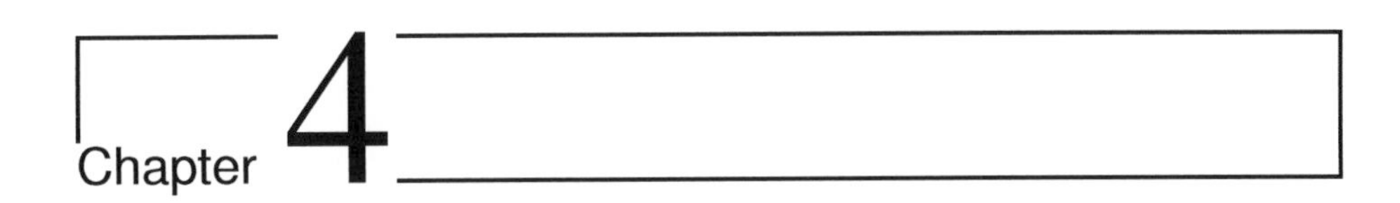

The Language of Models

I do not believe in things. I believe only in their relationships. — Georges Braque (1882-1963, cubist painter)

Mathematicians do not study objects, but relations among objects. — Henri Poincaré (1854-1912, mathematician)

One October I received a bill from the utility company: $52 for natural gas for the month. According to the bill, the average outdoor temperature during October was 49°F (9.5°C).

I had particular interest in this bill because two months earlier, in August, I replaced the old furnace in our house with a new, expensive, high-efficiency furnace that's supposed to be saving gas and money. This bill is the first one of the heating season and I want to know whether the new furnace is working.

To be able to answer such questions, I keep track of my monthly utility bills. The gasbill variableshows a lot of variation: from $3.42 to $240.90 per month. The 50% coverage interval is $21.40 to $117.90. The mean bill is $77.85. Judging from this, the $52 October bill is low. It looks like the new furnace is working!

Perhaps that conclusion is premature. My new furnace is just one factor that can influence the gas bill. Some others: the weather, the price of natural gas (which fluctuates strongly from season to season and year to year), the thermostat setting, whether there were windows or doors left open, how much gas we use for cooking, etc. Of all the factors, I think the weather and the price of natural gas are the most important. I'd like to take these into account when judging whether the $52 bill is high or low.

Rather than looking at the bill in dollars, it makes sense to look at the quantity of gas used. After all, the new furnace can only reduce the amount of gas used, it doesn't influence the price of gas. According to the utility bill, the usage in October was 65 ccf. ("ccf" means cubic feet, one way to measure the quantity of gas.) The variable ccf gives the historical values from past bills: the range is 0

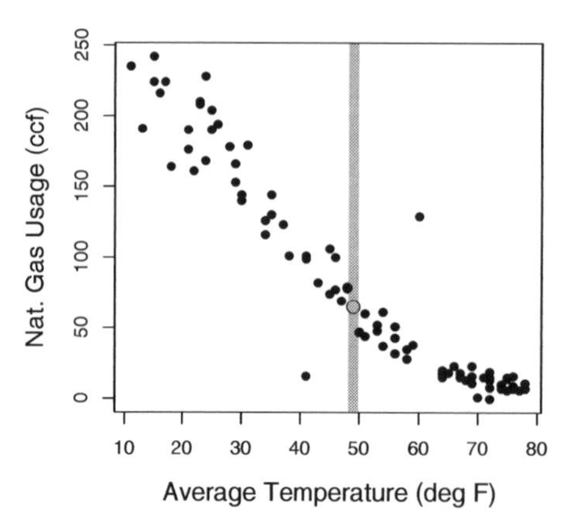

Figure 4.1: Monthly natural gas usage versus average outdoor temperature during the month.

to 242 ccf, so 65 ccf seems perfectly reasonable, but it's higher than the median monthly usage, which is 51 ccf. Still, this doesn't take into account the weather.

Now it is time to build a model: a representation of the utility data for the purpose of telling whether 65 ccf is low or high given the temperature. The variable temperature contains the average temperature during the billing period.

A simple graph of ccf versus temperature will suffice. (See Figure 4.1.) The open point shows the October bill (49 deg. and 65 ccf). The gray line indicates which points to look at for comparison, those from months near 49 degrees.

The graph suggests that 65 ccf is more or less what to expect for a month where the average temperature is 49 degrees. The new furnace doesn't seem to make much of a difference.

4.1 Models as Functions

Figure 4.1 gives a pretty good idea of the relationship between gas usage and temperature.

The concept of a **function** is very important when thinking about relationships. A function is a mathematical concept: the relationship between an **output** and one or more **inputs**. One way to talk about a function is that you plug in the inputs and receive back the output. For example, the formula $y = 3x + 7$ can be read as a function with input x and output y. Plug in a value of x and receive the output y. So, when x is 5, the output y is 22.

One way to represent a function is with a formula, but there are other ways as well, for example graphs and tables. Figure 4.2 shows a function representing the relationship between gas usage and temperature. The function is much simpler than the data. In the data, there is a scatter of usage levels at each tem-

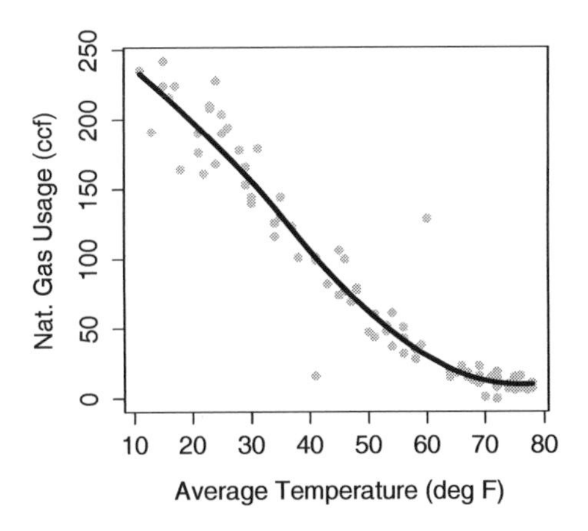

Figure 4.2: A model of natural gas usage versus outdoor temperature.

perature. But in the function there is only one output value for each input value.

Some vocabulary will help to describe how to represent relationships with functions.

- The **response variable** is the variable whose behavior or variation you are trying to understand. On a graph, the response variable is conventionally plotted on the vertical axis.

- The **explanatory variables** are the other variables that you want to use to explain the variation in the response. Figure 4.1, shows just one explanatory variable, temperature. It's plotted on the horizontal axis.

- **Conditioning on explanatory variables** means taking the value of the explanatory variables into account when looking at the response variables. When in Figure 4.1 you looked at the gas usage for those months with a temperature near 49°, you were conditioning gas usage on temperature.

- The **model value** is the output of a function. The function — called the **model function** — has been arranged to take the explanatory variables as inputs and return as output a typical value of the response variable. That is, the model function gives the typical value of the response variable *conditioning on* the explanatory variables. The function shown in Figure 4.2 is a model function. It gives the typical value of gas usage conditioned on the temperature. For instance, at 49°, the typical usage is 65 ccf. At 20°, the typical usage is much higher, about 200 ccf.

- The **residuals** show how far each case is for its model value. For example, one of the cases plotted in Figure 4.2 is a month where the temperature was 13° and the gas usage was 191 ccf. When the input is 13°, the model function gives an output of 228 ccf. So, for that case, the residual is $191 - 228 = -37$ ccf. Residuals are always "actual value minus model value."

Graphically, the residual for each case tells how far above the model function that case is. A negative residual means that the case is below the model function.

The idea of a function is fundamental to understanding statistical models. Whether the function is represented by a formula or a graph, the function takes one or more inputs and produces an output. In the statistical models in this book, that output is the model value, a "typical" or "ideal" value of the response variable at given levels of the inputs. The inputs are the values explanatory variables.

The model function describes how the typical value of the response variable depends on the explanatory variables. The output of the model function varies along with the explanatory variables. For instance, when temperature is low, the model value of gas usage is high. When temperature is high, the model value of gas usage is low. The idea of "depends on" is very important. In some fields such as economics, the term **dependent variable** is used instead of "response variable." Other phrases are used for this notion of "depends on," so you may hear statements such as these: "the value of the response *given* the explanatory variables," or "the value of the response *conditioned on* the explanatory variables."

The model function describes a relationship. If you plug in values for the explanatory variables for a given case, you get the model value for that case. The model value is usually different from one case to another, at least so long as the values of the explanatory variables are different. When two cases have exactly the same values of the explanatory values, they will have exactly the same model value even though the actual response value might be different for the two cases.

The residuals tell how each case differs from its model value. Both the model values and the residuals are important. The model values tell what's typical or average. The residuals tell how far from typical an individual case is likely to be. This might remind you of the mean and standard deviation.

As already said, models partition the variation in the response variable. Some of the variability is explained by the model, the remainder is unexplained. The model values capture the "deterministic" or "explained" part of the variability of the response variable from case to case. The residuals represent the "random" or "unexplained" part of the variability of the response variable.

4.2 Model Functions with Multiple Explanatory Variables

Historically, women tended to be paid less than men. To some extent, this reflected the division of jobs along sex lines and limited range of jobs that were open to women — secretarial, nursing, school teaching, etc. But often there was

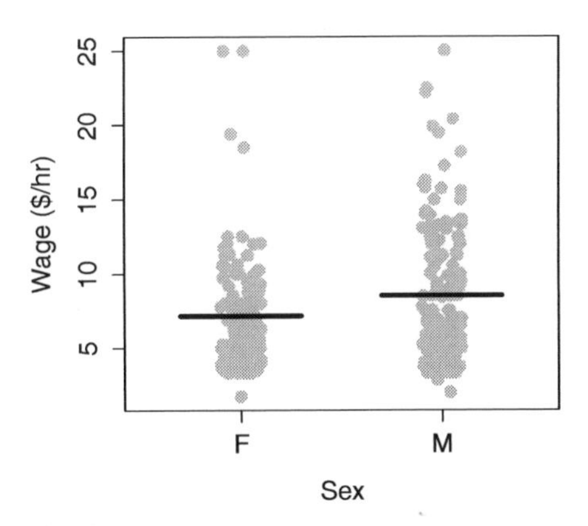

Figure 4.3: Hourly wages versus sex from the Current Population Survey data of 1985.

simple discrimination; an attitude that women's work wasn't as valuable or that women shouldn't be in the workplace. Over the past thirty or forty years, the situation is changing. Training and jobs that were once rarely available to women — police work, management, medicine, law, science — are now open to them.

Surveys consistently show that women tend to earn less than men, a "wage gap." To illustrate, consider data from one such survey, the Current Population Survey (CPS) from 1985. In the survey data, each case is one person. The variables are the person's hourly wages at the time of the survey, age, sex, marital status, the sector of the economy in which they work, etc.

One aspect of these data is displayed by plotting wage versus sex, as in Figure 4.3. The model plotted along with the data show that typical wages for men are higher than for women.

The situation is somewhat complex since the workforce reflected in the 1985 data is a mixture of people who were raised in the older system and those who were emerging in a more modern system. A woman's situation can depend strongly on when she was born. This is reflected in the data by the age variable.

There are other factors as well. The roles and burdens of women in family life remain much more traditional than their roles in the economy. Perhaps marital status ought to be taken into account. In fact, there are all sorts of variables that you might want to include — job type, race, location, etc.

A statistical model can include multiple explanatory variables, all at once. To illustrate, consider explaining wage using the worker's age, sex, and marital status.

In a typical graph of data, the vertical axis stands for the response variable and the horizontal axis for the explanatory variable. But what do you do when there is more than one explanatory variable? One approach, when some of the

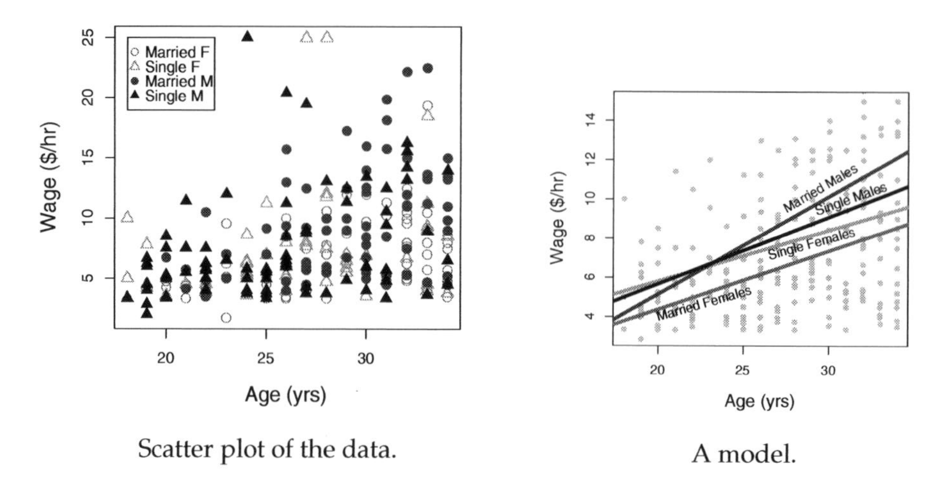

Scatter plot of the data. A model.

Figure 4.4: Hourly wages modeled by age, sex, and marital status.

explanatory variables are categorical, is to use differing symbols or colors to represent the differing levels of the categories. The left panel in Figure 4.4 shows wages versus age, sex, and marital status plotted out this way.

The first thing that might be evident from the scatter plot is that not very much is obvious from the data on their own. There does seem to be a slight increase in wages with age, but the cases are scattered all over the place.

The model, shown in the right panel of the figure, simplifies things. The relationships shown in the model are much clearer. You can see that wages tend to increase with age, and they do so differently for men and for women and differently for married people and single people.

Models can sometimes reveal patterns that are not evident in a graph of the data. This is a great advantage of modeling over simple visual inspection of data. There is a real risk, however, that a model is imposing structure that is not really there on the scatter of data, just as people imagine animal shapes in the stars. A skeptical approach is always warranted. Much of the second half of the book is about ways to judge whether the structure suggested by a model is justified by the data.

There is a tendency for those who first encounter models to fix attention on the clarity of the model and ignore the variation around the model. This is a mistake. Keep in mind the definition of statistics offered in the first chapter:

> *Statistics is the explanation of variation in the context of what remains unexplained.*

The scatter of the wage data around the model is very large; this is a very important part of the story. The scatter suggests that there might be other factors that account for large parts of person-to-person variability in wages, or perhaps just that randomness plays a big role.

Compare the broad scatter of wages to the rather tight way the gas usage in Figure 4.1 is modeled by average temperature. The wage model is explaining only a small part of the variation in wages, the gas-usage model explains a very large part of the variation in those data.

Adding more explanatory variables to a model can sometimes usefully reduce the size of the scatter around the model. If you included the worker's level of education, job classification, age, or years working in their present occupation, the unexplained scatter might be reduced. Even when there is a good explanation for the scatter, if the explanatory variables behind this explanation are not included in the model, the scatter due to them will appear as unexplained variation. (In the gas-usage data, it's likely that wind velocity, electricity usage that supplements gas-generated heat, and the amount of gas used for cooking and hot water would explain a lot of the scatter. But these potential explanatory variables were not measured.)

4.3 Reading a Model

There are two distinct ways that you can read a model.

Read out the model value. Plug in specific values for the explanatory variables and read out the resulting model value. For the model in Figure 4.1, an input temperature of 35° produces an output gas usage of 125 ccf. For the model in Figure 4.4 a single, 30-year old female has a model value of $8.10 per hour. (Remember, the model is based on data from 1985!)

Characterize the relationship described by the model. In contrast to reading out a model value for some specific values of the explanatory variables, here interest is in the overall relationship: how gas usage depends on temperature; how wages depend on sex or marital status or age.

Reading out the model value is useful when you want to make a prediction (What would the gas usage be if the temperature were 10°?) or when you want to compare the actual value of the response variable to what the model says is a typical value. (Is the gas usage in the 49° month lower than expected, perhaps due to my new furnace?).

Characterizing the relationship is useful when you want to make statements about broad patterns that go beyond individual cases. Is there really a connection between marital status and wage? Which way does it go?

The "shape" of the model function tells you about such broad relationships. Reading the shape from a graph of the model is not difficult.

For a quantitative explanatory variable, e.g., temperature or age, the model form is a continuous curve or line. An extremely important aspect of this curve is its slope. For the model of gas usage in Figure 4.2, the slope is down to the right: a negative slope. This means that as temperature increases, the gas usage goes down. In contrast, for the model of wages in Figure 4.4, the slope is up to the right: a positive slope. This means that as age increases, the wage goes up.

The slope is measured in the usual way: rise over run. The numerical size of the slope is a measure of the strength of the relationship, the sign tells which way the relationship goes. Some examples: For the gas usage model in Figure 4.2 in winter-like temperatures, the slope is about -4 ccf/degree. This means that gas usage can be expected to go down by 4 ccf for every degree of temperature increase. For the model of wages in Figure 4.4, the slope for single females is about 0.20 dollars-per-hour/year: for every year older a single female is, wages typically go up by 20 cents-per-hour.

Slopes have units. These are always the units of the response variable divided by the units of the explanatory variable. In the wage model, the response has units of dollars-per-hour while the explanatory variable `age` has units of years. Thus the units of the slope are dollars-per-hour/year.

For categorical variables, slopes don't apply. Instead, the pattern can be described in terms of *differences*. In the model where wage is explained only by sex, the difference between typical wages for males and females is 2.12 dollars per hour.

When there is more than one explanatory variable, there will be a distinct slope or difference associated with each.

When describing models, the words used can carry implications that go beyond what is justified by the model itself. Saying "the difference between typical wages" is pretty neutral: a description of the pattern. But consider this statement: "Typical wages go up by 20 cents per hour for every year of age." There's an implication here of **causation**, that as a person ages his or her wage will go up. That might in fact be true, but the data on which the model is based were not collected in a way to support that claim. Those data don't trace people as they age; they are not longitudinal data. Instead, the data are a snapshot of different people at different ages: cross-sectional data. It's dangerous to draw conclusions about changes over time from cross-sectional data of the sort in the CPS data set. Perhaps people's wages stay the same over time but that the people who were hired a long time ago tended to start at higher wages than the people who have just been hired.

Consider this statement: "A man's wage rises when he gets married." The model in Figure 4.4 is consistent with this statement; it shows that a married man's typical wage is higher than an unmarried man's typical wage. But does marriage cause a higher wage? It's possible that this is true, but that conclusion isn't justified from the data. There are other possible explanations for the link between marital status and wage. Perhaps the men who earn higher wages are more attractive candidates for marriage. It might not be that marriage causes higher wages but that higher wages cause marriage.

To draw conclusions about causation, it's important to collect data in an appropriate way. For instance, if you are interested in the effect of marriage on wages, you might want to collect data from individuals both before and after marriage and compare their change in wages to that over the same time period in individuals who don't marry. The strongest statements about causation require something more: that the condition be imposed experimentally, picking the people who are to get married at random. Such an experiment is hardly possible when it comes to marriage.

4.4 Choices in Model Design

The suitability of a model for its intended purpose depends on choices that the modeler makes. There are three fundamental choices:

1. The data.
2. The response variable.
3. The explanatory variables.

4.4.1 The Data

How were the data collected? Are they a random sample from a relevant sampling frame? Are they part of an experiment in which one or more variables were intentionally manipulated by the experimenter, or are they observational data? Are the relevant variables being measured? (This includes those that may not be directly of interest but which have a strong influence on the response.) Are the variables being measured in a meaningful way?

Unfortunately, work on building models often starts only after the data have been collected. Consulting statisticians often have a researcher approach them with a data set and ask for help. Regrettably, the answer is often, "It's too late. You needed to talk to me **before** you collected your data." Start thinking about your models while you are still planning your data collection.

When you are confronted with a situation where your data are not suitable, you need to be honest and realistic about the limitations of the conclusions you can draw. The issues involved will be discussed starting in Chapter 8.

4.4.2 The Response Variable

The appropriate choice of a response variable for a model is often obvious. The response variable should be the thing that you want to predict, or the thing whose variability you want to understand. Often, it is something that you think is the effect produced by some other cause.

For example, in examining the relationship between gas usage and outdoor temperature, it seems clear that gas usage should be the response: temperature is a major determinant of gas usage. But suppose that the modeler wanted to be able to measure outdoor temperature from the amount of gas used. Then it would make sense to take temperature as the response variable.

Similarly, wages make sense as a response variable when you are interested in how wages vary from person to person depending on traits such as age, experience, and so on. But suppose that a sociologist was interested in assessing the influence of income on personal choices such as marriage. Then the marital status might be a suitable response variable, and wage would be an explanatory variable.

Most of the modeling techniques in this book require that the response variable be quantitative. The main reason is that there are straightforward ways to measure variation in a quantitative variable and measuring variation is key to assessing the reliability of models. There are, however, methods for building models with a categorical response variable. (One of them, logistic regression, is the subject of Chapter 18.)

4.4.3 The Explanatory Variables

Choices of explanatory variables are richer and much more nuanced. Much of this book concerns ways to tell if an explanatory variable ought to be included in a model.

That said, some of the things that shape the choice of explanatory variables are obvious. Do you want to study sex-related differences in wage? Then sex had better be an explanatory variable. Is temperature a major determinant of the usage of natural gas? Then it makes sense to include it as an explanatory variable.

You will see situations where including an explanatory variable hurts the model, so it is important to be careful. (This will be discussed in Chapter 14.) A much more common mistake is to leave out explanatory variables. Unfortunately, few people learn the techniques for handling multiple explanatory variables and so your task will often need to go beyond modeling to include explaining how this is done.

When designing a model, you should think hard about what are potential explanatory variables and be prepared to include them in a model along with the variables that are of direct interest.

4.5 Model Terms

Once the modeler has selected explanatory variables, a choice must be made about **model terms**.

Notice that the models in the preceding examples have graphs of different shapes. The gas-usage model is a gentle curve, the wage-vs-sex model is just two values, and the more elaborate wage model is four lines with different shapes.

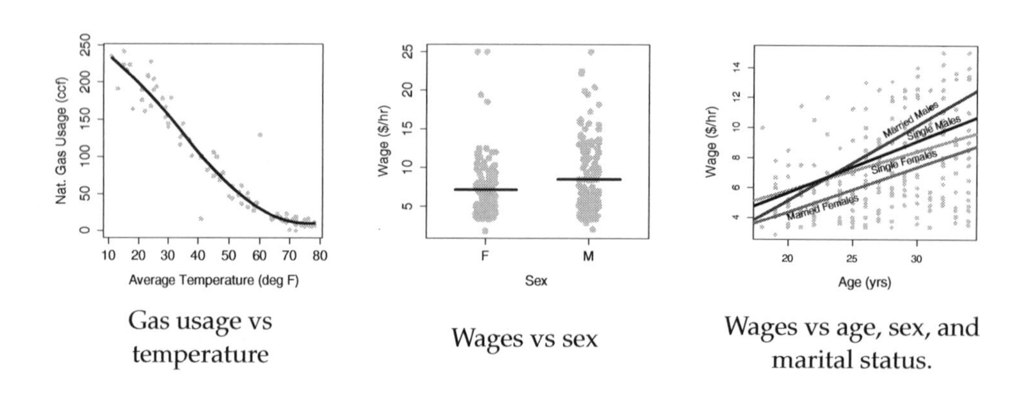

Gas usage vs temperature

Wages vs sex

Wages vs age, sex, and marital status.

The modeler determines the shape of the model through his or her choice of **model terms**. The basic idea of a model term is that explanatory variables can be included in a model in more than one way. Each kind of term describes a different way to include a variable in the model.

You need to learn to describe models using model terms for several reasons. First, you will communicate in this language with the computers that you will use to perform the calculations for models. Second, when there is more than one or two explanatory variables, it's hard to visualize the model function with a graph: knowing the language of model terms will help you "see" the shape of the function even when you can't graph it. Third, model terms are the way to talk about "parts" of models. In evaluating a model, statistical methods can be used to take the model apart and describe the contribution of each part. This analysis — the word "analysis" literally means to loosen apart — helps the modeler to decide which parts are worth keeping.

There are just a few basic kinds of models terms. They are:

1. the intercept term
2. main terms
3. interaction terms
4. transformation terms

Models almost always include the intercept term and a main term for each of the explanatory variables. (There are rare exceptions.) Transformation and interaction terms can be added to create more expressive or flexible shapes.

To form an understanding of how different kinds of terms contribute to the overall "shape" of the model function, it's best to look at the different shape functions that result from including different model terms. The next section illustrates this. Several differently shaped models are constructed of the same data plotted in Figure 4.5. The data are the record time (in seconds) for the 100 meter freestyle race along with the year in which the record was set and the sex of the swimmer. The response variable will be time, the explanatory variables will be year and sex.

Figure 4.5 shows some obvious patterns, seen most clearly in the plot of time versus year. The record time has been going down over the years. This is natural, since setting a new record means beating the time of the previous record. There's also a clear difference between the men's and women's records; men's record times are faster than women's, although the difference has decreased markedly over the years.

The following models may or may not reflect these patterns, depending on which model terms are included.

4.5.1 The Intercept Term (and no other terms)

The **intercept term** is included in almost every statistical model. The intercept term is a bit strange because it isn't something you measure; it isn't a variable.

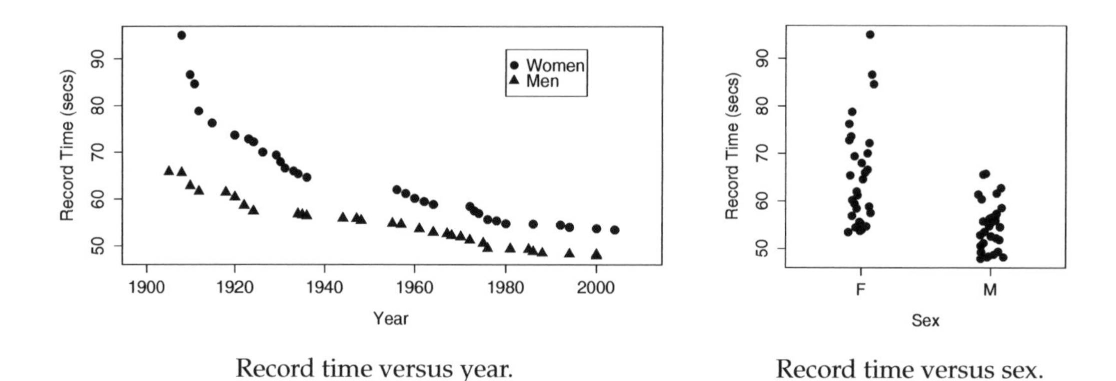

Record time versus year. Record time versus sex.

Figure 4.5: World record swimming times in the 100 meter freestyle.

(The term "intercept" will make sense when model formulas are introduced in the next chapter.)

The figure below shows the swimming data with a simple model consisting only of the intercept term.

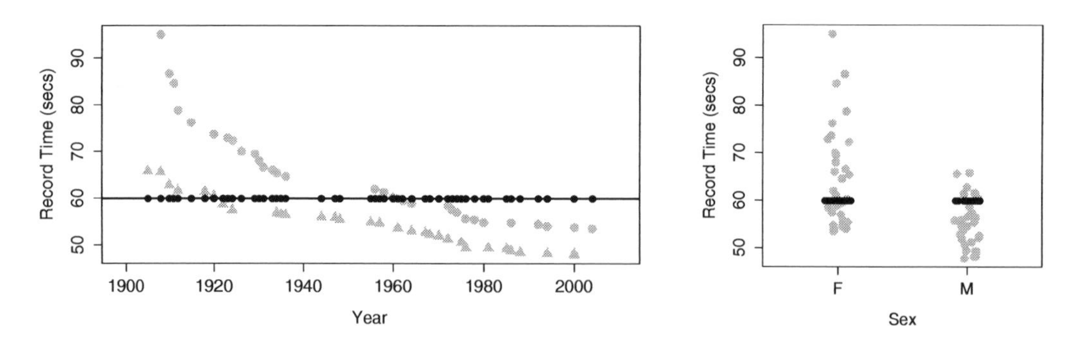

The model value for this model is exactly the same for every case. In order to create model variation from case to case, you would need to include at least one explanatory variables in the model.

4.5.2 Intercept and Main Terms

The most basic and common way to include an explanatory variable is as a main effect. Almost all models include the intercept term and a main term for each of the explanatory variables. The figures below show three different models each of this form:

The intercept and a main term from year. This produces model values that vary with year, but show no difference between the sexes. This is because sex has not been included in the model.

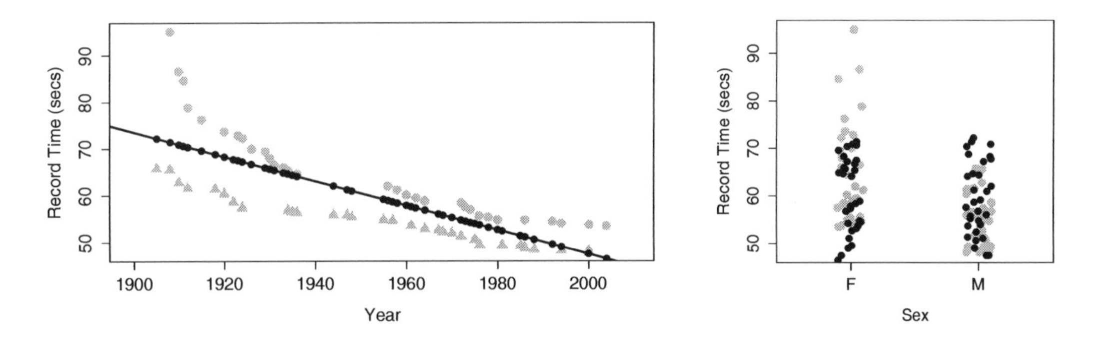

The model values have been plotted out as small black dots. The model pattern is evident in the left graph: swim time versus year. But in the right graph — swim time versus sex — it seems to be all scrambled. Don't be confused by this. The right-hand graph doesn't include year as a variable, so the dependence of the model values on year is not at all evident from that graph. Still, each of the model value dots in the left graph occurs in the right graph at exactly the same *vertical* coordinate.

The intercept and a main term from sex. This produces different model values for each level of sex.

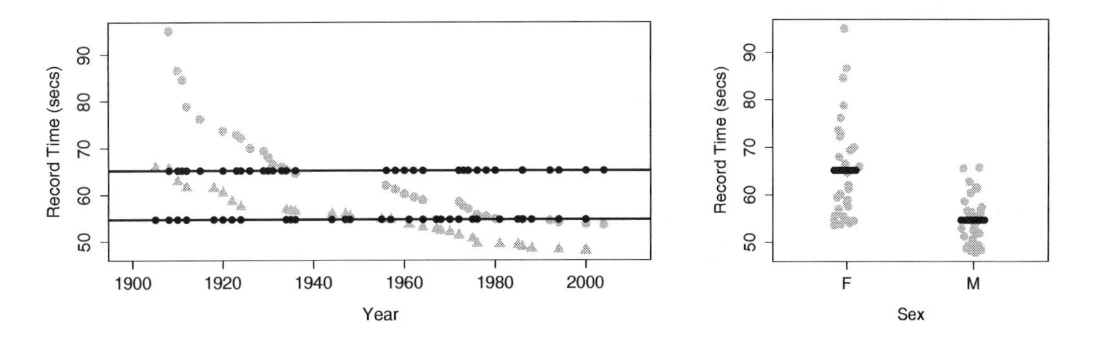

There is no model variation with year because year has not been included in the model.

The intercept and main terms from sex and from year. This model gives dependence on both sex and year.

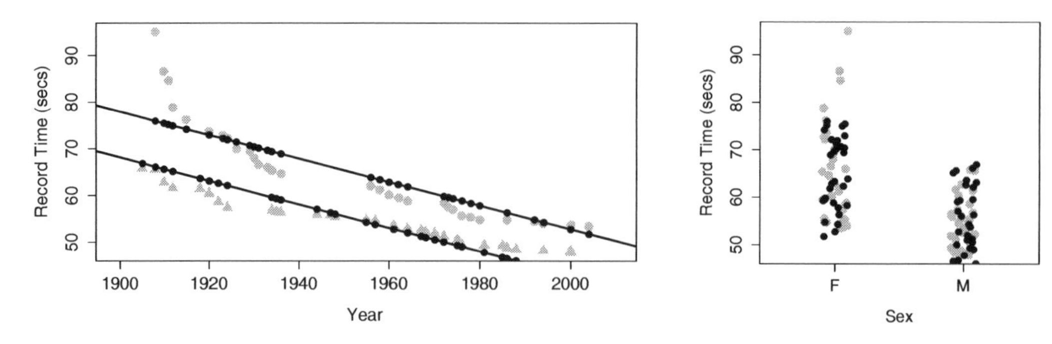

Note that the model values form two parallel lines in the graph of time versus year: one line for each sex.

4.5.3 Interaction Terms

Interaction terms combine two other terms, typically two main terms. An interaction term can describe how one explanatory variable modulates the role of another explanatory variable in modeling the relationship of both with the response variable.

In the graph, including the interaction term between sex and year produces a model with two non-parallel lines for time versus year. (The model also has the main terms for both sex and year and the intercept term, as always.)

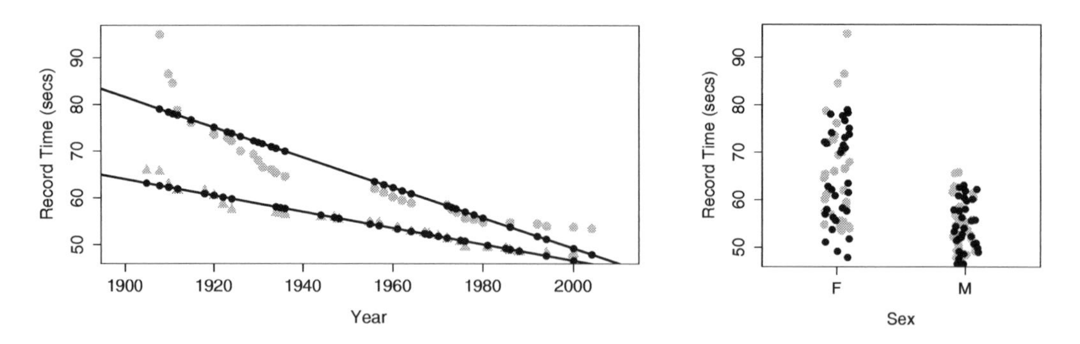

One way to think about the meaning of the interaction term in this model is that it describes how the effect of sex changes with year. Looking at the model, you can see how the difference between the sexes changes over the years; the difference is getting smaller. Without the interaction term, the model values would be two *parallel* lines; the difference between the lines wouldn't be able to change over the years.

Another, equivalent way to put things is that the interaction term describes how the effect of year changes with sex. The effect of year on the response is reflected by the slope of the model line. Looking at the model, you can see that the slope is different depending on sex: steeper for women than men.

For most people, it's surprising that one term — the interaction between sex and year — can describe both how the effect of year is modulated by sex, and how the effect of sex is modulated by year. But these are just two ways of looking at the same thing.

Aside. 4.1 Interaction terms and partial derivatives

The mathematically oriented reader may recall that one way to describe the effect of one variable on another is a partial derivative: the derivative of the response variable with respect to the explanatory variable. The interaction — how one explanatory variable modulates the effect of another on the response variable — corresponds to a mixed second-order partial derivative. Writing the response as z and the explanatory variables as x and y, the interaction corresponds to $\frac{\partial^2 z}{\partial x \partial y}$ which is exactly equal to $\frac{\partial^2 z}{\partial y \partial x}$. That is, the way that x modulates the effect of y on z is the same thing as the way that y modulates the effect of x on z.

A common misconception about interaction terms is that they describe how one explanatory variable affects another explanatory variable. Don't fall into this error. Model terms are always about how the response variable depends on the explanatory variables, not how explanatory variables depend on one another. An interaction term between two variables describes how two explanatory variables combine jointly to influence the response variable.

Once people learn about interaction terms, they are tempted to include them everywhere. After all, it's natural to think that world record swimming times would depend differently on year for women than for men. Of course wages might depend differently on age for men and women! Regretably, the uncritical use of interaction terms can lead to poor models. The problem is not the logic of interaction, the problem is in the data. As you will see in Chapter 11, interaction terms can introduce a problem called **multi-collinearity** which can reduce the reliability of models. Fortunately, it's not hard to detect multi-collinearity and to drop interaction terms if they are causing a problem. The model diagnostics that will be introduced in later chapters will make it possible to play safely with interaction terms.

4.5.4 Transformation Terms

A **transformation term** is a modification of another term using some mathematical transformation. Transformation terms only apply to quantitative variables. Some common transformations are x^2 or $\sqrt{x}$ or $\log x$, where the quantitative explantory variable is x.

A transformation term allows the model to have a dependence on x that is not a straight line. The graph shows a model that includes these terms: an intercept, main effects of sex and year, an interaction between sex and year, and a year-squared transformation term.

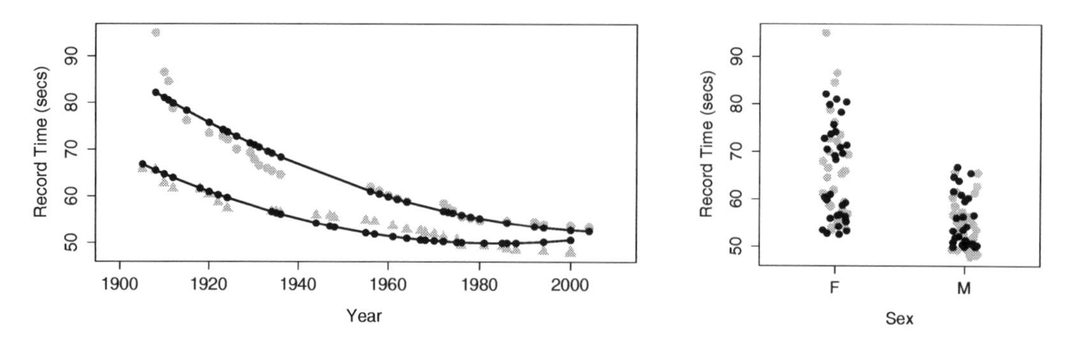

Adding in the year-squared term provides some curvature to the model function.

Aside. 4.2 Are swimmers slowing down?

Look carefully at the model with a year-squared transformation term. You may notice that, according to the model, world record times for men have been getting worse since about year 1990. This is, of course, nonsense. Records can't get worse. A new record is set only when an old record is beaten. The model doesn't know this common sense about records — the model terms allow the model to curve in a certain way and the model curves in exactly that way. What you probably want out of a model of world records is a slight curve that's constrained never to slope upward. There is no elementary way to do this. Indeed, it is an unresolved problem in statistics how best to include in a model addtional knowledge that you might have such as "world records can't get worse with time."

4.5.5 Main Effects without the Intercept.

It's possible to construct a model with main terms but no intercept terms. If the explanatory variables are all quantitative, this is almost always a mistake. The figure, which plots the model function for swim time modeled by age with no intercept term, shows why.

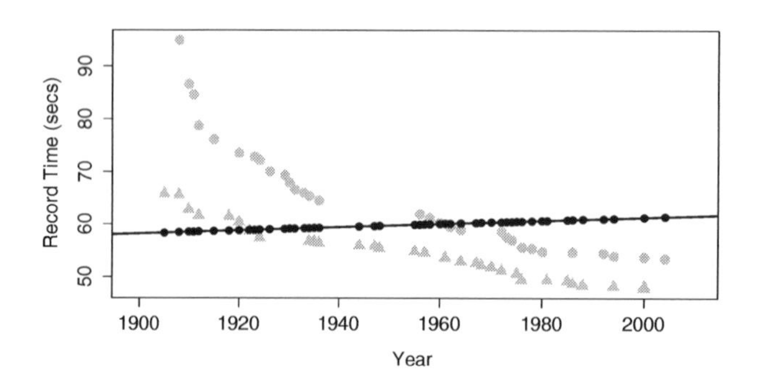

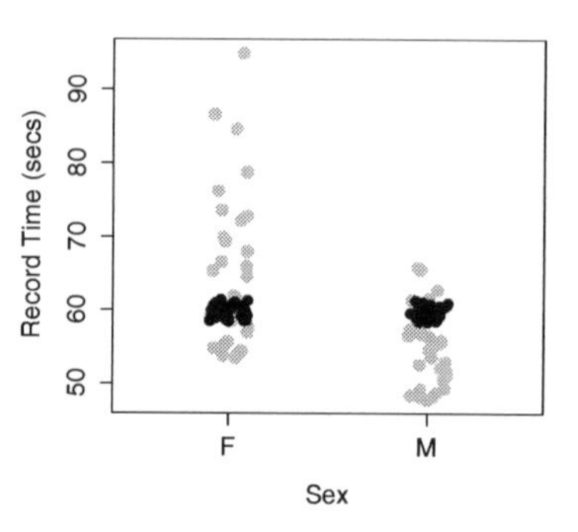

The model function is sloping slightly upward rather than falling as the data clearly indicate. This is because, without an intercept term, the model line is forced to go through the origin. The line is sloping upward so that it will show a time of zero in the hypothetical year zero! Silly. It's no wonder that the model function fails to look anything like the data.

Never leave out the intercept unless you have a very good reason. Indeed, statistical software typically includes the intercept term by default. You have to go out of your way to tell the software to exclude the intercept.

4.6 Standard Notation for Describing Model Design

There is a concise notation for specifying the choices made in a model design, that is, which is the response variable, what are the explanatory variables, and what model terms to use. This notation, introduced originally in [9], will be used throughout the rest of this book and you will use it in working with computers.

To illustrate, here is the notation for some of the models looked at earlier in this chapter:

- ccf ~ 1 + temperature
- wage ~ 1 + sex
- time ~ 1 + year + sex + year:sex

The ~ symbol (pronounced "tilde") divides each statement into two parts. On the left of the tilde is the name of the response variable. On the right is a list of model terms. When there is more than one model term, as is typically the case, the terms are separated by a + sign.

The examples show three types of model terms:

1. The symbol 1 stands for the intercept term.
2. A variable name (e.g., sex or temperature) stands for using that variable in a main term.
3. An interaction term is written as two names separated by a colon, for instance year:sex.

Although this notation looks like arithmetic or algebra, IT IS NOT. The plus sign does not mean arithmetic addition, it simply is the divider mark between terms. In English, one uses a comma to mark the divider as in "rock, paper, and scissors." The modeling notation uses + instead: "rock + paper + scissors." So, in the modeling notation 1 + age does NOT mean "arithmetically add 1 to the age." Instead, it means "two model terms: the intercept and age as a main term."

Similarly, don't confuse the tilde with an algebraic equal sign. The model statement is not an equation. So the statement wage ~ 1 + age does *not* mean "wage equals 1 plus age." Instead it means, "wage is the response variable and there are two model terms: the intercept and age as a main term."

As concise as the modeling notation is, it's used so much that they like to use some shorthand. Two main points will cover most of what you will do:

- You don't have to type the 1 term; it will be included by default. So, wage ~ age is the same thing as wage ~ 1 + age.

  On those very rare occasions when you might want to insist that there be no intercept term, you can indicate this with a minus sign: wage ~ age - 1.

- Almost always, when you include an interaction term between two variables, you will also include the main terms for those variables. The * sign can be used as shorthand. The model wage ~ 1 + sex + age + sex:age can be written simply as wage ~ sex * age.

4.7 Computational Technique

At the core of the language of modeling is the notation that uses the tilde character (~) to identify the response variable and the explanatory variables. This notation is incorporated into many of operators that you will use.

To illustrate the computer commands for modeling and graphically displaying relationships between variables, use the utilities data set:.

```
> utils = ISMdata("utilities.csv")
```

The examples make particular use of these variables

- ccf — the natural gas usage in cubic feet during the billing period.
- month — the month coded as 1 to 12 for January to December.
- temp — the average temperature during the billing period.

Another example uses Current Population Survey wage data:

```
> cps = ISMdata("cps.csv")
```

and focuses on the variables `wage`, `sex`, and `sector`.

4.7.1 Bi-variate Plots

The basic idea of a bi-variate (two variable) plot is to examine one variable as it relates to another. The conventional format is to plot the response variable on the vertical axis and an explanatory variable on the horizontal axis.

Quantitative Explanatory Variable

When the explanatory variable is quantitative, a scatter-plot is an appropriate graphical format. In the scatter plot, each case is a single point.

The basic computer operator for making scatter plots is `xyplot`:

```
> xyplot( ccf ~ temp, data=utils)
```

The first argument is a model formula written using the ~ modeling notation. This formula, `ccf` ~ `temp` is pronounced "ccf versus temperature."

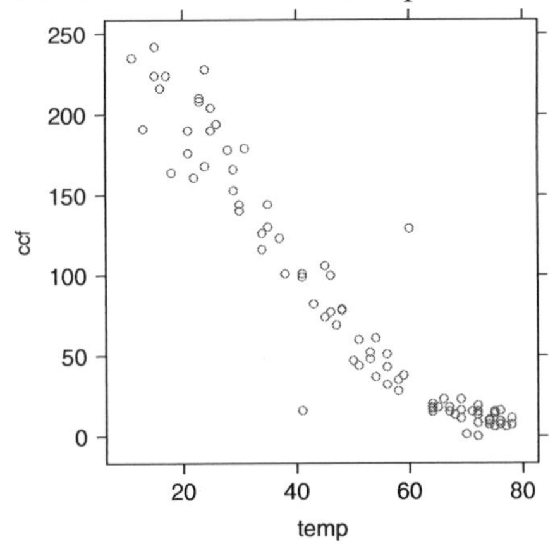

In order to keep the model notation concise, the model formula has left out the name of the data frame to which the variables belong. Instead, the frame is specified in the `data` argument. Since `data` has been set to be `utils`, the formula `ccf~temp` is effectively translated to `utils$ccf~utils$temp`.

You can specify the labels by hand, if you like. For example,

```
> xyplot( ccf ~ temp, data=utils,
    xlab="Temperature (deg F)",
    ylab="Natural Gas Usage (ccf)")
```

Categorical Explanatory Variable

When the explanatory variable is categorical, an appropriate format of display is the box-and-whiskers plot, made with the `bwplot` operator. Here, for example, is the `wage` versus `sex` from the Current Population Survey:

```
> bwplot( wage ~ sex, data=cps)
```

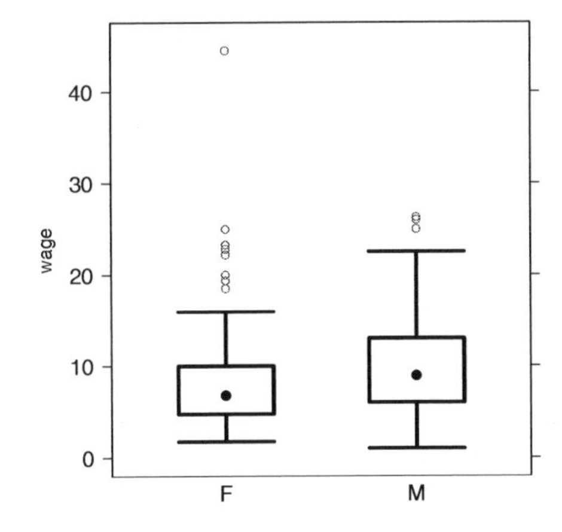

When there are many levels, and when the names of the levels are long, it can become hard to read the labels on the graph. An effective solution is to rotate the labels, perhaps by 45 degrees. Here is wage ~ sector

```
> bwplot(wage~sector, data=cps, scales=list(rot=45) )
```

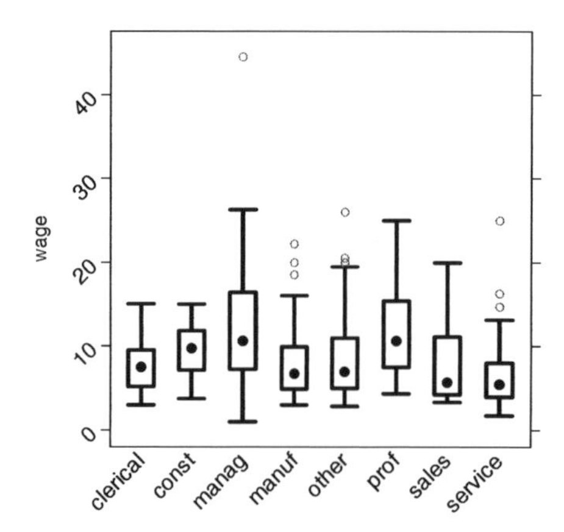

Admittedly, the argument `scales=list(rot=45)` is obscure. When you need to use it, just copy it from this example.

Notice that the outliers are setting the overall vertical scale for the graph and obscuring the detail at typical wage levels. You can use the `ylim` argument to set the scale of the y-axis however you want. For example:

```
>  bwplot(wage~sector, data=cps, scales=list(rot=45),
          ylim=c(0,30) )
```

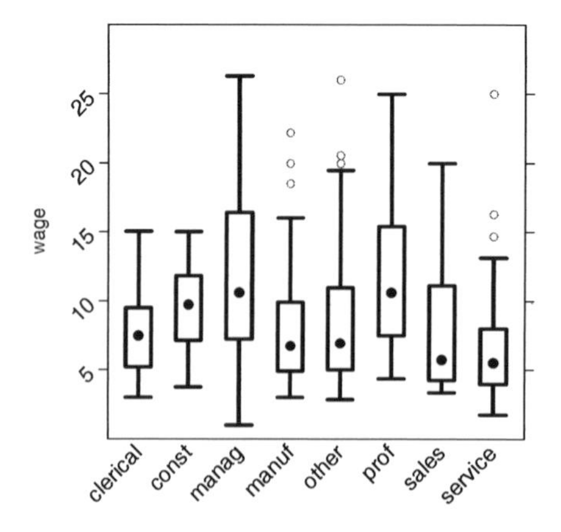

You can also make side-by-side density plots which show more detail than the box-and-whisker plots. For instance:

```
> densityplot( ~ wage, groups=sex, data=cps, auto.key=TRUE )
```

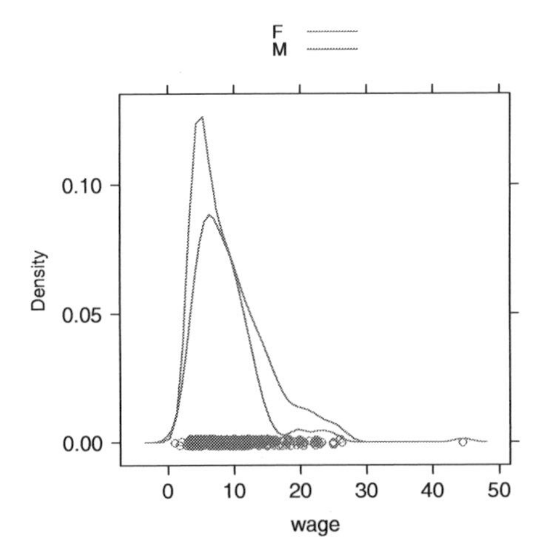

It seems a bit odd to have the notation ~wage — nothing is in the role of the response variable. This idiosyncratic notation perhaps is meant to reflect that wage is on the horizontal axis.

Multiple Explanatory Variables

The two-dimensional nature of paper or the computer screen lends itself well to displaying two variables: a response versus a single explanatory variable. Sometimes it is important to be able to add an additional explanatory variable. The graphics system gives a variety of options in this regard:

Coding the additional explanatory variable using color or symbol shapes. This is done by using the groups argument set to the name of the additional explanatory variable. For example:

```
> xyplot(wage ~ age, groups=sex, data=cps,
  auto.key=TRUE)
```

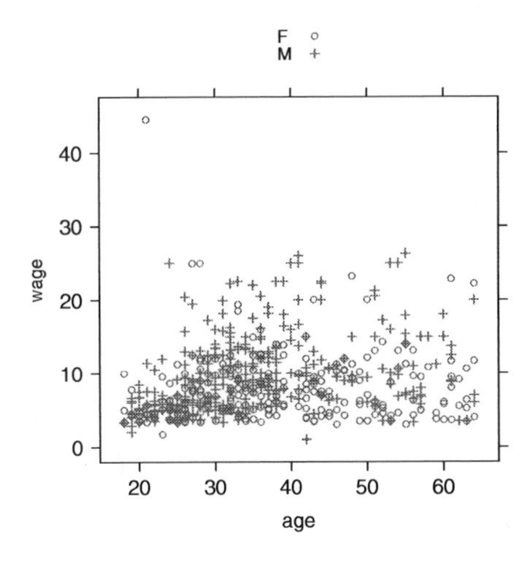

Splitting the plot into the groups defined by the additional explanatory variable. This is done by including the additional variable in the model formula using a | separator. For example: |)

```
> xyplot(wage ~ age | sex, data=cps)
```

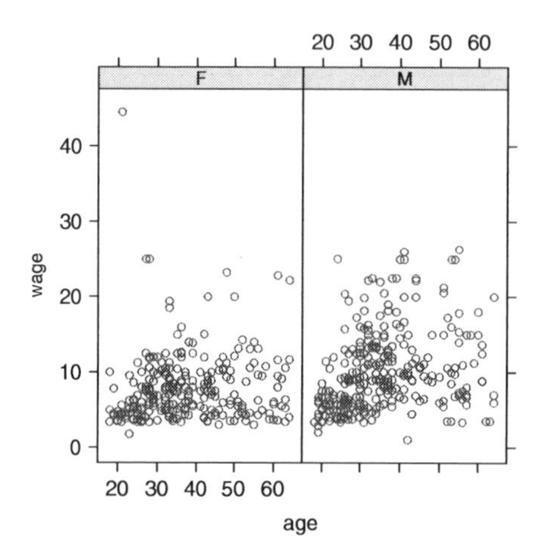

4.7.2 Fitting Models and Finding Model Values

The `lm` operator (short for "Linear Model") will translate a model design into fitted model values. It does this by "fitting" the model to data, a process that will be explained in later chapters. For now, focus on how to use `lm` to compute the fitted model values.

The `lm` operator uses the same model language as in the book. To illustrate, consider the world-record swim-times data :

```
> swim = ISMdata("swim100m.csv")
```

To construct the model time $\sim$ 1 for the swim data:

```
> mod1 = lm( time ~ 1, data=swim)
```

Here the model has been given a name, `mod1`, so that you can refer to it later. You can use any name you like, so long as it is valid in R.

Once the model has been constructed, the fitted values can be found using the `fitted` operator:

```
> fitted(mod1)
    1     2     3     4     5     6     7     8
59.92 59.92 59.92 59.92 59.92 59.92 59.92 59.92
    9    10    11    12    13    14    15    16
59.92 59.92 59.92 59.92 59.92 59.92 59.92 59.92
... and so on for 62 cases altogether.
```

There is an individual fitted model value for each case. Of course, in this model all the model values are exactly the same since the model time $\sim$ 1 treats all the cases as exactly the same.

In later chapters you'll see how to analyze the model values, make predictions from the model, and assess the contribution of each model term. For now, just look at the model values by plotting them out along with the data used. I'll plot out both the data values and the model values versus year just to emphasize that the model values are the same for every case:

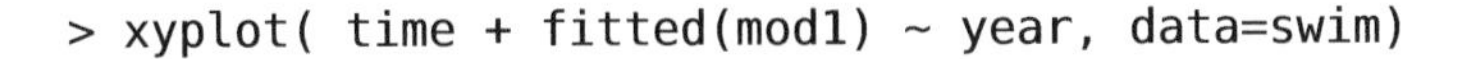

```
> xyplot( time + fitted(mod1) ~ year, data=swim)
```

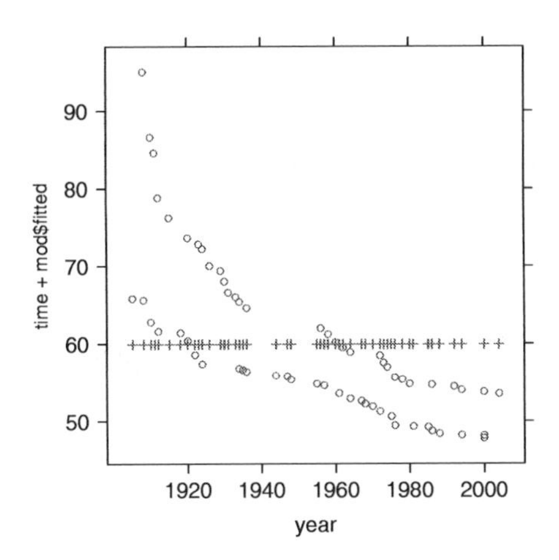

Pay careful attention to the syntax used in the above command. There are two quantities to the left of the ~. This is not part of the modeling language, where there is always a single response variable. Instead, it is a kind of shorthand, telling `xyplot` that it should plot out *both* of the quantities on the left side against the quantity on the right side. Of course, if you wanted to plot just the model values, without the actual data, you could specify the formula as `fitted(mod1)~year`.

Here are more interesting models:

```
mod2 = lm( time ~ 1+year, data=swim)
mod3 = lm( time ~ 1+sex, data=swim)
mod4 = lm( time ~ 1+sex+year, data=swim)
mod5 = lm( time ~ 1+year+sex+year:sex, data=swim)
```

You can, if you like, compare the fitted values from different models on one plot:

```
> xyplot( fitted(mod5) + fitted(mod3) ~ year, data=swim,
   auto.key=TRUE)
```

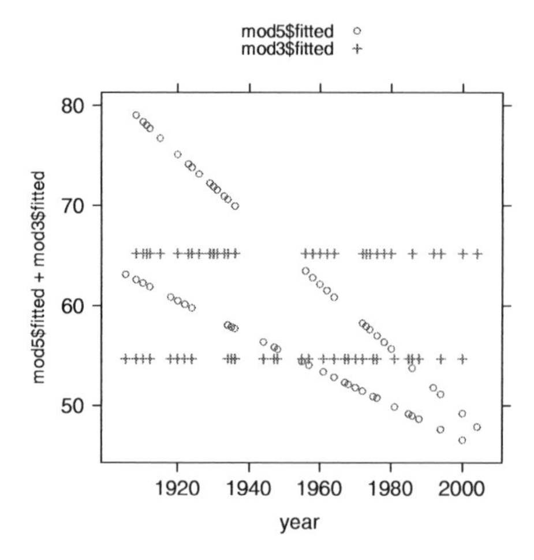

Shorthand notation for modeling

The intercept term is almost always included in models. For this reason, it's included by default even if you don't have the `1` term explicitly in the design of your model. For example, `mod2` could have been constructed with this statement:

```
mod2 = lm( time ~ year, data=swim )
```

There is also a shorthand for including both the main effects and the interaction terms. For example, `mod5` has terms for the main effects of `year` and `sex` and their interaction terms. This can be written:

```
mod5 = lm( time ~ year*sex, data=swim )
```

Suppressing the Intercept Term

You will rarely have to do it, but if you want to *exclude* the intercept term from a model, you use the notation `-1` in the model formula:

```
mod6 = lm( time ~ year-1, data=swim)
xyplot( time + fitted(mod6) ~ year, data=swim)
```

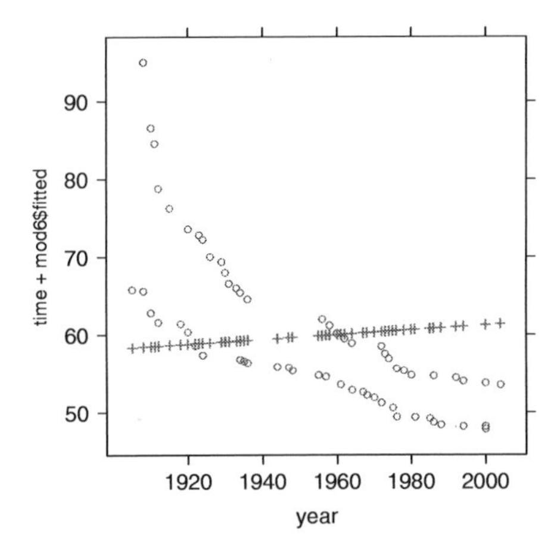

Interactions and Main Effects

Typically a model that includes an interaction term between two variables will include the main terms from those variables too. As a shorthand for this, the modeling language has a ∗ symbol. So, the formula `time~year+sex+year:sex` can also be written `time~year*sex`.

Transformation Terms

Transformation terms such as squares can also be included in the model formula. To mark the quantity clearly as a single term, it's best to wrap the term with `I()` as follows:

```
> mod7 = lm( time ~ year + I(year^2) + sex, data=swim)
```

Another way to accomplish this, for polynomials, is to use the operator `poly` as in the model formula `time~poly(year,2)+sex`.

Here's a plot of the result:

```
> xyplot( time + fitted(mod7) ~ year, data=swim)
```

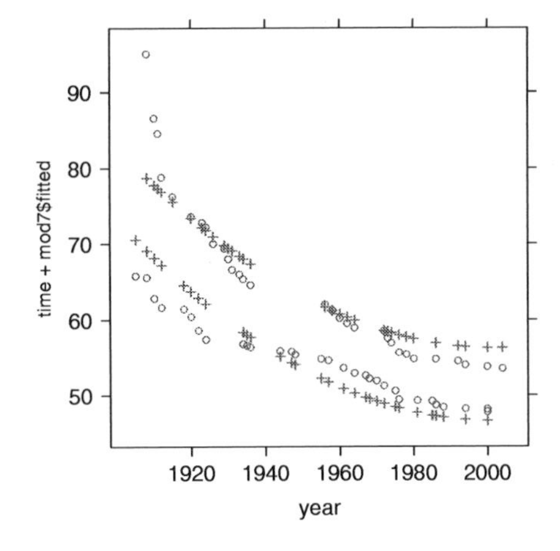

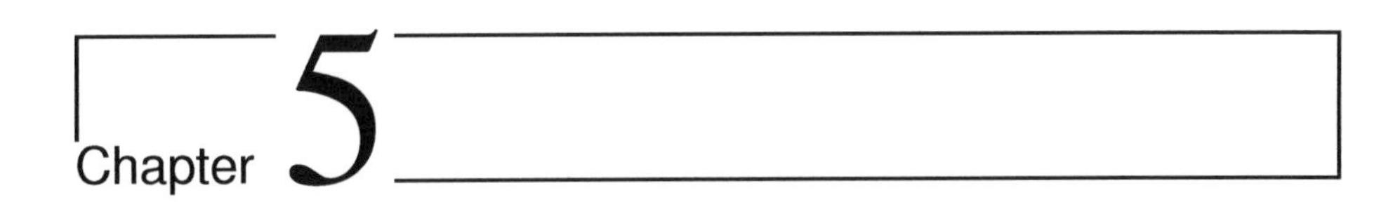

Model Formulas and Coefficients

> *All economical and practical wisdom is an extension or variation of the following arithmetical formula:* $2 + 2 = 4$. *Every philosophical proposition has the more general character of the expression* $a + b = c$. *We are mere operatives, empirics, and egotists, until we learn to think in letters instead of figures.* — Oliver Wendell Holmes (1841-1935)

The previous chapter presents models as graphs. The response variable is plotted on the vertical axis and one of the explanatory variables on the horizontal axis. Such a visual depiction of a model is extremely useful, when it can be done. The relationship between the response and explanatory variables can be seen as slopes or differences; interactions can be seen as differences between slopes; the match or mismatch between the data and the model can be visualized easily.

A graph is also a useful mode of communicating; many people have the graph-reading skills needed to interpret the sorts of models in the previous chapter, or, at least, to get an impression of what the models are about.

Presenting models via a graph is, however, very limiting. The sorts of models that can be graphed effectively have only one or two explanatory variables, whereas often models need many more explanatory variables. (There are models with thousands of explanatory variables, but even a model with only three or four explanatory variables can be impossible to graph in an understandable way.) Even when a model is simple enough to be graphed, it's helpful to be able to quantify the relationships. Such quantification becomes crucial when you want to characterize the reliability of a model or draw conclusions about the strength of the evidence to support the claim of a relationship shown by a model.

This chapter introduces important ways to present models in a non-graphical way as formulas and as coefficients.

5.1 The Linear Model Formula

Everyone who takes high-school algebra encounters this equation describing a straight-line relationship:

$$y = mx + b$$

The equation is so familiar to many people that they automatically make the following associations: x and y are the variables, m is the slope of the line, b is the y-intercept — the place the line crosses the y axis.

The straight-line equation is fundamental to statistical modeling, although the nomenclature is a little different. To illustrate, consider this model of the relationship between an adult's height and the height of his or her mother, based on Galton's height data:

Model Values height = 46.7 + 0.313 mother

This is a model represented not as a graph but as a **model formula**. Just reading the model in the same way as $y = mx + b$ shows that the intercept is 46.7 and the slope is 0.313. That is, if you make a graph of the model values of the response variable height (of the adult child) against the values of the explanatory variable mother, you will see a straight line with that slope and that intercept.

To find the model value given by the formula, just plug in numerical values for the explanatory variable. So, according to the formula, a mother who is 65 inches tall will have children with a typical height of $46.7 + 0.313 \times 65$, giving 67.05 inches.

You can also interpret the model in terms of the relationship between the child's height and the mother's height. Comparing two mothers who differ in height by one inch, their children would typically differ in height by 0.313 inches.

In the design language, the model is specified like this:

height ~ 1 + mother

There is a simple correspondence between the model design and the model formula. The model formula takes each of the terms in the model design and multiplies it by a number. The intercept term is multiplied by 46.7 and the mother term is multiplied by 0.313. Such numbers have a generic name: **model coefficients**. Instead of calling 0.313 the "slope," it's called the "coefficient on the term mother." Similarly, 46.7 could be the "coefficient on the intercept term," but it seems more natural just to call it the intercept.

Where did these precise values for the coefficients come from? A process called **fitting the model to the data** finds coefficients that bring the model values from the formula to correspond as closely as possible to the response values in the data. As a result, the coefficients that result from fitting a given model design will depend on the data to which the model is fitted. Chapter 6 describes the fitting process in more detail.

5.2 Linear Models with Multiple Terms

It's easy to generalize the linear model formula to include more than one explanatory variable or additional terms such as interactions. For each term, just add a new component to the formula. For example, suppose you want to model the child's height as a function of **both** the mother's and the father's height. As a model design, take

height $\sim$ 1 + mother + father

Here is the model formula for this model design fitted to Galton's data:

Model Values height = 22.3 + 0.283 mother + 0.380 father

As before, each of the terms has its own coefficient. This same pattern of terms and coefficients can be extended to include as many variables as you like.

Interaction terms also fit into this framework. Suppose you want to fit a model with an interaction between the father's height and the mother's height. The model design is

height $\sim$ 1 + mother + father + father:mother

For Galton's data, the formula that corresponds to this design is

Model Values height = 132.3 - 1.43 mother - 1.21 father + 0.0247 father $\times$ mother

The formula follows the same pattern as before: a coefficient multiplying the value of each term. So, to find the model value of height for a child whose mother is 65 inches tall and whose father is 67 inches tall, multiply the coefficient by the value of corresponding term: $132.3 - 1.43 \times 65 - 1.21 \times 67 + 0.0247 \times 65 \times 67$ giving 66.12 inches. The term father $\times$ mother may look a little odd, but it just means to multiply the mother's height by the father's height.

5.3 Formulas with Categorical Variables

Since quantitative variables are numbers, they can be reflected in a model formula in a natural, direct way: just multiply the value of the model term by the coefficient on that term.

Categorical variables are a little different. They don't have numerical values. It doesn't mean anything to multiply a category name by a coefficient.

In order to include a categorical variable in a model formula, a small translation is necessary. Suppose, for example, that you want to model the child's height by both its father's and mother's height and also the sex of the child. The model design is nothing new: just include a model term for sex.

Aside. 5.1 Interpreting Interaction Terms

The numerics of interaction terms are easy. You just have to remember to multiply the coefficient by the product of all the variables in the term. Multiply the coefficient time the values of all the variables in the interaction term. The meaning of interaction terms is somewhat more difficult. Many people initially mistake interaction terms to refer to a relationship between two variables. For instance, they would (wrongly) think that an interaction between mother's and father's heights means that, say, tall mothers tend to be married to tall fathers. Actually, interaction terms do not describe the relationship between the two variables, they are about *three* variables: how one explanatory variable modulates the effect of another on the response.

A model like height ~ 1+mother+father captures some of how a child's height varies with the height of either parent. The coefficient on mother (when fitting this model to Galton's data) was 0.283, indicating that an extra inch of the mother's height is associated with an extra 0.283 inches in the child. This coefficient on mother doesn't depend on the father's height; the model provides no room for it to do so.

Suppose that the relationship between the mother's height and her child's height is potentiated by the father's height. This would mean that if the father is very tall, then the mother has even more influence on the child's height than for a short father. That's an interaction.

To see this sort of effect in a model, you have to include an interaction term, as in the model height ~ 1+mother+father+mother:father. The coefficients from fitting this model to Galton's data allow you to compare what happens when the father is very short (say, 60 inches) to when the father is very tall (say, 75 inches). With a short father, an extra inch in the mother is associated with an extra 0.05 inches in the child. With a tall father, an extra inch in the mother is associated with an extra 0.42 inches in the child.

The interaction term can also be read the other way: the relationship between a father's height and the child's height is greater when the mother is taller.

Of course, this assumes that the model coefficients can be taken at face value as reliable. Later chapters will deal with how to evaluate the strength of the evidence for a model. It will turn out that Galton's data do not provide good evidence for an interaction between mother's and father's heights in determining the child's height.

height ~ 1 + mother + father + sex

The corresponding model formula for this design on the Galton data is

Model Values height = 15.3 + 0.322 mother + 0.406 father + 5.23 sex**M**

Interpret the quantitative terms in the ordinary way. The new term, 5.23 sex**M**, means "add 5.23 whenever the case has level M on sex." This is a very roundabout way of saying that males tend to be 5.23 inches taller than females, according to the model.

Another way to think about the meaning of sex**M** is that it is a new quantitative variable, called an **indicator variable**, that has the numeric value 1 when the case is level M in variable sex, and numeric value 0 otherwise.

A categorical variable will have one indicator variable for each level of the variable. Thus, a variable language with levels Chinese, English, French, German, Hindi, etc. has a separate indicator variable for each level.

For the sex variable, there are two levels: F and M. But notice that there is only one coefficient for sex, the one for sex**M**. Get used to this. Whenever there is an intercept term in a model, at least one of the indicator variables from any categorical variable will be left out. The level that is omitted is a **reference level**. Chapter 6 explains why things are done this way.

5.4 Model Coefficients Describe Relationships

Model coefficients describe the strength of relationships. Consider this model of height:

Model Values height = 15.3 + 0.322 mother + 0.406 father + 5.23 sex**M**

The coefficient 5.23 on sex**M** indicates that males are typically 5.23 inches taller than females (who are the reference group). Thus, the coefficient indicates the strength of the relationship between sex and height.

The coefficient 0.322 on the mother main term means that two mothers who differ in height by 1 inch will have children whose typical height differs by 0.322 inches. This means that there is indeed a relationship between the mother's height and the child's. If the coefficient were bigger, the relationship would be stronger.

The sign of a coefficient tells which way the relationship goes. If the coefficient on sex**M** had been −5.23, it would mean that males are typically shorter than females. If the coefficient on mother had been −0.322, then taller mothers would have shorter children.

When a variable is included in more than one model term, the individual coefficients can no longer be read directly as indicating how that variable contributes to the relationship. Instead, one needs to compare the model values under different settings of the explanatory variables. To see why, consider again the above formula. The model includes each variable only as a main term, so the result will be the same as you could read off directly from the coefficient. For instance, to see the strength of the relationship between mother's height and the child's height, pick values for the inputs, calculate the resulting model value, and then compare this to the model value for a slightly different value of mother's heights. Suppose that you pick sex as M, mother as 65 inches, and father as 68 inches. This results in a model value of $15.3 + 0.322 \times 65 + 0.406 \times 68 + 5.23 \times 1$ or 69.0744 inches. Now change the value for mother to 66; the resulting model value is different: 69.3959 inches. The difference between them is the typical height difference associated with a 1-inch difference in mother's height: 0.3215 inches, just what you already saw from the coefficient on mother.

Those familiar with calculus may recognize the connection to the *partial derivative* in this approach of comparing model values due to a change of inputs. Chapter 8 will consider the issues in more detail, particularly the matter of **holding constant** other variables or **adjusting** for other variables.

5.5 Model Values and Residuals

If you plug in the values of the explanatory variables from your data, the model formula gives, as an output, the model values. It's rarely the case that the model values are an exact match with the actual response variable in your data. The difference is the *residual.*

For any one case in the data, the residual is always defined in terms of the actual response variable and the model value that arises when that case's values for the explanatory variables are given as the input to the model formula. For instance, if you are modeling height, the residuals are

$$\text{height} = {}_{\text{Values}}^{\text{Model}}\ \text{height} + \text{residuals}$$

The residuals are always defined in terms of a particular data set and a particular model. Each case's residual would likely change if the model were altered or if another data set were used to fit the model.

5.6 Coefficients of Basic Model Designs

The presentation of a linear model as a set of coefficients is a compact shorthand for the complete model formula. It often happens that for some purposes, interest focuses on a single coefficient of interest. In order to help you to interpret coefficients correctly, it is helpful to see how they relate to some basic model designs. Often, the interpretation for more complicated designs is not too different.

To illustrate how basic model designs apply generally, I will use A to stand for a generic response variable, B to stand for a quantitative explanatory variable, and C for a categorical explanatory variable. As always, 1 refers to the intercept term.

Model A ∼ 1

The model A ∼ 1 is the simplest of all. There are no explanatory variables; the only term is the intercept. Think of this model as saying that all the cases are the same. In fitting the model, you are looking for a single value that is as close as possible to each of the values in A.

The coefficient from this model is the mean of A. The model values are the same for every case: the mean of all the samples. This is sometimes called the **grand mean** to distinguish it from group means, the mean of A for different groups.

Example 5.1: The mean height of all the cases in Galton's height data — the "grand mean" — is the coefficient on the model height ∼ 1. Fitting this model to Galton's data gives this coefficient:

Model Term	Coefficient
Intercept	66.7607

Thus, the mean height is 66.76 inches.

Example 5.2: The mean wage earned by all of the people in the Current Population Survey is given by the coefficient of the model wage ∼ 1. Fitting the model to the CPS gives:

Model Term	Coefficient
Intercept	8.96

Thus, the mean wage is $8.96 per hour.

Model A ∼ 1+G

The model A ∼ 1+G is also very simple. The categorical variable G can be thought of as dividing the data into groups, one group for each level of G. There is a separate model value for each of the groups.

The model values are the group-wise means: separate means of A for the cases in each group. The model coefficients, however, are not exactly these group-wise means. Instead, the coefficient of the intercept term is the mean of one group, which can be called the **reference group** or **reference level**. Each of the other coefficients is the *difference* between its group's mean and the mean of the reference group.

Example 5.3: Calculate the group-wise means of the heights of men and women in Galton's data by fitting the model height ~ sex.

Model Term	Coefficient
Intercept	64.11
sexM	5.12

The mean height for the reference group, women, is 64.11 inches. Men are taller by 5.12 inches. In the standard form of a model report, the identity of the reference group is not stated explicitly. You have to figure it out from which levels of the variable are missing.

By suppressing the intercept term, you change the meaning of the remaining coefficients; they become simple group-wise means rather than the difference of the mean from a reference group's mean. Here's the report from fitting the model height ~ sex - 1.

Model Term	Coefficient
sexF	64.11
sexM	69.23

It might seem obvious that this simple form is to be preferred, since you can just read off the means without doing any arithmetic on the coefficients. That can be the case sometimes, but almost always you will want to include the intercept term in the model. The reasons for this will become clearer when hypothesis testing is introduced in later chapters.

Example 5.4: Calculate group-wise means of wages in the different sectors of the economy by fitting the model wage ~ sector:

Model Term	Coefficient
Intercept	7.42
sector**const** 2.08	
sector**manag** 4.69	
sector**manuf** 0.61	
sector**other** 1.08	
sector**prof** 4.52	
sector**sales** 0.17	
sector**service** -0.89	

There are eight levels of the sector variable, so the model has 8 coefficients. The coefficient of the intercept term gives the group-wise mean of the reference group. The reference group is the one level that isn't listed explicitly in the other coefficients; it turns out to be the clerical sector. So, the mean wage of clerical workers is $7.42 per hour. The other coefficients give the *difference* between the mean of the reference group and the means of other groups. For example, workers in the construction sector make, on average $2.08 per hour more than clerical workers. Similarly, service sector works make 89 cents per hour *less* than clerical workers.

Model A ~ 1+B

Model A ~ 1+B is the basic straight line relationship. The two coefficients are the intercept and the slope of the line. The slope tells what change in A corresponds to a one-unit change in B.

Example 5.5: The model wage ~ 1+educ shows how workers wages are different for people with different amounts of education (as measured by years in school). Fitting this model to the Current Population Survey data gives the following coefficients:

Model Term	Coefficient
Intercept	-0.69
educ	0.74

According to this model, a one-year increase in the amount of education that a worker received is associated with a 74 cents per hour increase in wages. (Remember, these data are from 1985.)

It may seem odd that the intercept coefficient is negative. Nobody is paid a negative wage. Keep in mind that the intercept of a straight line $y = mx + b$ refers to the value of y when $x = 0$. The intercept coefficient -0.69, tells the typical wage for workers with zero years of education. There are no workers in the data set with zero years; only three workers have less than five years. In this data set, the intercept is an **extrapolation** outside of the range of the data.

Model A ~ 1 + G + B

The model A ~ 1 + G + B gives a straight-line relationship between A and B, but allows different lines for each group defined by G. The lines are different only in their intercepts; all of the lines have the same slope.

The coefficient labeled "intercept" is the intercept of the line for the reference group. The coefficients on the various levels of categorical variable G reflect how the intercepts of the lines from those groups differ from the reference group's intercept.

The coefficient on B gives the slope. Since all the lines have the same slope, a single coefficient will do the job.

Example 5.6: Wages versus educational level for the different sexes: wage ~ 1 + educ + sex The coefficients are

Model Term	Coefficient
Intercept -1.92773 sex**M**	2.27
educ	0.74

The educ coefficient tells how education is associated with wages. The sex**M** coefficient says that men tend to make $2.27 more an hour than women when comparing men and women with the same amount of education. Chapter 8 will explain why the inclusion of the educ term allows this comparison *at the same level* of education.

Note that the model A ~ 1 + G + B is the same as the model A ~ 1 + B + G. The order of model terms doesn't make a difference.

Model A ~ 1 + G + B + G:B

The model A ~ 1 + G + B + G:B is also a straight-line model, but now the different groups defined by G can have different slopes and different intercepts. (The interaction term G:B says how the slopes differ for the different group. That is, thinking of the slope as effect of B on A, the interaction term G:B tells how the effect of B is modulated by different levels of G.)

Example 5.7: Here is a model describing wages versus educational level, separately for the different sexes: wage ~ 1 + sex + educ + sex:educ

Model Term	Coefficient
Intercept	-3.10
sex**M**	4.20
educ	0.83
sex**M**:educ	-0.15

Interpreting these coefficients: For women, an extra year of education is associated with an increase of wages of 83 cents per hour. For men, the relationship is weaker: an increase in education of one year is associated with only an increase of wages of only $0.831 - 0.148$ or 68 cents per hour.

5.7 Coefficients have Units

A common convention is to write down coefficients and model formulas without being explicit about the units of variables and coefficients. This convention is unfortunate. Although leaving out the units leads to neater tables and simpler-looking formulas, the units are fundamental to interpreting the coefficients. Ignoring the units can mislead severely.

To illustrate how units come into things, consider the model design wages ~ 1 + educ + sex. Fitting this model design to the Current Population Survey gives this model formula:

Model Values wage = -1.93 + 0.742 educ + 2.27 sex**M**

A first glance at this formula might suggest that sex is more strongly related than educ to wage. After all, the coefficient on educ is much smaller than the coefficient on sexM. But this interpretation is invalid, since it doesn't take into account the units of the variables or the coefficients.

The response variable, wage, has units of dollars-per-hour. The explanatory variable educ has units of years. The explanatory variable sex is categorical and has no units; the indicator variables for sex are just zeros and ones: pure numbers with no units.

The coefficients have the units needed to transform the quantity that they multiply into the units of the response variable. So, the coefficient on educ has units of "dollars-per-hour per year." This sounds very strange at first, but remember that the coefficient will multiply a quantity that has units of years, so the product will be in units of dollars-per-hour, just like the wage variable. In this formula, the units of the intercept coefficient and the coefficient on sexM both have units of dollars-per-hour, because they multiply something with no units and need to produce a result in terms of dollars-per-hour.

A person who compares 0.742 dollars-per-hour per year with 2.27 dollars per hour is comparing apples and oranges, or, in the metric-system equivalent, comparing meters and kilograms. If the people collecting the data had decided to measure education in months rather than in years, the coefficient would have been a measly $0.742/12 = 0.0618$ even though the relationship between education and wages would have been exactly the same.

Aside. 5.2 Comparing Coefficients

If you want to compare the two coefficients, say the coefficient on educ and the coefficient on sexM, you have to put them on a common footing. It's not always obvious how to do this, because the coefficient have different units. So you have to be clever.

One approach that might work here is to find the number of years of education that produces a similar size effect of education on wages as is seen with sexM. The answer turns out to be about 3 years. (To see this, note that the wage gain associated with sexM, 2.27 dollars per hour to the wage gain associated with three years of education, $3 \times 0.742 = 2.21$ dollars per hour.)

Thus, according to the model, being a male is equivalent in wage gains to an increase of 3 years in education.

5.8 Untangling Explanatory Variables

One of the advantages of using formulas to describe models is the way they facilitate using multiple explanatory variables. Many people assume that they can study relationships one variable at a time, even when they know there are influences from multiple factors. Underlying this assumption is a belief that influences add up in a simple way. Indeed model formulas without interaction terms actually do simply add up the contributions of each variable.

But this does not mean that an explanatory variable can be considered in isolation from other explanatory variables. There is something else going on that makes it important to consider the explanatory variables not one at a time but simultaneously, at the same time.

In many situations, the explanatory variables are themselves related to one another. As a result, variables can to some extent stand for one another. An effect attributed to one variable might equally well be assigned to some other variable.

Due to the relationships between explanatory variables, you need to untangle them from one another. The way this is done is to use the variables together in a model, rather than in isolation. The way the tangling shows up is in the way the coefficient on a variable will change when another variable is added to the model or taken away from the model. That is, model coefficients on a variable tend to depend on the context set by other variables in the model.

To illustrate, consider this criticism of spending on public education in the United States from a respected political essayist:

> *The 10 states with the lowest per pupil spending included four — North Dakota, South Dakota, Tennessee, Utah — among the 10 states with the top SAT scores. Only one of the 10 states with the highest per pupil expenditures — Wisconsin — was among the 10 states with the highest SAT scores. New Jersey has the highest per pupil expenditures, an astonishing $10,561, which teachers' unions elsewhere try to use as a negotiating benchmark. New Jersey's rank regarding SAT scores? Thirty-ninth... The fact that the quality of schools... [fails to correlate] with education appropriations will have no effect on the teacher unions' insistence that money is the crucial variable.* — George F. Will, (September 12, 1993), "Meaningless Money Factor," The Washington Post, C7. Quoted in [10].

The response variable here is the score on a standardized test taken by many students finishing high school: the SAT. The explanatory variable — there is only one — is the level of school spending. But even though the essayist implies that spending is *not* the "crucial variable," he doesn't include any other variable in the analysis. In part, this is because the method of analysis — pointing to individual cases to illustrate a trend — doesn't allow the simultaneous consideration of multiple explanatory variables. (Another flaw with the informal method of comparing cases is that it fails to quantify the strength of the effect: Just how much negative influence does spending have on performance? Or is the claim that there is no connection between spending and performance.)

DATA FILE
`sat.csv`

You can confirm the claims in the essay by modeling. The datasetcontains state-by-state information from the mid-1990s on *per capita* yearly school expenditures in thousands of dollars, average statewide SAT scores, average teachers' salaries, and other variables.[10]

The analysis in the essay corresponds to a simple model: sat $\sim$ 1 + expend. Fitting the model to the state-by-state data gives a model formula,

$$\text{sat}_{\text{Model Values}} = 1089 - 20.9\ \text{expend}$$

The formula is consistent with the claim made in the essay; the coefficient on expend is negative. According to the model, an increase in expenditures by \$1000 per capita is associated with a 21 point decrease in the SAT score. That's not very good news for people who think society should be spending more on schools: the schools that spend the least per capita have the highest average SAT scores. (A 21 point decrease in the SAT doesn't mean much for an individual student, but as an average over tens of thousands of students, it's a pretty big deal.)

Perhaps expenditures is the wrong thing to look at — you might be studying administrative inefficiency or even corruption. Better to look at teachers' salaries. Here's the model sat $\sim$ 1 + salary, where salary is the average annual salary of public school teachers in \$1000s:

$$\underset{\text{Values}}{\text{Model}}\ \text{sat} = 1159 - 5.54\ \text{salary}$$

The essay's claim is still supported. Higher salaries are associated with lower average SAT scores! But maybe states with high salaries manage to pay well because they overcrowd classrooms. So, look at the average student/teacher ratio in each state: sat $\sim$ 1 + ratio

$$\underset{\text{Values}}{\text{Model}}\ \text{sat} = 921 + 2.68\ \text{ratio}$$

Finally, a positive coefficient! That means ... larger classes are associated with higher SAT scores.

All this goes against the conventional wisdom that holds that higher spending, higher salaries, and smaller classes will be associated with better performance.

At this point, many advocates for more spending, higher salaries, and smaller classes will explain that you can't measure the quality of an education with a standardized test, that the relationship between a student and a teacher is too complicated to be quantified, that students should be educated as complete human beings and not as test-taking machines, and so on. Perhaps.

Whatever the criticisms of standardized tests, they have the great benefit of allowing comparisons across different conditions: different states, different curricula, etc. If there is a problem with the tests, it isn't standardization itself but the material that is on the test. Absent a criticism of that material, rejections of standardized testing ought to be treated with considerable skepticism.

But there is something wrong with using the SAT as a test, even if the content of the test is good. What's wrong is that the test isn't required for all students. Depending on the state, a larger or smaller fraction of students will take the SAT. In states where very few students take the SAT, those students who do are the ones bound for out-of-state colleges, in other words, the high performers.

What's more, the states which spend the least on education tend to have the fewest students who take the SAT. That is, the fraction of students taking the SAT is entangled with expenditures and other explanatory variables.

To untangle the variables, they have to be included simultaneously in a model. That is, in addition to expend or salary or ratio, the model needs to take into account the fraction of students who take the SAT (variable frac).

Here are three models that attempt to untangle the influences of frac and the spending-related variables:

$$\text{sat}_{\text{Model Values}} = 994 + 12.29\ \text{expend} - 2.85\ \text{frac}$$

$$\text{sat}_{\text{Model Values}} = 988 + 2.18\ \text{salary} - 2.78\ \text{frac}$$

$$\text{sat}_{\text{Model Values}} = 1119 - 3.73\ \text{ratio} - 2.55\ \text{frac}$$

In all three models, including frac has completely altered the relationship between performance and the principal explanatory variable of interest, be it expend, salary, or ratio. Not only is the coefficient different, it is different in sign.

Chapter 8 discusses why adding frac to the model can be interpreted as an attempt to examine the other variables while holding frac constant, as if you compared only states with similar values of frac.

The situation seen here, where adding a new explanatory variable (e.g., frac) changes the sign of the coefficient on another variable (e.g., expend, salary, ratio) is called **Simpson's paradox**.

Simpson's Paradox is an extreme version of a common situation: that the coefficient on an explanatory variable can depend on what other explanatory variables have been included in the model. In other words, the role of an explanatory variable can depend, sometimes strongly, on the context set by other explanatory variables. You can't look at explanatory variables in isolation; you have to interpret them in context.

There is nothing magical about Simpson's Paradox or the dependence of model coefficients on context. It appears paradoxical only when the details of model fitting are hidden in a black box of software. You will open the black box in Chapter 11, which will help you to understand when and why Simpson's Paradox occurs and how to anticipate it.

But which is the right model? What's the right context? Do expend, salary, and ratio have a positive role in school performance as the second set of models indicate, or should you believe the first set of models? This is an important question and one that comes up often in statistical modeling. At one level, the answer is that you need to be aware that context matters. At another level, you should always check to see if your conclusions would be altered by including or excluding some other explanatory variable. At a still higher level, the choice of which variables to include or exclude needs to be related to the modeler's ideas about what causes what.

5.9 Why Linear Models?

Many people are uncomfortable with using linear models to describe potentially complicated relationships. The process seems a bit unnatural: specify the model terms, fit the model to the data, get the coefficients. How do you know that a model fit in this way will give a realistic match to the data? Coefficients seem an overly abstract way to describe a relationship. Modeling without graphing seems like dancing in the dark; it's nice to be able to see your partner.

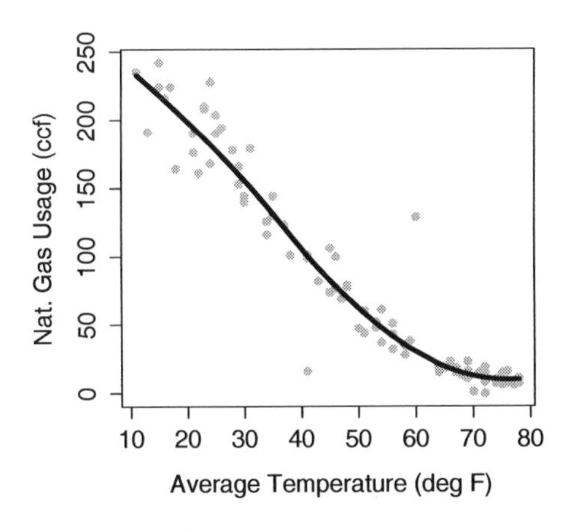

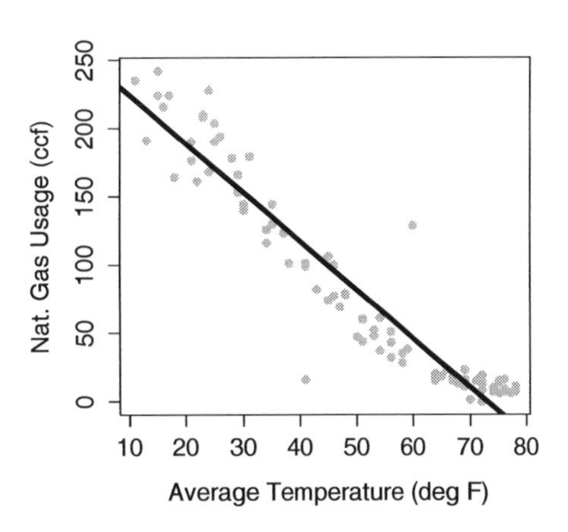

Figure 5.1: Two models of natural gas usage versus outdoor temperature.

Return to the example of world record swim times plotted in Figure 4.5 on page 78. There's a clear curvature to the relationship between record time and year. Without graphing the data, how would you have known whether to put in a transformation term? How would you know that you should use sex as an explanatory variable? But when you do graph the data, you can see easily that something is wrong with a straight-line model like time $\sim$ 1 + year.

However, the advantages of graphing are obvious only in retrospect, once you have found a suitable graph that is informative. Why graph world record time against year? Why not graph it versus body weight of the swimmer or the latitude of the pool in which the record was broken?

Researchers decide to collect some variables and not others based on their knowledge and understanding of the system under study. People know, for example, that world records can only get better over the years. This may seem utterly obvious but it is nonetheless a bit of expert knowledge. It is expert knowledge that makes year an obvious explanatory variable. In general, expert knowledge comes from experience and education rather than data. It's common knowledge, for example, that in many sports, separate world records are kept for men and women. This feature of the system, obvious to experts, makes sex a sensible choice as an explanatory variable.

When people use a graph to look at how well a model fits data, they tend to look for a satisfyingly snug replication of the details. In Chapter 4 a model was shown of the relationship between monthly natural gas usage in a home and the average outdoor temperature during the month. Figure 5.1 shows that model along with a straight-line model. Which do you prefer? Look at the model in the left panel. That's a nice looking model! The somewhat irregular curve seems like a natural shape, not a rigid and artificial straight line like the model in the right panel.

Yet how much better is the curved model than the straight-line model? The straight-line model captures an important part of the relationship between natural gas usage and outdoor temperature: that colder temperatures lead to more usage. The overall slopes of the two models are very similar. The biggest discrepancy comes at warm temperatures, above 65°F. But people who know about home heating can tell you that above 65°F, homes don't need to be heated. At those temperatures gas usage is due only to cooking and water heating: a very different mechanism. To study heating, you should use only data for fairly low temperatures, say below 65°F. To study two different mechanisms — heating at low temperatures and no heating at higher temperatures — you should perhaps construct two different models, perhaps by including an interaction between temperature as a quantitative variable and a categorical variable that indicates whether the temperature is above 65°.

Humans are powerful pattern recognition machines. People can easily pick out faces in a crowd, but they can also pick out faces in a cloud or on the moon. The downside to using human criteria to judge how well a model fits data is the risk that you will see patterns that aren't warranted by the data.

To avoid this problem, you can use formal measures of how well a model fits the data, measures based on the size of residuals and that take into account chance variations in shape. Much of the rest of the book is devoted to such measures.

With the formal measures of fit, a modeler has available a strategy for finding effective models. First, fit a model that matches, perhaps roughly, what you know about the system. The straight-line model in the right panel of Figure 5.1 is a good example of this. Check how good the fit is. Then, try refining the model, adding detail by including curvy transformation terms. Check the fit again and see if the improvement in the fit goes beyond what would be expected from change.

A reasonable strategy is to start with model designs that include only main terms for your explanatory variables (and, of course, the intercept term, which is to be included in almost every model). Then add in some interaction terms, see if they improve the fit. Finally, you can try transformation terms to capture more detail.

Most of the time you will find that the crucial decision is which explanatory variables to include. Since it's difficult to graph relationships with multiple explanatory variables, the benefits of making a snug fit to a single variable are illusory.

Often the relationship between an explanatory variable and a response is rather loose. People, as good as they are at recognizing patterns, aren't effective at combining lots of cases to draw conclusions of overall patterns. People can have trouble seeing for forest for the trees. Figure 5.2 shows Galton's height data: child's height plotted against mother's height. It's hard to see anything more than a vague relationship in the cloud of data, but it turns out that there is sufficient data here to justify a claim of a pretty precise relationship. Use your human skills to decide which of the two models in Figure 5.2 is better. Pretty hard to decide, isn't it! Decisions like this are where the formal methods of

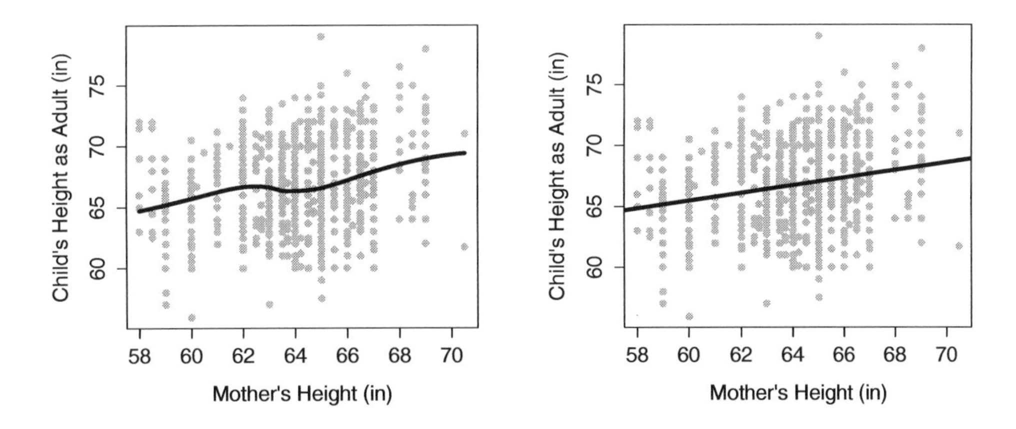

Figure 5.2: Two models of child's height as an adult versus mother's height.

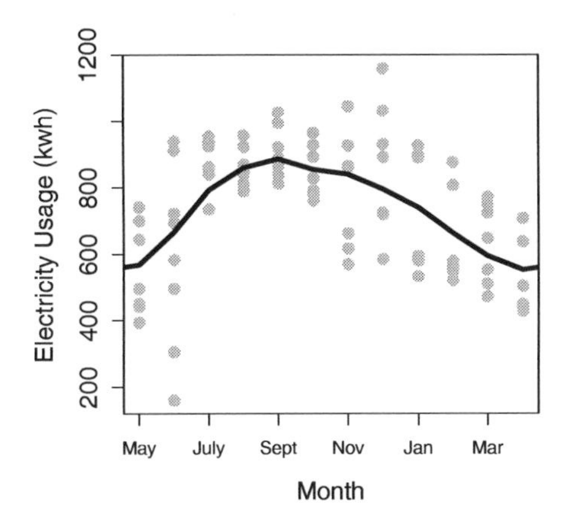

Figure 5.3: Electricity usage versus Month has a relationship that can't be captured by a straight-line model.

linear models pay off. Start with the straight-line terms then see if elaboration is warranted.

There are, however, some situations when you can anticipate that straight-line model terms will not do the job. For example, consider electricity use in a house that is heated electrically in the winter and cooled electrically in the summer. The relationship between electricity and temperature can be expected to be V-shaped — heavy use both for very cold temperatures when the heat is on and for very warm temperatures when air conditioning is in use. Or consider the relationship between college grades and participation in extra-curricular activities such as school sports, performances, the newspaper, etc. There's reason to believe that some participation in extra-curricular activities is associated with higher grades. This might be because students who are doing well have the confidence to spend time on extra-curriculars. But there's also reason to think that very heavy participation takes away from the time students need to study. So the overall relationship between grades and participation might be Λ-shaped. Relationships that have a V- or Λ-shape won't be effectively captured by straight-line models; transformation terms and interaction terms will be required.

5.10 Computational Technique

The `lm` operator finds model coefficients. To illustrate, here's a pair of statements that read in a data frame and fit a model to it:

```
> swim = ISMdata("swim100m.csv")
> mod = lm( time ~ year + sex, data=swim)
```

The first argument to `lm` is a model design, the second is the data frame.

The object created by `lm` — here given the name `mod` — contains a variety of information about the model. To access the coefficients themselves, use the `coef` operator applied to the model:

```
> coef(mod)
(Intercept)        year        sexM
   555.7168     -0.2515     -9.7980
```

As shorthand to display the coefficients, just type the name of the object that is storing the model:

```
> mod

Call:
lm(formula = time ~ year + sex, data = swim)

Coefficients:
(Intercept)         year          sexM
    555.717       -0.251        -9.798
```

A more detailed report can be gotten with the `summary` operator. This gives additional statistical information that will be used in later chapters:

```
> summary(mod)
Coefficients:
            Estimate Std. Error t value Pr(>|t|)
(Intercept) 555.7168    33.7999   16.44  < 2e-16
year         -0.2515     0.0173  -14.52  < 2e-16
sexM         -9.7980     1.0129   -9.67  8.8e-14
```

From time to time in the exercises, you will be asked to calculate model values "by hand." This is accomplished by multiplying the coefficients by the appropriate values and adding them up. For example, the model value for a male swimmer in 2010 would be:

```
> 555.7 - 0.2515*2010 - 9.798
[1] 40.39
```

Notice that the "value" used to multiply the intercept is always 1, and the "value" used for a categorical level is either 0 or 1 depending on whether there is a match with the level. In this example, since the swimmer in question was male, the value of sex**M** is 1. If the swimmer had been female, the value for sex**M** would have been 0.

When a model includes interaction terms, the interaction coefficients need to be multiplied by all the values involved in the interaction. For example, here is a model with an interaction between year and sex:

```
> mod2 = lm( time ~ year*sex, data=swim)
> coef(mod2)
(Intercept)        year       sexM   year:sexM
   697.3012     -0.3240  -302.4638      0.1499
> 697.3 -0.3240*2010 - 302.5 +0.1499*2010
[1] 44.86
```

The `year:sexM` coefficient is being multiplied by the year (2010) and the value of sex**M**, which is 1 for this male swimmer.

5.10.1 Other Useful Operators

cross will combine two categorical variables into a single variable. For example, in the Current Population Survey data, the variable sex has levels F and M, while the variable race has levels W and NW. Crossing the two variables combines them; the new variable has four levels: F.NW, M.NW, F.W, M.W:

```
> cross(cps$sex,cps$race)
 [1] M.W  M.W  F.W  F.W  M.W  F.W  F.W  M.W  M.W
[10] F.W  M.W  M.W  M.W  M.W  M.W  M.W  M.W  M.NW
[19] M.NW M.W  F.NW M.NW F.W  F.W  M.NW F.W  F.W
     ... and so on.
Levels: F.NW M.NW F.W M.W
```

as.factor will convert a quantitative variable to a categorical variable. This is useful when a quantity like month has been coded as a number, say 1 for January and 2 for February, etc. but you do not want models to treat it as such.

To illustrate, consider two different models of the usage temperature versus month:

```
utils = ISMdata("utilities.csv")
mod1 = lm( temp ~ month, data=utils)
mod2 = lm( temp ~ as.factor(month), data=utils)
```

DATA FILE
utilities.csv

Here are the graphs of those models:

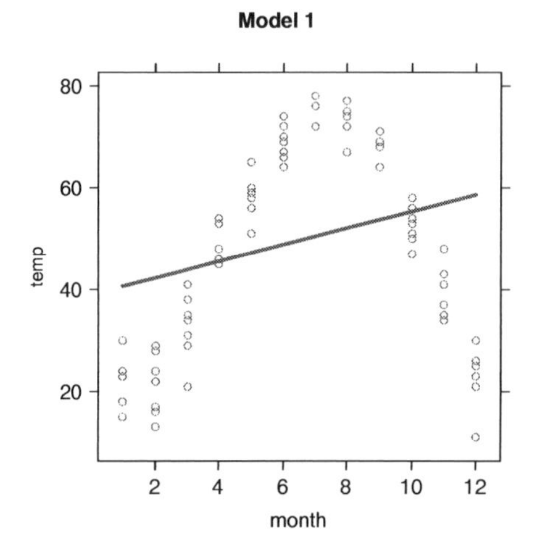

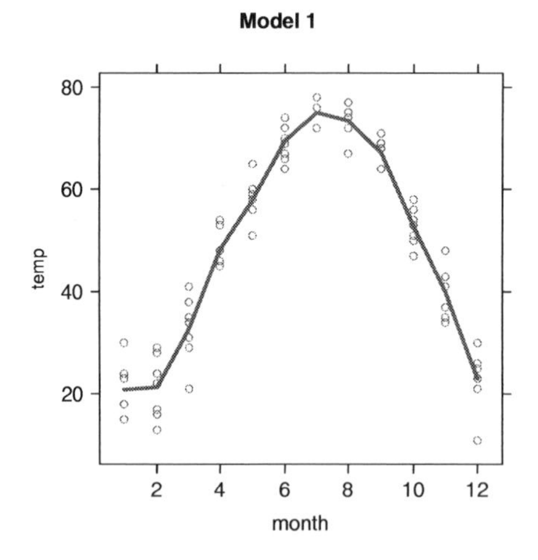

In the first model, month is treated quantitatively, so the model term month produces a straight-line relationship that does not correspond well to the data.

In the second model, month is treated categorically, allowing a more complicated model relationship.

Chapter 6 Fitting Models to Data

With four parameters, I can fit an elephant; with five I can make it wiggle its trunk. — John von Neumann

The purpose of models is not to fit the data but to sharpen the questions. — Samuel Karlin

The selection of variables and terms is part of the modeler's art. It is a creative process informed by your goals, your understanding of how the system you are studying works, and the data you can collect. Later chapters will deal with how to evaluate the success of your creation as well as general principles for designing effective models that capture the salient aspects of your system.

This chapter is about how to find the coefficients. The goal is to find the "best" coefficients — to fit the model to the data. The outcome of fitting is the coefficients that capture as well as possible the variation in the response variable.

Fitting a model — once you have specified the design — is an entirely automatic process that requires no human decision making. It's ideally suited to computers and, in practice, you will always use the computer to find the best fit. Nevertheless, it's important to understand the way in which the computer finds the coefficients. Much of the logic of interpreting models is based on how firmly the coefficients are tied to the data, to what extent the data dictate precise values of the coefficients. The fitting process reveals this. In addition, depending on your data and your model design, you may introduce ambiguities into your model that make the model less reliable than it could be. These ambiguities, which stem from **redundancy**, are revealed by the fitting process.

6.1 The Least Squares Criterion

Fitting a model is somewhat like buying clothes. You choose the kind of clothing that you want: the cut, style, and color. There are usually several or many different items of this kind, and you have to pick the one that matches your body appropriately. In this analogy, the style of clothing is the model design — you choose this. Your body shape is like the data — you're pretty much stuck with what you've got.

There is a system of sizes of clothing: the waist, hip, and bust sizes, inseam, sleeve length, neck circumference, foot length, etc. By analogy, these are the model coefficients. When you are in the fitting room of a store, you try on different items of clothing with a range of values of the coefficients. Perhaps you bring in pants with waist sizes 31 through 33 and leg sizes 29 and 30. For each of the pairs of pants that you try on, you get a sense of the overall fit, although this judgment can be subjective. By trying on all the possible sizes, you can find the item that gives the best overall match to your body shape.

Similarly, in fitting a model, the computer tries different values for the model coefficients. Unlike clothing, however, there is a very simple, entirely objective criterion for judging the overall fit called the **least squares** criterion. To illustrate, consider the following very simple data set of age and height in children:

age (years)	height (cm)
5	80
7	100
10	110

As the model design, take height $\sim$ 1 + age. To highlight the link between the model formula and the data, here's a way to write the model formula that includes not just the names of the variables, but the actual column of data for each variable. Each of these columns is called a **vector**. Note that the intercept vector is just a column of 1s.

$$\text{Model Values height} = c_1 \underset{\text{intercept}}{\begin{bmatrix} 1 \\ 1 \\ 1 \end{bmatrix}} + c_2 \underset{\text{age}}{\begin{bmatrix} 5 \\ 7 \\ 10 \end{bmatrix}} .$$

The goal of fitting the model is to find the best numerical values for the two coefficients c_1 and c_2.

To start, just guess something: $c_1 = 5$ and $c_2 = 10$. Just a guess. Plugging in these guesses to the model formula gives a model value of height for each case:

$$\text{Model Values height} = 5 \underset{\text{intercept}}{\begin{bmatrix} 1 \\ 1 \\ 1 \end{bmatrix}} + 10 \underset{\text{age}}{\begin{bmatrix} 5 \\ 7 \\ 10 \end{bmatrix}} = \begin{bmatrix} 55 \\ 75 \\ 105 \end{bmatrix} .$$

Arithmetic with vectors is done in the ordinary way. Coefficients multiply each element of the vector, so that $10\begin{bmatrix} 5 \\ 7 \\ 10 \end{bmatrix}$ gives $\begin{bmatrix} 50 \\ 70 \\ 100 \end{bmatrix}$. To add or subtract two vectors means to add or subtract the corresponding elements: $\begin{bmatrix} 5 \\ 5 \\ 5 \end{bmatrix} + \begin{bmatrix} 50 \\ 70 \\ 100 \end{bmatrix}$ gives $\begin{bmatrix} 55 \\ 75 \\ 105 \end{bmatrix}$.

The residuals tell you how the response values (that is, the actual heights) differ from the model value of height. For this model, the residuals are

$$\overset{\text{height}}{\begin{bmatrix} 80 \\ 100 \\ 110 \end{bmatrix}} - \overset{\text{Model height}}{\begin{bmatrix} 55 \\ 77 \\ 105 \end{bmatrix}} = \overset{\text{residuals}}{\begin{bmatrix} 25 \\ 23 \\ 5 \end{bmatrix}}.$$

The smaller the residuals, the better the model values match the actual response values.

Try another guess of the coefficients, say $c_1 = 40$ and $c_2 = 8$. Plugging in these guesses ...

$$\text{Model Values } \text{height} = 40 \overset{\text{intercept}}{\begin{bmatrix} 1 \\ 1 \\ 1 \end{bmatrix}} + 8 \overset{\text{age}}{\begin{bmatrix} 5 \\ 7 \\ 10 \end{bmatrix}} = \overset{\text{Model height}}{\begin{bmatrix} 80 \\ 96 \\ 120 \end{bmatrix}}$$

The residuals from the new coefficients are

$$\text{residuals} = \begin{bmatrix} 80 \\ 100 \\ 110 \end{bmatrix} - \begin{bmatrix} 80 \\ 96 \\ 120 \end{bmatrix} = \begin{bmatrix} 0 \\ 4 \\ -10 \end{bmatrix}.$$

Which of the two model formulas is better? For the first case — the 5-year old who is 80 cm tall — the second formula is exactly right, but the first formula understates the height by 25 cm. Similarly, the second formula is better for the second case. But for the third case, the residual from the first formula is only 5 cm, while the second formula gives a bigger residual: -10 cm.

It is not uncommon in comparing two formulas to find, as in this model, one formula is better for some cases and the other formula is better for other cases. This is not so different from clothing, where one item may be too snug in the waist but just right in the leg, and another item may be perfect in the waist but too long in the leg.

What's needed is a way to judge the *overall* fit. In modeling, the overall fit of a model formula is measured by a single number: the sum of squares of the residuals. For the first formula, this is $25^2 + 23^2 + 5^2 = 1179$. For the second formula, the sum of squares of the residuals is $0^2 + 4^2 + (-10)^2 = 116$. By this measure, the second formula gives a much better fit.

To find the **best** fit, just keep trying different formulas until one is found that gives the least sum of square residuals: in short, the **least squares**. The process isn't nearly so laborious as it might seem. There are systematic ways to find the coefficients that give the least sum of square residuals, just as there are systematic ways to find clothing that fits without trying on every item in the store. The details of these systematic methods are not important here. Later chapters explore the geometry of fitting, making it easy to see how a computer can quickly find the coefficients that give the best fitting model formula.

Using the least squares criterion, a computer will quickly find the best coefficients for the height vs age model to be $c_1 = 54.211$ and $c_2 = 5.789$. Plugging in these coefficients to the model formula produces fitted model values of $\begin{bmatrix} 83.158 \\ 94.737 \\ 112.105 \end{bmatrix}$, residuals of $\begin{bmatrix} -3.158 \\ 5.263 \\ -2.105 \end{bmatrix}$, and therefore a sum of square residuals of 42.105, clearly much better than either previous two model formulas.

How do you know that the coefficients that the computer gives are indeed the best? You can try varying them in any way whatsoever. Whatever change you make, you will discover that the sum of square residuals is never going to be less than it was for the coefficients that the computer provided.

The smallest possible sum of square residuals is zero. This can occur only when the model formula gives the response values exactly, that is, when the fitted model values are an exact match with the response values. Because models typically do not capture all of the variation in the response variable, usually you do not see a zero sum of square residuals. When you do, however, it may well be a sign that there is something wrong, that your model is too detailed.

It's important to keep in mind that the coefficients found using the least squares criterion are the best possible but only in a fairly narrow sense. They are the best *given* the design of the model and *given* the data used for fitting. If the data change (for example, you collect new cases), or if you change the model design, the best coefficients will likely change as well. It makes sense to compare the sum of square residuals for two model formulas only when the two formulas have the same design and are fitted with the same data. So don't try to use the sum of square residuals to decide which of two different designs are better. Tools to do that will be introduced in later chapters.

Why the sum of square residuals? Why not some other criterion? The answers are in part mathematical, in part statistical, and in part historical.

In the late 18th century, three different criteria were competing, each of which made sense:

1. The least squares criterion, which seeks to minimize the sum of square residuals.

2. The least absolute value criterion. That is, rather than the sum of square residuals, one looks at the sum of absolute values of the residuals.

3. Make the absolute value of the residual as small as possible for the single worst case.

All of these criteria make sense in the following ways. Each tries to make the residuals small, that is, to make the fitted model values match closely to the response values. Each treats a positive residual in the same way as a negative residual; a mismatch is a mismatch regardless of whether the fitted values are too high or too low. The first and second criteria each combine all of the cases in determining the best fit.

The third criterion is hardly ever used because it allows one or two cases to dominate the fitting process — only the two most extreme residuals count in evaluating the model. Think of the third criterion as a dead end in the historical evolution of statistical practice.

Computationally, the least squares criterion leads to simpler procedures than the least absolute value criterion. In the 18th century, when computing was done by hand, this was a very important issue. It's no longer so, and there can be good statistical reasons to favor a least absolute value criterion when it is thought that the data may contain outliers.

The least squares criterion is best justified when variables have a bell-shaped distribution. Insofar as this is the situation — and it often is — the least squares criterion is arguably better than least absolute value.

Another key advantage of the least squares criterion is in interpretation, as you'll see in the next section.

6.2 Partitioning Variation

A model is intended to explain the variation in the response variable using the variation in the explanatory variables. It's helpful in quantifying the success of this to be able to measure how much variation there is in the response variable, how much has been explained, and how much remains unexplained. This is a partitioning of variation into parts.

You might think that such a partitioning would always be possible and would always make sense, but in fact it depends on how one chooses to measure variation. To illustrate, suppose you and your friend earn $50 by door-to-door bagel deliveries. You decide to partition this: you keep $30 and your friend gets $20. (Perhaps you worked harder than your friend.) Whichever way you partition the money between you and your friend, the total amount will always equal exactly the sum of the amounts that you and your friend get. Obvious.

But suppose you decided — bizarrely, admittedly — to measure money not in the ordinary sense but as "square root dollars." The two of you start out with $\sqrt{\$50} = 7.07$ square-root dollars. After splitting it up, You have $\sqrt{\$30} = 5.48$ square dollars and your friend gets $\sqrt{\$20} = 4.47$ square-root dollars. Notice that the partitioning doesn't work: $7.07 \neq 5.48 + 4.47$. That's one reason why it's natural to stick with counting money in an ordinary way, and not with silly square-root dollars.

Now consider how variation in the response variable is broken into parts by a model, partitioned into the variation in the fitted model values and the variation in the residuals.

One way to measure the variation is with the standard deviation. (See Chapter 3.) The table below gives the standard deviations of the response variable, the fitted model values, and the residuals for the best-fitting model (found by the computer) for height versus age:

	Source	Standard Deviation
	Fitted Model values	14.570
+	Residuals	4.588
$\neq$	Response Variable (height)	15.275

A little arithmetic shows that the partitioning does not work: $15.275 \neq 14.570 + 4.588$.

The problem is that measuring variability with a standard deviation is very much like using "square-root dollars."

If you want the partitioning to work, you have to measure variability in the right way. One of the consequences of using the least squares criterion to fit models is that there is actually a simple way to measure variability that does produce a meaningful partitioning: measure variability with the variance.

	Source	Variance
	Fitted Model values	212.28
+	Residuals	21.05
=	Response Variable (height)	233.33

Now the partitioning works: $233.33 = 212.28 + 21.05$.

The variance — and its square root, the standard deviation — measure deviation from a central value. Although the standard deviation has nicer units, the variance is best from a deeper point of view: it provides a natural partitioning of variation.

Another sensible choice is the sum of squares, which also permits the partitioning of the response variable into parts: the fitted model values and the residuals.

	Source	Sum of Squares
	Fitted Model values	28457.89
+	Residuals	42.11
=	Response Variable (height)	28500.00

Again, the partitioning works: $28500.00 = 28457.89 + 42.11$.

Perhaps this use of squares — the variance, the sum of squares — reminds you of another kind of partitioning. The **Pythagorean theorem** says that the length of the hypotenuse of a right triangle and the lengths of the two other sides are related like this: $A^2 = B^2 + C^2$. This is a kind of partitioning: the square length of the hypotenuse can be partitioned into two other square lengths: those of the legs.

In later chapters you'll see that there is a close connection between the right-triangle geometry, the least-squares criterion for fitting, and the partitioning of variation.

Example 6.1: Residuals in Global Temperature In studying human-induced global climate change, one issue is the extent to which fluctuations in global temperatures reflect naturally occurring climate variability. A strategy for addressing this is to model global temperature using as explanatory variables measurements of year-to-year natural variability in some climate systems.

Figure 6.1 shows a record of global mean temperature from the late 1800s through this century from Thompson et al. [11]. The global mean temperature shows a slow increase of the sort described by the phrase "global warming." But there is also a sharp dip in the middle. Might this be due to short term fluctuations, such as those due to the El Niño/Southern Oscillation (ENSO) or the "cold oceans-warm land" (COWL) system in the Northern Hemisphere?

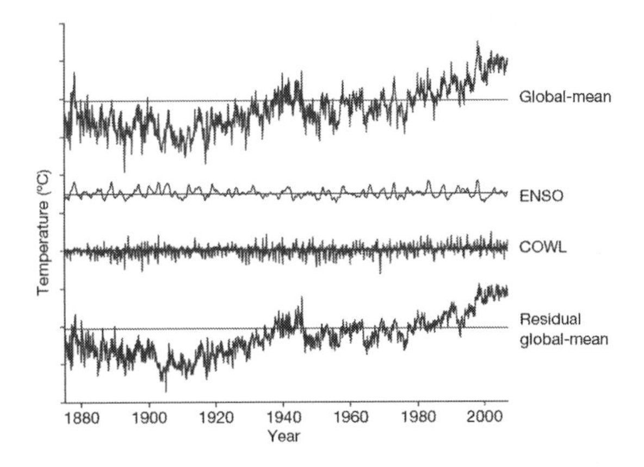

Figure 6.1: Global mean temperatures (top) partitioned into variation associated with natural climate variation ("ENSO" and "COWL") and residual variation (at the bottom). From Thompson *et al.* [11]

In the figure, the global mean temperature data has been partitioned into three components: (1) variation associated with ENSO, (2) variation associated with COWL, and (3) the residuals. The residuals show a long-term warming trend, but they also show a sharp short-term discontinuity in 1945 when the residuals suddenly drop. This suggests that the drop in 1945 is not due to ENSO or COWL.

Thompson *et al.* investigated the 1945 discontinuity, in part by considering additional explanatory variables such as timing of large volcanic eruptions. They traced the 1945 discontinuity to an interesting form of sampling bias. During World War II, the fraction of sea-surface measurements made by US ships increased steadily, up to about 80% of all measurements in August 1945. Then there was a sudden drop to about 50%, as measurements from UK ships increased. US ships tended to make measurements from engine-room intakes; these are biased warmer than the ambient sea water. UK ships tended to use uninsulated buckets to make measurements; these are biased cold. When the mix of measurements shifted from the US to the UK, so did the bias in the measurements.

Residuals are not necessarily random or impossible to explain, they are just the part of the variation in your response variable that is not explained by your model.

6.3 Redundancy

Data sets often contain variables that are closely related. For instance, consider the following data on prices of used cars (the Ford Taurus model), the year the

DATA FILE
used-fords.csv

car was built, and the number of miles the car has been driven.

Price	Year	Mileage
14997	2008	22613
8995	2007	53771
7990	2006	36050
18990	2008	25896

... and so on.

Although the year and mileage variables are different, they are related: older cars typically have higher mileage than newer cars. A group of students thought to see how price depends on year and mileage. After collecting a sample of 635 cars advertised on the Internet, they fit a model price ~ mileage + year. The coefficients came out to be:

price ~ mileage + year

Term	Coefficient	Units
Intercept	-1136311.500	dollars
mileage	-0.069	dollars/mile
year	573.500	dollars/year

The coefficient on mileage is straightforward enough; according to the model the price of these cars typically decreases by 6.9 cents per mile. But the intercept seems strange: very big and negative. Of course this is because the intercept tells the model value at zero mileage in the year zero! They checked the plausibility of the model by plugging in values for a 2006 car with 50,000 miles: $-1136312 - 0.069 \times 50000 + 2006 \times 573.50$ giving $10679.50 — a reasonable sounding price for such a car.

The students thought that the coefficients would make more sense if they were presented in terms of the age of the car rather than the model year. Since their data was from 2009, they calculated the age by taking $2009 -$ year. Then they fit the model price ~ mileage + age.

price ~ mileage + age

Term	Coefficient	Units
Intercept	15850.000	dollars
mileage	-0.069	dollars/mile
age	-573.500	dollars/year

The mileage coefficient is unchanged, but now the intercept term looks much more like a real price: the price of a hypothetical brand new car with zero miles. The age coefficient makes sense; typically the cars decrease in price by $573.50 per year of age. Again, the students calculated the model for a 2006 car (thus age 3 years in 2009) with 50,000 miles: $10679.50, the same as before.

Surprised that two different sets of coefficients could give the same model value, the students computed the fitted model value for all the cars in their data set for both models. Exactly the same for every car.

It makes sense that the two models should give the same model values. They are based on exactly the same information. The only difference is that in one model the age is given by the model year of the car, in the other by the age itself.

The students decided to experiment. "What happens if we include both age and year in the model?" The started with the model price ~ mileage + year + age. The software responded by giving coefficients for the first three model terms — the same intercept, mileage and year coefficients as in the first table — but reported the age coefficient as "NA," not available. Why?

Actually, there is no mathematical reason why the computer could not have given coefficients for all four model terms. For example, any of the following sets of coefficients would give exactly the same model values for *any* inputs of year and mileage, with age being appropriately calculated from year.

Term	Set 1	Set 2	Set 3	Set 4
Intercept	15850.000	-185050.000	-734511.500	-1136311.500
mileage	-0.069	-0.069	-0.069	0.069
year	0.000	100.000	373.500	573.500
age	-573.500	-473.500	-200.000	0.000

The overall pattern is that the coefficient on age minus the coefficient on year has to be -573.5. This works because age and year are basically the same thing since age $= 2009 -$ year.

The dual identity of age and year is **redundancy**. It arises whenever an explanatory model vector can be modeled exactly in terms of the other explanatory vectors. Here, age can be computed by 2009 times the intercept plus -1 times the year.

The problem with redundancy is that it creates ambiguity. Which of the four sets of coefficients in the table above is the right one? Whenever there is redundancy, there is no unique set of coefficients that gives the best fit of the model to the data.

In order to avoid this ambiguity, modeling software is written to spot any redundancy and drop the redundant model vectors from the model. One possibility would be to report a coefficient of zero for the redundant vector. But this could be misleading. For instance, the user might interpret the zero coefficient on Set 1 above as meaning that year isn't associated with price. But that would be wrong; year is a very important determinant of price, it's just that all the relationship is represented by the coefficient on age in Set 1. Rather than report of a coefficient of zero, it's helpful when software reports the redundancy by displaying a NA. This makes it clear that it is redundancy that is the issue and not that a coefficient is genuinely zero.

Example 6.2: Almost redundant

When model vectors are redundant, modeling software can spot the situation and do the right thing. But sometimes, the redundancy is only approximate. In such cases it's important for you to be aware of the potential for problems.

In collecting the Current Population Survey wage data, interviewers asked people their age and the number of years of education they had. From this, they computed the number of years of experience. They presumed that kids spend six years before starting school. So a 40-year old with 12 years of education would have 22 years of experience: $6 + 12 + 22 = 40$. This creates redundancy of experience with age and education. Ordinarily, the computer could identify such a situation. However, there is a mistake in the CPS data. Case 350 is an 18-year old woman with 16 years of education. This seems hard to believe since very few people start their education at age 2. Presumably, either the age or education were mis-recorded. But either way, this small mistake in one case means that the redundancy among experience, age, and education is not exact. It also means that standard software doesn't spot the problem when all three of these variables are included in a model. As a result, coefficients on these variables are highly unreliable.

The situation of approximate redundancy is called **multi-collinearity**. With multi-collinearity, as opposed to exact redundancy, there is a unique least squares fit of the model to the data. However, this uniqueness is hardly a blessing. What breaks the tie to produce a unique best fitting set of coefficients are the small deviations from exact redundancy. These are often just a matter of random noise, arithmetic round-off, or mistakes as with case 350 in the Current Population Survey data. As a result, the choice of the winning set of coefficients is to some extent arbitrary and potentially misleading.

The costs of multi-collinearity can be measured when fitting a model. (See section 14.7.) When the costs of multi-collinearity are too high, the modeler often chooses to take terms out of the model.

6.4 Computational Technique

Using the `lm` software is mainly a matter of familiarity with the model design language. Computing the fitted model values and the residuals is done with the `fitted` and `resid`. These operators take a model as an input. To illustrate:

```
swim = ISMdata("swim100m.csv")
mod1 = lm( time ~ year + sex, data=swim)
```

DATA FILE
swim100m.csv

Once you have constucted the model, you can use `fitted` and `resid`:

```
> fitted(mod1)
    1     2     3     4     5     6     7     8
66.88 66.13 65.62 65.12 63.61 63.11 62.61 62.10
 ... and so on ...
   55    56    57    58    59    60    61    62
58.82 58.32 57.82 56.31 54.80 54.30 52.79 51.78
> resid(mod1)
     1      2      3      4      5      6
-1.081 -0.526 -2.823 -3.520 -2.212 -2.709
  ... and so on.
```

Sometimes it's helpful to look for outliers in the residuals, and to plot the residuals versus the fitted model values or versus explanatory variables. For instance:

```
> bwplot( as.numeric(resid(mod1)) )
> subset(swim, outlier(resid(mod1)))
   year time sex
32 1908 95.0   F
33 1910 86.6   F
34 1911 84.6   F
> xyplot( resid(mod1) ~ fitted(mod1) )
> xyplot( resid(mod1) ~ year, data=swim)
```

(Technical note: The `as.numeric` operator translates the residuals to a format that `bwplot` will work with.)

6.4.1 Sums of Squares

Computations can be performed on the fitted model values and the residuals, just like any other quantity:

```
> mean( fitted(mod1))
[1] 59.92
> var( resid(mod1))
[1] 15.34
> sd( resid(mod1))
[1] 3.917
> summary( resid(mod1))
 Min.  1st Qu.  Median    Mean   3rd Qu.   Max.
-4.70  -2.70    -0.597  2.98e-16  1.28     19.1
```

Sums of squares are very important in statistics. Here's how to calculate them for the response values, the fitted model values, and the residuals:

```
> sum( swim$time^2 )
[1] 228635
> sum( fitted(mod1)^2 )
[1] 227699
> sum( resid(mod1)^2 )
[1] 935.8
```

The partitioning of variation by models is seen by the way the sum of squares of the fitted and the residuals add up to the sum of squares of the response:

```
> 227699 + 935.8
[1] 228635
```

Don't forget the squaring stage of the operation! The sum of the residuals (without squaring) is very different from the sum of squares of the residuals:

```
> sum( resid(mod1) )
[1] 1.849e-14
> sum( resid(mod1)^2 )
[1] 935.8
```

Take care in reading numbers formatted like `1.849e-14`. The notation stands for 1.849×10^{-14}. That number, 0.00000000000001849, is effectively zero compared the residuals themselves!

6.4.2 Redundancy

The `lm` operator will automatically detect redundancy and deal with it by leaving the redundant terms out of the model.

To see how redundancy is handled, here is an example with a constructed redundant variable in the swimming dataset. The following statement adds a new variable to the dataframe counting how many years after the end of World War II each record was established:

```
> swim$afterwar = swim$year - 1945
```

Here is a model that doesn't involve redundancy

```
> mod1 = lm( time ~ year + sex, data=swim)
> coef(mod1)
(Intercept)        year        sexM
   555.7168     -0.2515     -9.7980
```

When the redundant variable is added in, `lm` successfully detects the redundancy and handles it. This is indicated by a coefficient of NA on the redundant variable.

```
> mod2 = lm( time ~ year + sex + afterwar, data=swim)
> coef(mod2)
(Intercept)        year        sexM    afterwar
   555.7168     -0.2515     -9.7980          NA
```

In the absence of redundancy, the model coefficients don't depend on the order in which the model terms are specified. But this is not the case when there is redundancy, since any redundancy is blamed on the later variables. For instance, here afterwar has been put first in the explanatory terms, so `lm` identifies year as the redundant variable:

```
> mod3 = lm( time ~ afterwar + year + sex, data=swim)
> coef(mod3)
(Intercept)    afterwar        year        sexM
    66.6199     -0.2515          NA     -9.7980
```

Even though the coefficients are different, the fitted model values and the residuals are exactly the same (to within computer round-off) regardless of the order of the model terms.

```
> fitted(mod2)
    1     2     3     4     5     6
66.88 66.13 65.62 65.12 63.61 63.11 and so on.
> fitted(mod3)
    1     2     3     4     5     6
66.88 66.13 65.62 65.12 63.61 63.11 and so on.
```

Note that whenever you use a categorical variable and an intercept term in a model, there is a redundancy. This is not shown explicitly. For example, here is a model with no intercept term, and both levels of the categorical variable sex show up with coefficients:

```
> lm( time ~ sex - 1, data=swim)
Coefficients:
sexF  sexM
65.2  54.7
```

If the intercept term is included (as it is by default unless -1 is used in the model formula), one of the levels is simply dropped in the report:

```
> lm( time ~ sex, data=swim)
Coefficients:
(Intercept)          sexM
       65.2         -10.5
```

Remember that this coefficient report implicitly involves a redundancy. If the software had been designed differently, the report might look like this:

```
(Intercept)     sexF       sexM
       65.2       NA      -10.5
```

6.4.3 Technical Notes: Missing Data

Occasionally, you may have a data frame with missing data. The lm program will handle this sensibly by excluding those cases where one or more of the variables used in the model are missing. Those cases with missing data will show up in the lists of residuals and fitted values as NAs. (This depends on the setting of the na.action argument to lm. In the ISM library, this has been set to na.exclude.)

The NAs can cause a problem in calculations involving the residuals or the fitted values. For instance, suppose you have constructed a model called mymodel fitted to a data frame with missing data. Here is the sum of squares of the fitted values:

```
> sum( fitted(mymodel)^2 )
[1] NA
```

Any NAs in the fitted values propagate through to the overall sum.

There are a couple of ways to deal with this:

```
> sum( fitted(mymodel)^2, na.rm=TRUE )
[1] 683434.3
> sum( na.omit(fitted(mymodel))^2 )
[1] 683434.3
```

You can also delete any cases with missing data from the data frame *before* fitting the model:

```
> cleaned = na.omit(swim)
```

Be careful, however, since this will exclude any cases that have any missing variables, even if those variables are not involved in the model.

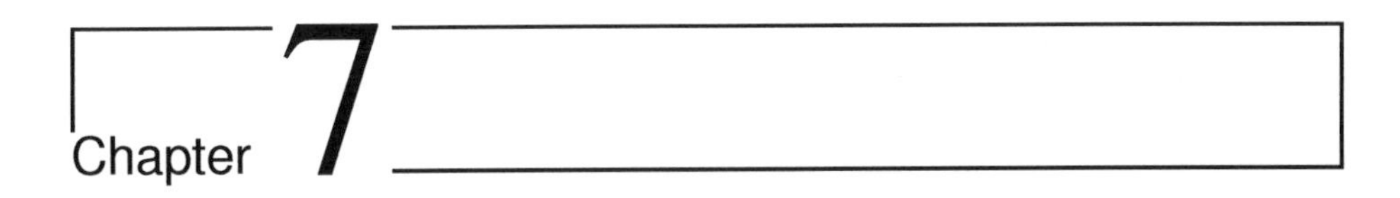

Chapter 7

Measuring Correlation

The invalid assumption that correlation implies cause is probably among the two or three most serious and common errors of human reasoning. — Stephen Jay Gould

"Co-relation or correlation of structure" is a phrase much used ... but I am not aware of any previous attempt to define it clearly, to trace its mode of action in detail, or to show how to measure its degree.' — Francis Galton, 1888

The last chapter described how the variance of the response variable can be divided into two parts: the variance of the fitted model values and the variance of the residuals. This partitioning is at the heart of a statistical model; the more of the variation that's accounted for by the model, the less is left in the residuals.

Because of the partitioning, an effective way to summarize a model is the proportion of the total variation in the response variable that is accounted for by the model. This description is called the R^2 (**"R-Squared"**) of the model. It is a ratio:

$$R^2 = \frac{\text{variance of fitted model values}}{\text{variance of response values}}.$$

Another name for R^2 is the **coefficient of determination**, but this is not a coefficient in the same sense used to refer to a multiplier in a model formula.

7.1 Properties of R^2

R^2 has a nice property that makes it easy to interpret: its value is always between zero and one. When $R^2 = 0$, the model accounts for none of the variance of the response values: the model is useless. When $R^2 = 1$, the model captures all

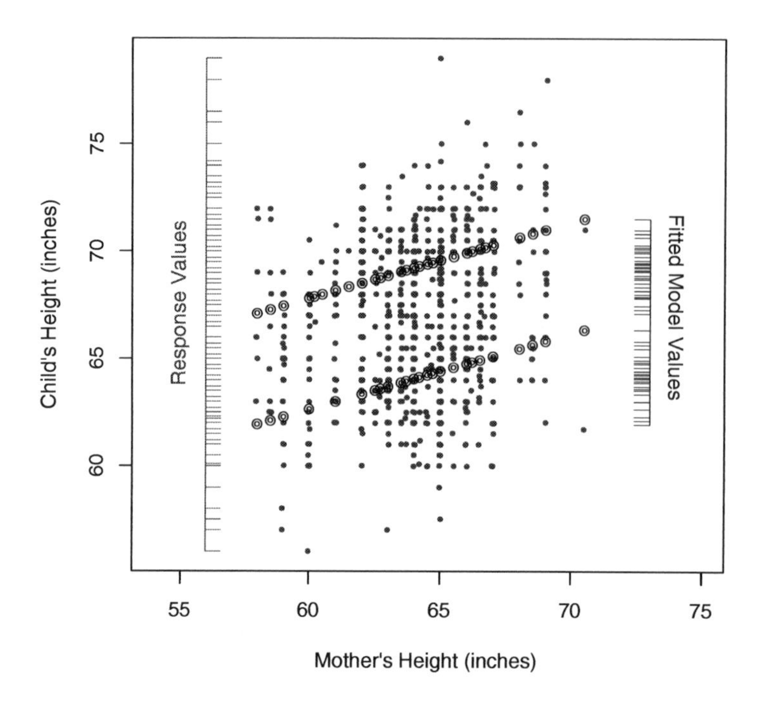

Figure 7.1: The small dots show child's height versus mother's height from Galton's data. The bigger symbols show the fitted model values from the linear model height ~ 1 + mother + sex that includes both mother's height and child's sex as explanatory variables.

of the variance of the response values: the model values are exactly on target. Typically, R^2 falls somewhere between zero and one, meaning that the model accounts for some, but not all, of the variance in the response values.

The history of R^2 can be traced to a paper[12] presented to the Royal Society in 1888 by Francis Galton. It is fitting to illustrate R^2 with some of Galton's data: measurements of the heights of parents and their children.

DATA FILE
galton.csv

Figure 7.1 shows a scatter plot of the children's height versus mother's height. Also plotted are the fitted model values for the model height ~ mother + sex. The model values show up as two parallel lines, one for males and one for females. The slope of the lines shows how the child's height typically varies with the mother's height.

Two rug plots have been added to the scatter plot in order to show the distribution of the response values (child's height) and the fitted model values (the output of the model). The rug plots are positioned vertically because each of them relates to the response variable height, which is plotted on the vertical axis.

The spread of values in each rug plot gives a visual indication of the variation. It's evident that the variation in the response variable is larger than the variation in the fitted model values. The variance quantifies this. For height, the variance is 12.84 square-inches. (Recall that the units of the variance are always the square of the units of the variable.) The fitted model values have a variance of 7.21 square-inches. The model captures a little more than half of the variance; the R^2 statistic is

$$R^2 = \frac{7.21}{12.84} = 0.56$$

One nice feature of R^2 is that the hard-to-interpret units of variance disappear — square-inches is a strange way to describe height! Since R^2 is the ratio of the response value variance to the fitted model value variance, the units cancel out. R^2 is unitless.

Example 7.1: Quantifying the capture of variation Back in Chapter 4, a model of natural gas usage and a model of wages were presented. It was claimed that the natural gas model captured more of the variation in natural gas usage than the wage model captured in the hourly wages of workers. At one level, such a claim is absurd. How can you compare natural gas usage to hourly wages? If nothing else, they have completely different units.

R^2 is what enables this to be done. For any model, R^2 describes the fraction of the variance in the response variable that is accounted for by the model. Here are the R^2 for the two models:

- wage ~ sex*married: $R^2 = 0.07$
- ccf ~ temp: $R^2 = 0.91$

Sex and marital status explain only a small fraction of the variation in wage. Temperature explains the large majority of the variation in natural gas usage (ccf).

7.2 Simple Correlation

The idea Galton introduced in 1888 is that the relationship between two variables — "co-relation" as he phrased it — can be described with a single number. This is now called the **correlation coefficient**, written r and often simply called "r".

The correlation coefficient is still very widely used. Part of the reason for this is that the single number describes an important mathematical aspect of the relationship between two variables: the extent to which they are aligned or **collinear**.

The correlation coefficient r is the square root of R^2 for the simple straight-line model: response ~ explanatory variable. As with all square roots, there is a choice of negative or positive. The sign of the r is chosen to indicate the slope of the model line.

It may seem very limiting, in retrospect, to define correlation in terms of such a simple model. But Galton was motivated by a wonderful symmetry involved in the correlation coefficient: the R^2 from the straight-line model of A versus B is the same as that from the straight-line model of B versus A. In this sense, the correlation coefficient describes the relationship between the two variables and not merely the dependence of one variable on another. Of course, a more modern perspective on relationships accepts that additional variables can be involved as well as interaction and transformation terms.

Example 7.2: Relationships without Correlation Consider the how the temperature varies over the course of the year. In most places, temperature changes predictably from month to month: a very strong relationship. Figure 7.2 shows data for several years of the average temperature in St. Paul, Minnesota (in the north-central US). Temperature is low in the winter months (January, February, December, months 1, 2, and 12) and high in the summer months (June, July, August, months 6, 7, and 8).

The Figure also shows a very simple straight-line model: `temperature` ~ `month`. This model doesn't capture very well the relationship between the two variables. Still, like any model, it has an R^2. Since the response values have a variance of 409.3 deg.F^2 and the fitted model values have a variance of 25.6 deg.F^2 the R^2 value works out to be $R^2 = \frac{25.6}{409.3} = 0.0624$.

As a way to describe the overall relationship between variables, r is limited. Although r is often described as measuring the relationship between two variables, it's more accurate to say that r quantifies a particular model rather than "the relationship" between two variables.

But not all relationships are linear. The simple correlation coefficient r does not reflect any nonlinear relationship in the data.

If you want to describe a relationship, you need to choose a model appropriately. For the temperature versus month data, the straight line model isn't appropriate. A nonlinear model, as in Figure 7.3, does a much better job. The R^2 of this model is 0.946. This value confirms the very strong relationship between month and temperature.

Example 7.3: R versus R^2 When working with the correlation coefficient or the coefficient of determination, it's important to keep in mind whether what's being presented is R or R^2.

To illustrate, consider the relationship between standardized college admissions tests and performance in college. A widely used college admissions test in the US is the SAT, administered by the College Board. The first-year grade-point average is also widely used to reflect a student's performance in college.

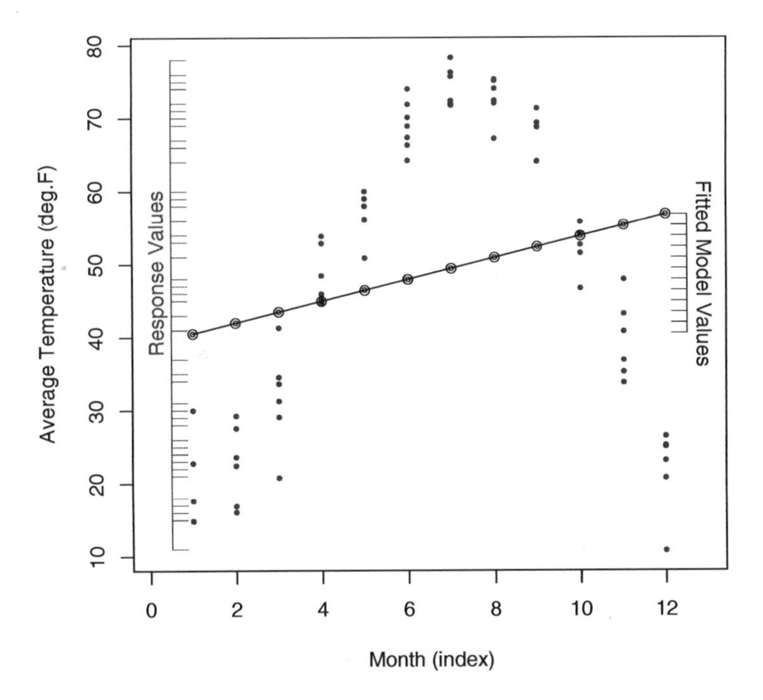

Figure 7.2: Temperature versus month in St. Paul, Minnesota, from the utilities dataset and a linear model of the data. The variance of the fitted model values is much smaller than the variance of the response values, producing a low $R^2 = 0.0624$.

The College Board publishes its studies that relate students' SAT scores to their first-year GPA.[13] The reports present results in terms of R — the square-root of R^2. A typical value is about $R = 0.40$.

Translating this to R^2 gives $0.40^2 = 0.16$. Seen as R^2, the connection between SAT scores and first-year GPA does not seem nearly as strong! Of course, $R = 0.40$ and $R^2 = 0.16$ are exactly the same thing. Just make sure that when you are reading about a study, you know which one you're dealing with. (In Chapter 10, correlation will be presented in terms of angles and measured in degrees. This helps to avoid ambiguities or mis-interpretations.)

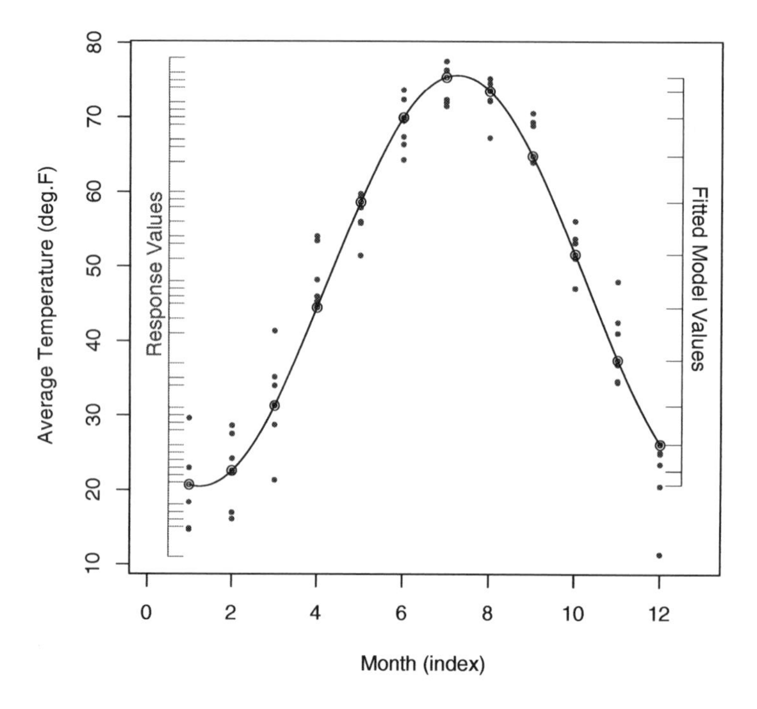

Figure 7.3: Temperature versus month fitted to a nonlinear model. $R^2 = 0.946$. The variance of the fitted model values is almost as large as the variance of the response values.

Whether an R^2 of 0.16 is "large" depends on your purpose. It's low enough to indicate that SAT provides little ability to predict GPA for an individual student. But 0.16 is high enough — as you'll see in Chapter 16 — to establish that there is indeed a link between SAT score and college performance when averaged over lots of students. Consequently, there can be some legitimate disagreement about how to interpret the R^2. In reporting on the 2008 College Board report, the New York *Times* headlined, "Study Finds Little Benefit in New SAT" (June 18, 2008) while the College Board press release on the previous day announced, "SAT Studies Show Test's Strength in Predicting College Success." Regretably, neither the news reports nor the press release stated the R^2.

7.3 Nested Models

Models are often built up in stages; start with an existing model and then add a new model term. This process is important in deciding whether the new term contributes to the explanation of the reponse variable. If the new, extended model is substantially better than that old one, there's reason to think that the new term is contributing.

You can use R^2 to measure how much better the new model is than the old model. But be careful. It turns out that whenever you add a new term to a model, R^2 can increase. It's never the case that the new term will cause R^2 to go down. Even a random variable, one that's utterly unrelated to the response variable, can cause an increase in R^2 when it's added to a model.

This raises an important question: When is an increase in R^2 big enough to justify concluding that the added explanatory term is genuinely contributing to the model? Answering this question will require concepts that will be introduced in later chapters.

An important idea is that of **nested models**. Consider two models: a "small" one and a "large" one that includes all of the explanatory terms in the small one. The small model is said to be nested in the large model.

To illustrate, consider a series of models of the Galton height data. Each model will take child's height as the response variable, but the different models will include different variables and model terms constructed from these variables.

Model A height ~ 1

Model B height ~ 1 + mother

Model C height ~ 1 + mother + sex

Model D height ~ 1 + mother + sex + mother:sex

Model B includes all of the explanatory terms that are in Model A. So, Model A is nested in Model B. Similarly, A and B are both nested in Model C. Model D is even more comprehensive; A, B, and C are all nested in D.

When one model is nested inside another, the variance of the fitted model values from the smaller model will be less than the variance of the fitted model values from the larger model. The same applies to R^2, since R^2 is proportional to the variance of the fitted model values.

In the four nested models listed above, for example, the increase in fitted variance and R^2 is this:

Model	A	B	C	D
Variance of model values	0.00	0.52	7.21	7.22
R^2	0.00	0.0407	0.5618	0.5621

Notice that the variance of the fitted model values for Model A is exactly zero. This is because Model A has only the interept term 1 as an explanatory term. This simple model treats all cases as the same and so there is no variation at all in the fitted model values: the variance and R^2 are zero.

Example 7.4: R^2 Out of the Headlines

Consider this headline from the *BBC News*: "Finger length 'key to aggression.'" The length of a man's fingers can reveal how physically aggressive he is. Or, from *Time* magazine, this summary of the same story: "The shorter a man's index finger is relative to his ring finger, the more aggressive he is likely to be (this doesn't apply to women)."

These news reports were based on a study examining the relationship between finger-length ratios and aggressiveness.[14] It has been noticed for years that men tend to have shorter index fingers compared to ring fingers, while women tend the opposite way. The study authors hypothesized that a more "masculine finger ratio" might be associated with other masculine traits, such as aggressiveness. The association was found and reported in the study: $R^2 = 0.04$. The underlying data are shown in this figure from the original report:

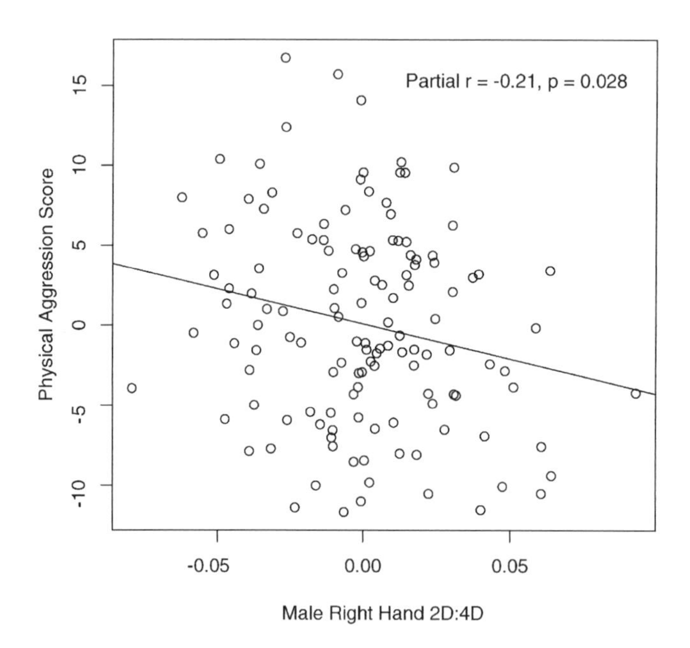

As always, R^2 gives the fraction of the variance in the response variable that is accounted for by the model. This means that of all the variation in aggressiveness among the men involved in the study, 4% was accounted for, leaving 96% unexplained by the models. This suggests that finger length is no big deal.

Headline words like "key" and "reveal" are vague but certainly suggest a strong relationship. The news media got it wrong.

7.4 Toward Statistical Models

The models that you create by fitting a model design to data are certainly **empirical** models; they reflect the observations and measurements contained in your data. But they are not yet fully statistical. What will bring statistics into play is the interpretation of the model to assess *how much evidence* the data provide in support of the model.

The interpretation of the strength of evidence is a subtle logical exercise that will require some new tools. The following chapters introduce those tools: ways of thinking about uncertainty, ways to measure how surprising a measurement or coefficient is, ways to quantify your expectations and uncertainties.

The purpose of this section is to give a very brief summary, so that you can anticipate and prepare for what will be coming.

First, it's very important to distinguish between the **strength of a relationship** and the **strength of evidence** for that relationship. Saying that a relationship between two variables is strong is sensible when a change in one of the variables is associated with a large change in the other variable. Drawing on Galton's data on the heights of adult children for an example, one can say that the relationship between the mother's height and the child's adult height is strong if, for instance, taller mothers tend to have much taller children than shorter mothers do. Similarly, the relationship between a person's sex and height is strong if the different sexes tend to have substantially different heights.

For simple models, a convenient way to describe the strength of a relationship is by the model coefficients. A bigger coefficient — whether it be positive or negative — indicates a stronger relationship. (See Chapter 5.) A vanishingly weak relationship is indicated by a coefficient that is near zero.

The strength of the evidence is a very different thing and involves both **how much data there is** and **how well the data match the model**. Both matters are important.

To illustrate, imagine that you have just two cases in your data, as shown in Figure 7.4A. With just two points in the scatter plot, a straight-line model will fit the data perfectly — just draw the line that connects the two points. As a result, r for this model is perfect: 1.00. However, with just two cases, the strength of the evidence is nil. The reason is that a model that fits just as well as this one (a perfect fit!) can be constructed for *any data whatsoever*. The slope of the line does indeed depend on the specifics of the data, but a perfect fit will result for fitting a line to any data with just two cases, even data that are just random junk. You can always connect two points with a line.

Now imagine that you have ten cases, as in Figure 7.4B. The ten points don't all fit on the best-fitting line — the model doesn't fit the data perfectly. $r = 0.41$. Yet the evidence for the model relationship in B is somewhat stronger than in A. The reason is that the ten points are all pretty close to the line. Some new methods will be needed before a precise definition can be given for "pretty close." For now, "pretty close" means closer than you would expect if the data had been random junk without meaning.

With more data, the evidence for the relationship can be stronger still. Figure 7.4C shows a situation with fifty cases. Although hardly any of the cases fall exactly on the line, the cases are close enough and there are so many of them that the evidence for the model is very strong.

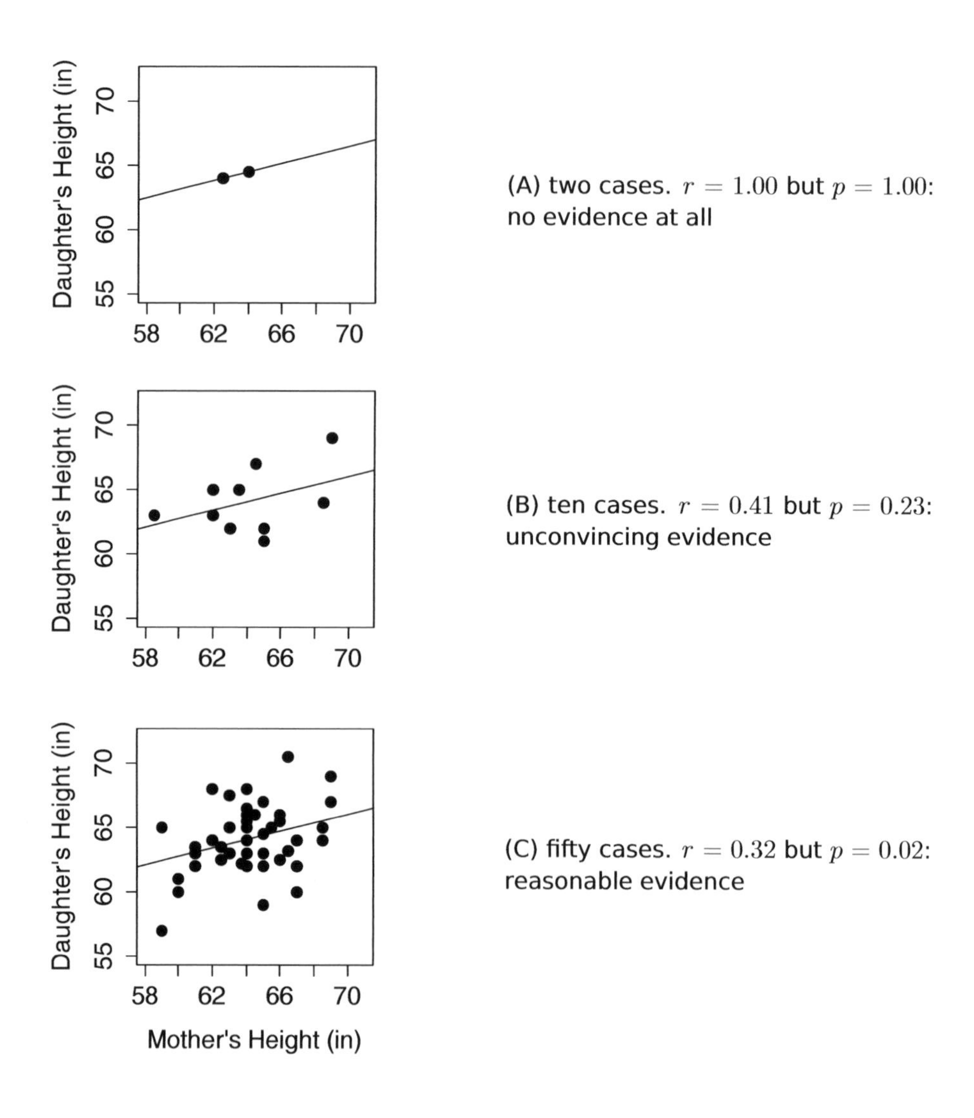

Figure 7.4: A model reflecting the relationship between a woman's height as an adult and the height of the woman's mother.

One common way to quantify the strength of evidence is with a **p-value**. This is a number between 0 and 1 that measures the surprise at seeing how close the data are to the model under the assumption that the data are really just random. A small p-value — a number near zero — indicates great surprise and therefore strong evidence that the data are not random, that they do inform the model. Later chapters will introduce and explain p-values, which have a subtle and delicate interpretation. For the data in Figure 7.4 panel A, the p-value is 1 — the weakest possible. In panel B, the p-value is 0.23, a value typically considered to reflect unconvincing evidence. In panel C, the p-value is 0.02 — a level taken as convincing in the science literature.

Note that even though the strength of evidence is very different from panel A to B to C, the models fit to the data in each of the three panels show almost exactly relationship between the mother's height and her daughter's height. Strength of a relationship is not the same thing as strength of evidence. Models become statistical models when you use the data to evaluate the strength of the evidence.

7.5 Computational Technique

The coefficient of determination, R^2, compares the variation in the response variable to the variation in the fitted model value. It can be calculated as a ratio of variances:

```
> swim = ISMdata("swim100m.csv")
> mod1 = lm( time ~ year + sex, data=swim )
> var(fitted(mod1))/var(swim$time)
[1] 0.844
```

R^2 is so widely used to describe how well a model captures the variation in the response that it is a standard part of a regression report:

```
> summary(mod1)
Coefficients:
            Estimate Std. Error t value  Pr(>|t|)
(Intercept) 555.7168    33.7999   16.44   < 2e-16
year         -0.2515     0.0173  -14.52   < 2e-16
sexM         -9.7980     1.0129   -9.67   8.8e-14

Residual standard error: 3.98 on 59 degrees of freedom
Multiple R-Squared: 0.844, Adjusted R-squared: 0.839
F-statistic:  160 on 2 and 59 DF,  p-value: <2e-16
```

7.5.1 Correlation coefficient, r

The `cor` operator will compute the correlation coefficient r between two variables. To illustrate, here is r between `age` and `wage` in the Current Population

Survey data.

```
> cps = ISMdata("cps.csv")
> cor( cps$wage, cps$age)
[1] 0.1770
```

r is closely to R^2 of the straight-line model. Here is R^2 of the model wage ~ 1+age:

```
> mod2 = lm( wage ~ 1 + age, data=cps )
> var(fitted(mod2))/var(cps$wage)
[1] 0.03132
```

Just take the square-root to find r from R^2:

```
> sqrt(0.03132)
[1] 0.1770
```

Since the variance is the square of the standard deviation, the above could have been calculated in a more streamlined way:

```
> sd( fitted( mod2)) / sd( cps$wage )
[1] 0.1770
```

Or, you can use the `cor` operator.

The correlation coefficient r summarizes only the simple linear model A ~ B+1 where B is quantitative. But the coefficient of determination, R^2, summarizes any model; it is much more useful.

7.5.2 Typical Size of Residuals

The typical size of a residual gives an indication of how the actual response values are scattered around the fitted model values. One way to estimate this size is with the standard deviation:

```
> sd( resid(mod1) )
[1] 3.917
```

When models are to be used for prediction, the typical size of a residual provides a simple guide to how precise an individual prediction will be. Suppose, for example, that you want to guess the hourly wage of a 30-year old female working in the management sector. You have constructed this model:

DATA FILE
cps.csv

```
> mod3 = lm( wage ~ age + sex + sector, data=cps )
> summary(mod3)
Coefficients:
              Estimate Std. Error t value Pr(>|t|)
(Intercept)    4.071     0.7762     5.24  2.3e-07
age            0.079     0.0168     4.70  3.4e-06
```

```
sexM            2.150     0.4379    4.93  1.1e-06
sectorconst     0.215     1.1590    0.19   0.853
sectormanag     4.245     0.7814    5.43  8.5e-08
sectormanuf    -0.352     0.7368   -0.48   0.633
 ... and so on for the other sectors.

Residual standard error: 4.5 on 524 degrees of freedom
Multiple R-Squared: 0.245,        Adjusted R-squared: 0.232
F-statistic: 18.9 on 9 and 524 DF,  p-value: <2e-16
```

Taking into account that the prediction is to be made for a female in the management sector, the prediction would be

```
> 4.071 + 0.079*30 + 2.150*0 + 0.215*0 + 4.245*1 + -0.352*0
[1] 10.69
```

(Keep in mind that these data are from the 1980s!)

The size of a typical residual from this model is

```
> sd( resid(mod3))
[1] 4.464
```

indicating how different from the prediction of $10.69 a typical individual's wage is likely to be. It turns out that a better estimate of the size of a typical residual is the **residual standard error**. This is a little bigger than the standard deviation of the residuals and is given in the regression report. (It's 4.5 in the above report.)

In section 14.5 you will see that in assessing the likely residual from a prediction, it's important to take into account the uncertainty in the coefficients themselves.

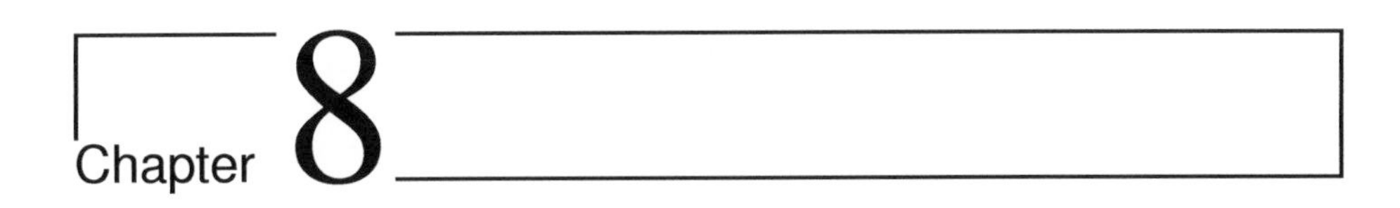

Total and Partial Relationships

I do not feel obliged to believe that the same God who has endowed us with sense, reason, and intellect has intended us to forgo their use. —Galileo Galilei

One of the most important ideas in science is **experiment**. In a simple, ideal form of an experiment, you cause one explanatory factor to vary and, holding all the other conditions constant, observe the result on some other variable. A famous story of such an experiment involves Galileo Galilei (1564-1642) dropping balls of different masses but equal diameter from the Leaning Tower of Pisa.[1] Would a heavy ball fall faster than a light ball, as theorized by Aristotle 2000 years previously? The quantity that Galileo varied was the weight of the ball, the quantity he observed was how fast the balls fell, the conditions he held constant were the height of the fall and the diameter of the balls. The experimental method of dropping balls side by side also holds constant the atmospheric conditions: temperature, humidity, wind, air density, etc.

Today, Galileo's experiment seems obvious. But not at the time. In the history of science, Galileo's work was a landmark: he put *observation* at the fore, rather than the beliefs passed down from authority. Aristotle's ancient theory, still considered authoritative in Galileo's time, was that heavier objects fall faster.

The ideal of "holding all other conditions constant" is not always so simple as with dropping balls from a tower. Consider an experiment to test the effect of a blood-pressure drug. Take two groups of people, give the people in one group the drug and give nothing to the other group. Observe how blood pressure changes in the two groups. The factor being caused to vary is whether or not a person gets the drug. But what is being held constant? Presumably the researcher took care to make the two groups as similar as possible: similar med-

[1] The picturesque story of balls dropped from the Tower of Pisa may not be true. Galileo did record experiments done by rolling balls down ramps.

ical conditions and histories, similar weights, similar ages. But "similar" is not "constant."

For non-experimentalists — people who study data collected through observation, without doing an experiment — a central question is whether there is a way to mimic "holding all other conditions constant." For example, suppose you observe the academic performance of students, some taught in large classes and some in small classes, some taught by well-paid teachers and some taught by poorly-paid teachers, some coming from families with positive parental involvement and some not, and so on. Is there a way to analyze data so that you can separate the influences of these different factors, examining one factor while, through analysis if not through experiment, holding the others constant?

In this chapter you'll see how models can be used to examine data as if some variables were being held constant. Perhaps the most important message of the chapter is that there is no point hiding your head in the sand; simply ignoring a variable is not at all the same thing as holding that variable constant. By including multiple variables in a model you make it possible to interpret that model in terms of holding the variables constant. But there is no methodological magic at work here. The results of modeling can be misleading if the model does not reflect the reality of what is happening in the system under study. Understanding how and when models can be used effectively, and when they can be misleading, will be a major theme of the remainder of the book.

8.1 Total and Partial Relationships

The common phrase "all other things being equal" is an important qualifier in describing relationships. To illustrate: A simple claim in economics is that a high price for a commodity reduces the demand. For example increasing the price of heating fuel will reduce demand as people turn down thermostats in order to save money. But the claim can be considered obvious only with the qualifier *all other things being equal*. For instance, the fuel price might have increased because winter weather has increased the demand for heating compared to summer. Thus, higher prices may be associated with higher demand. Unless you hold other variables constant — e.g., weather conditions — increased price may not in fact be associated with lower demand.

In fields such as economics, the Latin equivalent of "all other things being equal" is sometimes used: **ceteris paribus**. So, the economics claim would be, "higher prices are associated with lower demand, *ceteris paribus*."

Although the phrase "all other things being equal" has a logical simplicity, it's impractical to implement "all." Instead of the blanket "all other things," it's helpful to be able to consider just "some other things" to be held constant, being explicit about what those things are. Other phrases along these lines are "taking into account ..." and "controlling for ...." Such phrases apply when you want to examine the relationship between two variables, but there is one or more additional variables that may be coming into play. The additional variables are sometimes called **covariates** or **confounders**.

A covariate is just an ordinary variable. The use of the word "covariate" rather than "variable" highlights the interest in holding this variable constant, to indicate that it's not a variable of primary interest.

As an example of a covariate, consider the following news report:

> **Heart Surgery Drug Carries High Risk, Study Says.** A drug widely used to prevent excessive bleeding during heart surgery appears to raise the risk of dying in the five years afterward by nearly 50 percent, an international study found.
>
> The researchers said replacing the drug — aprotinin, sold by Bayer under the brand name Trasylol — with other, cheaper drugs for a year would prevent 10,000 deaths worldwide over the next five years.
>
> Bayer said in a statement that the findings are unreliable because Trasylol tends to be used in more complex operations, and the researchers' statistical analysis did not fully account for the complexity of the surgery cases.
>
> The study followed 3,876 patients who had heart bypass surgery at 62 medical centers in 16 nations. Researchers compared patients who received aprotinin to patients who got other drugs or no antibleeding drugs. Over five years, 20.8 percent of the aprotinin patients died, versus 12.7 percent of the patients who received no antibleeding drug.
>
> When researchers adjusted for other factors, they found that patients who got Trasylol ran a 48 percent higher risk of dying in the five years afterward.
>
> The other drugs, both cheaper generics, did not raise the risk of death significantly.
>
> The study was not a randomized trial, meaning that it did not randomly assign patients to get aprotinin or not. In their analysis, the researchers took into account how sick patients were before surgery, but they acknowledged that some factors they did not account for may have contributed to the extra deaths. - Carla K. Johnson, Associated Press, 7 Feb. 2007

The report involves several variables. Of primary interest is the relationship between (1) the risk of dying after surgery and (2) the drug used to prevent excessive bleeding during surgery. Also potentially important are (3) the complexity of the surgical operation and (4) how sick the patients were before surgery. Bayer disputes the published results of the relationship between (1) and (2) holding (4) constant, saying that it's also important to hold variable (3) constant.

The term **partial relationship** describes a relationship with one or more covariates being held constant. A useful thing to know in economics might be the partial relationship between fuel price and demand with weather conditions being held constant. Similarly, it's a partial relationship when the article refers to the effect of the drug on patient outcome in those patients with a similar complexity of operation.

In contrast to a partial relationship where certain variables are being held constant, there is also a **total relationship**: how an explanatory variable is related to a response variable *letting those other explanatory variables change as they will.* (The corresponding Latin phrase is **mutatis mutandis.**)

In the aprotinin drug example, the total relationship involves a death rate of 20.8 percent of patients who got aprotinin, versus 12.7 percent for others. This implies an increase in the death rate by a factor of 1.64. When the researchers looked at a partial relationship (holding constant the patient sickness before the operation), the death rate was seen to increase by less: a factor of 1.48. Judging from this difference of the partial and total relationships, the sicker patients tended to be given aprotinin.

Here's another illustration of the difference between partial and total relationships. I was once involved in a budget committee that recommended employee health benefits for the college at which I work. At the time, college employees who belonged to the college's insurance plan received a generous subsidy for their health insurance costs. Employees who did not belong to the plan received no subsidy but were instead given a moderate monthly cash payment. After the stock-market crashed in year 2000, the college needed to cut budgets. As part of this, it was proposed to eliminate the cash payment to the employees who did not belong to the insurance plan. This proposal was supported by a claim that this would save money without reducing health benefits. I argued that this claim was about a partial relationship: how expenditures would change assuming that the number of people belonging to the insurance plan remained constant. I thought that this partial relationship was irrelevant; the loss of the cash payment would cause some employees, who currently received health benefits through their spouse's health plan, to switch to the college's health plan. Thus, the total relationship between the cash payment and expenditures might be the opposite of the partial relationship: the savings from the moderate cash payment would trigger a much larger expenditure by the college.

Perhaps it seems obvious that one should be concerned with the "big picture," the total relationship between variables. If eliminating the cash payment increases expenditures overall, it makes no sense to focus exclusively on the narrow savings from the suspending the payment itself. On the other hand, in the aprotinin drug example, for understanding the impact of the drug itself it seems important to take into account how sick the various patients were and how complex the surgical operations. There's no point ascribing damage to aprotinin that might instead be the result of complicated operations or the patient's condition.

Whether you wish to study a partial or a total relationship is largely up to you and the context of your work. But certainly you need to know which relationship you are studying.

Consider another example of partial versus total relationships. In 1962, naturalist author Rachel Carson published *Silent Spring*,[15] a powerful indictment of the widespread use of pesticides such as DDT. Carson pointed out the links between DDT and dropping populations of birds such as the bald eagle. She also speculated that pesticides were the cause of a recent increase in the number of human cancer cases. The book's publication was instrumental in the eventual banning of DDT.

The increase in deaths from cancer over time is a total relationship between cancer deaths and time. It's relevant to consider a partial relationship between the number of cancer deaths and time, holding the population constant. This partial relationship can be indicated by a death *rate*: say, the number of cancer deaths per 100,000 people. It seems obvious that the covariate of population size ought to be held constant. But there are still other covariates to be held constant. The decades before *Silent Spring* had seen a strong decrease in deaths at young ages from other non-cancer diseases which now were under greater control. It had also seen a strong increase in smoking. When adjusting for for these covariates, the death rate from cancer was actually falling, not increasing as Carson claimed.[16]

As you'll see, the interpretation of a model as describing total or partial relationships depends on which explanatory variables have been included in the model.

Example 8.1: Used Car Prices Figure 8.1 shows a scatter plot of the price of used Honda Accords versus the number of miles each car has been driven. The graph shows a pretty compelling relationship: the more miles that a car goes, the lower the price. This can be summarized by a simple linear model: price ~ mileage. Fitting such a model gives this model formula

$$\underset{\text{Values}}{\text{Model}}\ \text{price} = 20770 - 0.10\,\text{mileage}.$$

Keeping in mind the units of the variables, the price of these Honda Accords typically falls by about 10 cents per mile driven. Think of that as the cost of the wear and tear of driving: depreciation.

As the cars are being driven, other things are happening to them. They are wearing out, they are being involved in minor accidents, and they are getting older. The relationship shown in Figure 8.1 takes none of these into account. As mileage changes, the other variables such as age are changing as they will: a total relationship.

In contrast to the total relationship, the *partial* relationship between price and mileage holding age constant tells you something different than the total relationship. The partial relationship would be relevant, for instance, if you were interested in the cost of driving a car. This cost includes gasoline, insurance, repairs, and depreciation of the car's value. The car will age whether or not you drive it; the extra depreciation due to driving it will be indicated by the partial relationship between price and mileage holding age constant.

The most intuitive way to hold age constant is to look at the relationship between price and mileage for a subset of the cars; include only those cars of a given age. This is shown in Figure 8.2. The cars have been divided into age groups (less than 2 years old, between 3 and 4 years old, etc.) and the data for each group has been plotted separately together with the best fitting linear model for the cars in that group. From the figure it's easy to see that the slope of the fitted models for each group is shallower than the slope fitted to all the cars

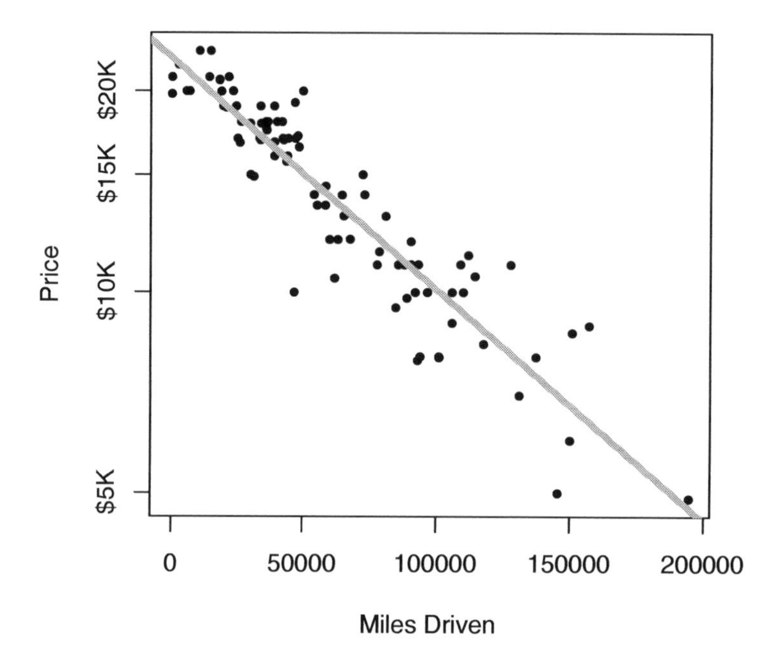

Figure 8.1: The price of used cars falls with increasing miles driven. The gray diagonal line shows the best fitting linear model. Price falls by about $10,000 for 100,000 miles, or 10 cents per mile driven.

combined. Instead of the price falling by about 10 cents per mile as it does for all the cars combined, within the 4-8 year old group the price decrease is only about 7 cents per mile, and only 3 cents per mile for cars older than 8 years.

By looking at the different age groups individually, you are holding age approximately constant in your model. The relationship you find in this way between price and mileage is a partial relationship. Of course, there are other variables that you didn't hold constant. So, to be precise, you should describe the relationship you found as a partial relationship *with respect to age*.

8.1.1 Models and Partial Relationships

Models make it easy to estimate the partial relationship between a response variable and an explanatory variable, holding one or more covariates constant.

The first step is to fit a model using both the explanatory variable and the covariates that you want to hold constant. For example, to find the partial relationship between car price and miles driven, holding age constant, fit the

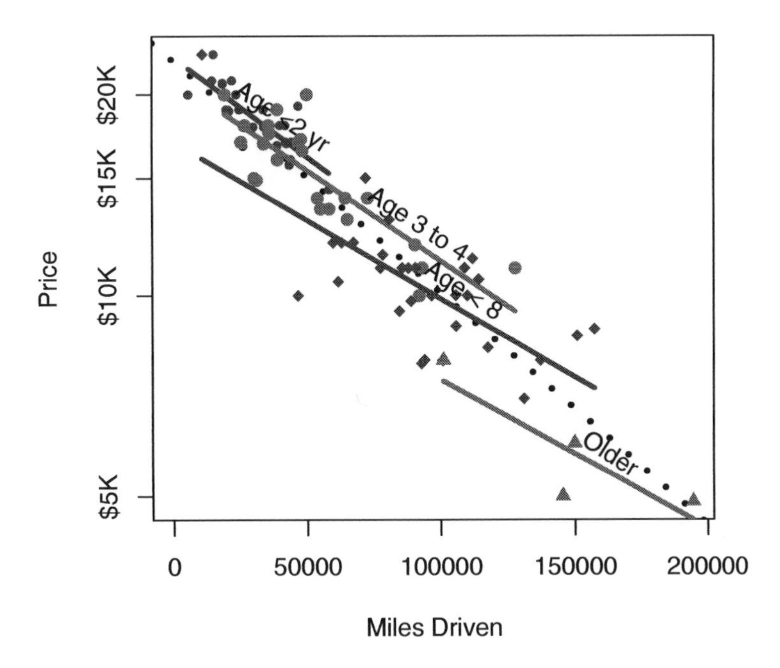

Figure 8.2: The relationship between price and mileage for cars in different age groups, indicating the partial relationship between price and mileage holding age constant. (The dotted diagonal line shows the total relationship between price and age, from Figure 8.1.)

model price ~ mileage+age. For the car-price data from Figure 8.1, this gives the model formula

$$\text{Model Values price} = 21330 - 0.077\,\text{mileage} - 538\,\text{age}$$

The second step is to interpret this model as a partial relationship between price and mileage holding age constant. A simple way to do this is to plug in some particular value for age, say 1 year. With this value plugged in, the formula for price as a function of mileage becomes

$$\text{Model Values price} = 21330 - 0.077\,\text{mileage} - 538 \times 1 = 20792 - 0.077\,\text{mileage}$$

The partial relationship is that price goes down by 0.077 dollars per mile, holding age constant.

Note the use of the phrase "estimate the partial relationship" in the first paragraph of this section. The model you fit creates a representation of the system you are studying that incorporates both the variable of interest and the covariates in explaining the response values. In this mathematical representation, it's easy to hold the covariates constant. If you don't include the covariate in the model, you can't hold it constant and so you can't estimate the partial relationship between the response and the variable of interest while holding the covariate constant. But even when you do include the covariates in your model, there is a legitimate question of whether your model is a faithful reflection of reality; holding a covariate constant in a model is not the same thing as holding it constant in the real world. These issues, which revolve around the idea of the causal relationship between the covariate and the response, are discussed in Chapter 20.

Aside. 8.1 Partial change and partial derivatives

If you are familiar with calculus and **partial derivatives**, you may notice that this rate is the partial derivative of price with respect to mileage. Using partial derivatives allows one to interpret more complicated models relatively easily. For example, you might have chosen to include in the model the interaction term between mileage and age. This would give a model with four terms:

Model Values $$\text{price} = 22140 - 0.094\,\text{mileage} - 750\,\text{age} + 0.0034\,\text{mileage} \times \text{age}$$

For this model, the partial relationship between price and mileage is not just the coefficient on mileage. Instead it is the partial derivative of price with respect to mileage, or:

$$\frac{\partial \text{price}}{\partial \text{mileage}} = -0.094 + 0.0034 \text{age}$$

Keeping in mind the units of the variables, this means that for a new car (age $= 0$), the price declines by 0.094 — 9.4 cents per mile. But for a 10-year old car, the decline is less rapid: $-0.094 + 10 \times 0.0034 = -0.060$ — only six cents a mile.

8.1.2 Adjustment

The table shows data on professional and sales employees of a large mid-western US trucking company: the annual earnings in 2007, sex, age, job title, how many years of employment with the company. Data such as these are sometimes used to establish whether or not employers discriminate on the basis of sex.

sex	earnings	age	title	hiredyears
M	35000	25	PROGRAMMER	0
F	36800	62	CLAIMS ADJUSTER	5
F	25000	34	RECRUITER	1
M	45000	44	CLAIMS ADJUSTER	0
M	40000	30	PROGRAMMER	5

... and so on for 129 cases altogether.

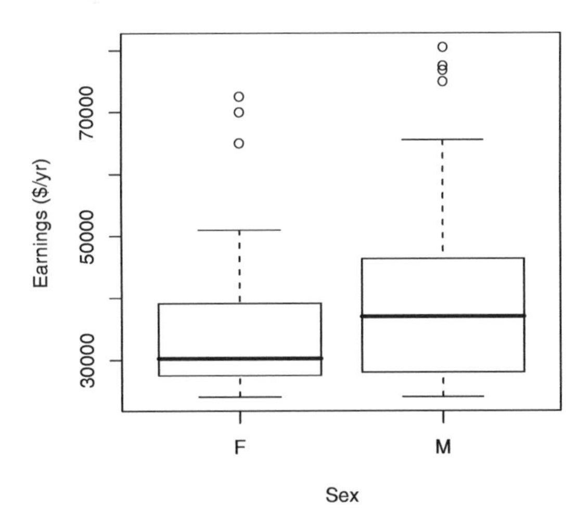

Figure 8.3: The distribution of annual earnings broken down by sex for professional and sales employees of a trucking company.

A boxplot reveals a clear pattern: men are being paid more than women. (See Figure 8.3.) Fitting the model earnings $\sim$ sex indicates the average difference in earnings between men and women:

$$\text{earnings} = 35501 + 4735\,\text{sexM}.$$

Since earnings are in dollars per year, Men are being paid, on average, $4735 more per year than women. This difference reflects the *total* relationship between earnings and sex, letting other variables change as they will.

Notice from the boxplot that even within the male or female groups, there is considerable variability in annual earnings from person to person. Evidently, there is something other than sex that influences the wages.

An important question is whether you should be interested in the total relationship between earnings and sex, or the partial relationship, holding other variables constant. This is a difficult issue. Clearly there are some legitimate reasons to pay people differently, for example different levels of skill or experience or different job descriptions, but it's always possible that these legitimate factors are being used to mask discrimination.

For the moment, take as a covariate something that can stand in as a proxy for experience: the employee's age. Unlike job title, age is hardly something that can be manipulated to hide discrimination. Figure 8.4 shows the employees earnings plotted against age. Also shown are the model lines of wages against age, fitted separately for men and women. It's evident that for both men and women, earnings tend to increase with age. The formulas for the two lines are:

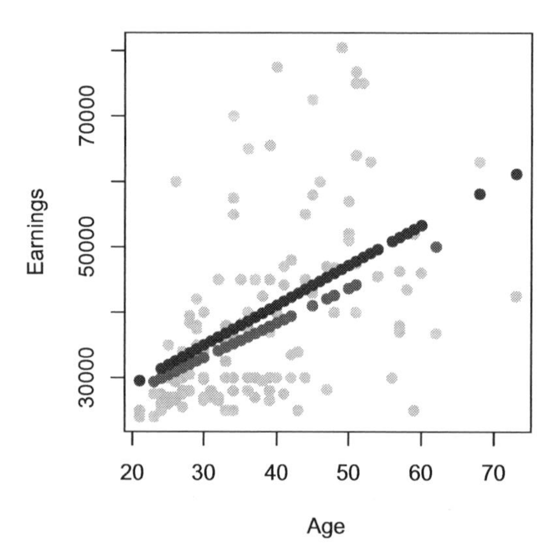

Figure 8.4: Annual earnings versus age. The lines show fitted models made separately for men and women.

For Men:	$\underset{\text{Values}}{\text{Model}}$ Earnings $=$ 16735 + 609 age
For Women	$\underset{\text{Values}}{\text{Model}}$ Earnings $=$ 17178 + 530 age

From the graph, you can see the partial relationship between earnings and sex, holding age constant. Pick an age, say 30 years. At 30 years, according to the model, the difference in annual earnings is \$1931, with men making more. At 40 years of age, the difference between the sexes is even more (\$2722), at 20 years of age, the difference is less (\$1140). All of these partial differences (holding age constant) are substantially less than the difference when age is not taken into account (\$4735).

One way to summarize the differences in earnings between men and women is to answer this question: How would the earnings of the women have been different if the women were men? Of course you can't know all the changes that might occur if the women were somehow magically transformed, but you can use the model to calculate the change assuming that all other variables except sex are held constant. This process is called **adjustment**.

To find the women's wages adjusted as if they were men, take the data for the women and plug it into the model formula for men. The difference between the earnings of men and women, adjusting for age, is \$2293. Differences in age appear to account for more than half of the overall earnings difference between men and women.

Of course, before you draw any conclusions, you need to know how precise these coefficients are. For instance, it's a different story if the sex difference is

$2293±10 or if it is $2293±5000. In the later case, it would be sensible to conclude only that the data leave the matter of wage difference undecided. Later chapters in this book describe how to characterize the precision of an estimate.

Another key matter is that of causation. $2293 indicates a difference, but doesn't say completely where the difference comes from. By adjusting for age, the model disposes of the possibility that the earnings difference reflects differences in the typical ages of male and female workers. It remains to be found out whether the earnings difference might be due to different skill sets, discrimination, or other factors.

8.1.3 Simpson's Paradox

Sometimes the total relationship is the opposite of the partial relationship. This is **Simpson's paradox**.

One of the most famous examples involves graduate admissions at the University of California in Berkeley. It was observed that graduate admission rates were lower for women than for men overall. This reflects the total relationship between admissions and sex. But, on a department-by-department basis, admissions rates for women were consistently as high or higher than for men. The partial relationship, taking into account the differences between departments, was utterly different from the total relationship.

Example 8.2: Do Life Vests Save Lives? According to legend, during a government reform in the 1970s the US Coast Guard was asked to justify its various regulations. One of these regulations is that ships must carry life preservers for crew and passengers. This hardly seems a controversial regulation, but following the general orders for review, the Guard undertook a small study of the effectiveness of life vests.

The study involved looking at the reports of man-overboard incidents in large ships. These reports include information about whether the victim was wearing a life vest and whether the victim drowned.

A clerk was assigned to page through report sheets, perhaps instructed to spend a day examining the first 500 reports in the files. He would have constructed a tally sheet counting how many incidents fit into the various categories.

Aggregated Counts

	Wearing Vest	No Vest
Survived	100	71
Drowned	300	29
Total	400	100

The table says that 75% of those wearing a vest drowned whereas only 29% of those drowned who were not wearing a vest. That is, wearing a vest is not effective; perhaps they are even harmful! At least, that's the story when you look at the total relationship between wearing a vest and drowning.

Some insightful researcher decided to break down the data depending on the weather situation: a partial relationship, holding weather conditions constant. In foul weather, sailors tend to wear life vests. In fair weather they don't. But falling overboard in foul weather is particularly dangerous since you may quickly be lost from view even if you stay afloat. Once lost, you are unlikely to be found.

Fair Weather

	Wearing Vest	No Vest
Survived	19	70
Drowned	1	10
Total	20	80

Foul Weather

	Wearing Vest	No Vest
Survived	81	1
Drowned	299	19
Total	380	20

The new tables merely disaggregate the counts in the top table. If you add the fair- and foul-weather tables together, you get the aggregated table.

The disaggregated tables show that $19/20 = 95\%$ of people survived who were wearing a life vest in fair weather, compared to $70/80 = 87.5\%$ of those who were not wearing a life vest. So, in fair weather, a life vest seems to help. In foul weather, $81/380 = 21\%$ of people survived who were wearing a vest, compared to only $1/20 = 5\%$ of those not wearing a vest. So, in foul weather also it helps to wear a life vest.

This is Simpson's Paradox: a total relationship indicates one thing, the partial relationship indicates exactly the opposite.

DATA FILE
sat.csv

Example 8.3: SAT Scores and School Spending Chapter 5 (page 104) showed some models relating school expenditures to SAT scores. The model sat $\sim$ 1 + expend produced a negative coefficient on expend, suggesting that higher expenditures are associated with lower test scores. Including another variable, the fraction of students who take the SAT (variable frac) reversed this relationship.

The model sat $\sim$ 1 + expend + frac attempts to capture how SAT scores depend both on expend and frac. In interpreting the model, you can look at how the SAT scores would change with expend while holding frac constant. That is, from the model formula, you can study the partial relationship between SAT and expend while holding frac constant.

The example also looked at a couple of other fiscally related variables: student-teacher ratio and teachers' salary. The total relationship between each of the fiscal variables and SAT was negative — for instance, higher salaries were associated with lower SAT scores. But the partial relationship, holding frac constant, was the opposite: Simpson's Paradox.

For a moment, take at face value the idea that higher teacher salaries and smaller class sizes are associated with better SAT scores as indicated by the following models:

Model Values sat = 988 + 2.18 salary - 2.78 frac
Model Values sat = 1119 - 3.73 ratio - 2.55 frac

In thinking about the impact of an intervention — changing teachers' salaries or changing the student-teacher ratio — it's important to think about what other things will be changing during the intervention. For example, one of the ways to make student-teacher ratio smaller is to hire more teachers. This is easier if salaries are held low. Similarly, salaries can be raised if fewer teachers are hired: increasing class size is one way to do this. So, salaries and student-teacher ratio are in conflict with each other.

If you want to anticipate what might be the effect of a change in teacher salary while holding student-teacher ratio constant, then you should include ratio in the model along with salary (and frac, whose dominating influence remains confounded with the other variables if it is left out of the model):

Model Values sat = 1058 + 2.55 salary - 4.64 ratio - 2.91 frac

Comparing this model to the previous ones gives some indication of the trade-off between salaries and student-teacher ratio. When `ratio` is included along with salary, the salary coefficient is somewhat bigger: 2.55 versus 2.18. This suggests that if salary is increased while holding constant the student-teacher ratio, salary has a stronger relationship with SAT scores than if salary is increased while allowing student-teacher ratio to vary in the way it usually does, accordingly.

Of course, you still need to have some way to determine whether the precision in the estimate of the coefficients is adequate to judge whether the detected difference in the salary coefficient is real — 2.18 in one model and 2.55 in the other. Such issues are introduced in Chapter 14.

8.2 Explicitly Holding Covariates Constant

The distinction between explanatory variables and covariates is in the modeler's mind. When it comes to fitting a model, both sorts of variables are considered on an equal basis when calculating the residuals and choosing the best fitting model to produce a model function. The way that you choose to interpret and analyze the model function is what determines whether you are examining partial change or total change.

The intuitive way to hold a covariate constant is to do just that. Experimentalists arrange their experimental conditions so that the covariates are the same. Think of Galileo using balls of the same diameter and varying only the mass. In a clinical trial of a new drug, perhaps you would test the drug only on women so that you don't have to worry about the covariate sex.

When you are not doing an experiment but rather working with observational data, you can hold a covariate constant by throwing out data. Do you want to see the partial relationship between price and mileage while holding age constant? Then restrict your analysis to cars that are all the same age, say 3 years old. Want to know the relationship between breath-holding time and body size holding sex constant? Then study the relationship in just women or in just men.

Dividing data up into groups of similar cases is an intuitive way to study partial relationships. It can be effective, but it is not a very efficient way to use data.

The problem with dividing up data into groups is that the individual groups may not have many cases. For example, for the used cars shown in Figure 8.2 there are only a dozen or so cases in each of the groups. To get even this number of cases, the groups had to cover more than one year of age. For instance, the group labeled "age < 8" includes cars that are 5, 6, 7, and 8 years old. It would have been nice to be able to consider six-year old cars separately from seven-year old cars, but this would have left me with very few cases in either the six- or seven-year old group.

At the same time, it seems reasonable to think that 5- and 7-year old cars have something to say about 6-year old cars; you would expect the relationship between price and mileage to shift gradually with age. For instance, the relationship for 6-year old cars should be intermediate to the relationships for 5- and for 7-year old cars.

Modeling provides a powerful and efficient way to study partial relationships that does not require studying separate subsets of data. Just include multiple explanatory variables in the model. Whenever you fit a model with multiple explanatory variables, the model gives you information about the partial relationship between the response and each explanatory variable *with respect to each of the other explanatory variables.*

8.3 Adjustment and Truth

It's tempting to think that including covariates in a model is a way to reach the truth: a model that describes how the real world works, a model that can correctly anticipate the consequences of interventions such as medical treatments or changes in policy, etc. This overstates the power of models.

A model design — the response variable and explanatory terms — is a statement of a hypothesis about how the world works. If this hypothesis happens to be right, then under ideal conditions the coefficients from the fitted model will approximate how the real world works. But if the hypothesis is wrong, for example if an important covariate has been left out, then the coefficients may not correctly describe how the world works.

In certain situations — the idealized **experiment** — researchers can create a world in which their modeling hypothesis is correct. In such situations there can be good reason to take the model results as indicating how the world works. For

Aside. 8.2 Divide and Be Conquered!

Efficiency starts to be a major issue when there are many covariates. Consider a study of the partial relationship between lung capacity and smoking, holding constant all these covariates: sex, body size, smoking status, age, physical fitness. There are two sexes and perhaps three or more levels of body size (e.g., small, medium, large). You might divide age into five different groups (e.g., pre-teens, teens, young adults, middle aged, elderly) and physical fitness into three levels (e.g., never exercise, sometimes, often). Taking the variables altogether, there are now $2 \times 3 \times 5 \times 3 = 90$ groups. It's very inefficient to treat these 90 groups completely separately, as if none of the groups had anything to say about the others. A model of the form lung capacity ~ body size + sex + smoking status + age + fitness can not only do the job more efficiently, but avoids the need to divide quantitative variables such as body size or age into categories.

To illustrate, consider this news report:

> Higher vitamin D intake has been associated with a significantly reduced risk of pancreatic cancer, according to a study released last week.
>
> Researchers combined data from two prospective studies that included 46,771 men ages 40 to 75 and 75,427 women ages 38 to 65. They identified 365 cases of pancreatic cancer over 16 years.
>
> Before their cancer was detected, subjects filled out dietary questionnaires, including information on vitamin supplements, and researchers calculated vitamin D intake. After statistically adjusting for [that is, holding constant] age, smoking, level of physical activity, intake of calcium and retinol and other factors, the association between vitamin D intake and reduced risk of pancreated cancer was still significant.
>
> Compared with people who consumed less than 150 units of vitamin D a day, those who consumed more than 600 units reduced their risk by 41 percent. - New York Times, 19 Sept. 2006, p. D6.

There are more than 125,000 cases in this study, but only 365 of them developed pancreatic cancer. If those 365 cases had been scattered around dozens or hundreds of groups and analyzed separately, there would be so little data in each group that no pattern would be discernible.

this reason, the results from studies based on experiments are generally taken as more reliable than results from non-experimental studies. But even when an experiment has been done, the situation may not be ideal; experimental subjects don't always do what the experimenter tells them to and uncontrolled influences can sometimes remain at play.

It's appropriate to show some humility about models and recognize that they can be no better than the assumptions that go into them. Useful object lessons are given by the episodes where conclusions from modeling (with careful adjustment for covariates) can be compared to experimental results. Some examples (from [17]):

- Does it help to use telephone canvassing to get out the vote? Models suggest it does, but experiments indicate otherwise.

- Is a diet rich in vitamins, fruits, vegetables and low in fat protective against cancer, heart disease or cognitive decline? Models suggest yes, but experiments generally do not.

Presumably what is happening in situations such as these is that an important covariate or other model terms have been left out of the models.

8.4 Computational Technique

8.4.1 Adjustment

There are two basic approaches to adjusting for covariates. Conceptually, the simplest one is to hold the covariates constant at some level when collecting data or by extracting a subset of data which holds those covariates constant. The other approach is to include the covariates in your models.

For example, suppose you want to study the differences in the wages of male and females. The very simple model wage ~ sex might give some insight, but it attributes to sex effects that might actually be due to level of education, age, or the sector of the economy in which the person works. Here's the result from the simple model:

DATA FILE
cps.csv

```
> cps = ISMdata("cps.csv")
> mod0 = lm( wage ~ sex, data=cps)
> summary(mod0)

 Coefficients:
             Estimate Std. Error t value Pr(>|t|)
 (Intercept)    7.879      0.322   24.50  < 2e-16
 sexM           2.116      0.437    4.84  1.7e-06

 Residual standard error: 5.03 on 532 degrees of freedom
 Multiple R-Squared: 0.0422, Adjusted R-squared: 0.0404
```

The coefficients indicate that a typical male makes $2.12 more per hour than a typical female. (Notice that $R^2 = 0.0422$ is very small: sex explains hardly any of the person-to-person variability in wage.)

By including the variables age, educ, and sector in the model, you can adjust for these variables:

```
> mod1 = lm( wage ~ age + sex + educ + sector, data=cps)
> summary(mod1)
```

```
Coefficients:
              Estimate Std. Error t value Pr(>|t|)
(Intercept)    -4.6941     1.5378   -3.05  0.00238
age             0.1022     0.0166    6.17  1.4e-09
sexM            1.9417     0.4228    4.59  5.5e-06
educ            0.6156     0.0944    6.52  1.6e-10
sectorconst     1.4355     1.1312    1.27  0.20500
... and so on for the other sectors.

Residual standard error: 4.33 on 523 degrees of freedom
Multiple R-Squared: 0.302,  Adjusted R-squared: 0.289
F-statistic: 22.6 on 10 and 523 DF, p-value: <2e-16
```

The adjusted difference between the sexes is $1.94 per hour. (The $R^2 = 0.30$ from this model is considerably larger than for mod1, but still a lot of the person-to-person variation in wages has not be captured.)

It would be wrong to claim that simply including a covariate in a model guarantees that an appropriate adjustment has been made. The effectiveness of the adjustment depends on whether the model design is appropriate, for instance whether appropriate interaction terms have been included. However, it's certainly the case that if you **don't** include the covariate in the model, you have **not** adjusted for it.

The other approach is to subsample the data so that the levels of the covariates are approximately constant. For example, here is a subset that considers workers between the ages of 30 and 35 with between 10 to 12 years of education and working in the sales sector of the economy:

```
> small = subset(cps, age <=35 & age >= 30 &
                      educ>=10 & educ <=12 &
                      sector=="sales" )
```

The choice of these particular levels of age, educ, and sector is arbitrary, but you need to choose some level if you want to hold the covariates appproximately constant.

The subset of the data can be used to fit a simple model:

```
> mod4 = lm( wage ~ sex, data=small)
> summary(mod4)
```

```
Coefficients:
            Estimate Std. Error t value Pr(>|t|)
(Intercept)    4.500      0.500     9.0     0.07
sexM           4.500      0.866     5.2     0.12
```

At first glance, there might seem to be nothing wrong with this approach and, indeed, for very large data sets it can be effective. In this case, however, there are only 3 cases that satisfy the various criteria: two women and one man.

```
> table( small$sex )
F M
2 1
```

So, the $4.50 difference between the sexes and wages depends entirely on the data from a single male! (Chapter 14 describes how to assess the precision of model coefficients. This one works out to be 4.50 ± 11.00 — not at all precise.)

8.4.2 Calculating Model Values

The `fitted` operator will produce the fitted model values from a model. Of course, the model formula describes a relationship between inputs and outputs. The inputs used by `fitted` are those in the data frame used for fitting the model.

You will sometimes want to calculate the model values for different inputs. For example, you might want to calculate a partial difference by computing the fitted model values for two hypothetical cases that differ with respect to only one variable.

DATA FILE
kidsfeet.csv

To illustrate, consider a model that tries to explain foot width as a function of some other variables for the 4th graders in the data set on kids' feet: :

```
> feet = ISMdata("kidsfeet.csv")
> mod = lm(width ~ length + sex, data=feet)
> mod
Coefficients:
(Intercept)    length      sexG
     3.6412    0.2210   -0.2325
```

This model is simple enough that you can see the relationships directly from the coefficients. For every additional cm of foot length, a child's foot width is typically greater by 0.22 cm. Girls feet are typically 0.23 cm narrower than boys feet.

Now to calculate model values for new inputs. The general idea is to plug in the new inputs into the formula and compute the output. For example, suppose you want to find the model values for Nancy, a girl whose foot is 26 cm long. Plug in the values for `sex` and `length` into the model formula, using the coefficients that were calculated by `lm`:

```
> 3.64117 + 0.22103*26 - 0.23252
[1] 9.1553
```

Note that the sexG term was included because this kid was a girl. If the new kid had been a boy, the calculation would be 3.6412+0.2210*26.

When models are complicated, perhaps including many explanatory variables, interaction terms, etc., it's difficult to do the calculation by hand. Instead, you can use R to perform the calculation for you. The only difficult part is preparing the inputs in the right format. This involves creating a new data frame with exactly the same variable names as in the original data used to fit the model and then entering the input values that you want.

First, you need to create a new data frame with the new input values.

One way to do this is to use spreadsheet software. You need to make sure that the variable names are spelled exactly as in the original data frame used to fit the model. If there are categorical variables involved, you need to be careful to use exactly the same spellings for the levels of those variables. For example, for the variable sex in the data used in this example, the levels are B and G — you have to use those and not M or F or f or "boy" or anything else.

Once you have your new data frame, read it into R in the conventional way. For example, if your new frame is in a file called nancy.csv, then read it in with the read.csv or ISMdata operator.

```
> nancy = read.csv("nancy.csv")
> predict(mod, newdata=nancy)
[1] 9.1553
```

As an alternative to using a spreadsheet program, you may prefer to create the new data frame entirely within R. To do this, start by creating a new data frame with space to enter the new data. An easy way to do this is to copy one (or several) cases from the original data frame. Here's one way to make a new data frame called nancy and holding just one case:

```
> nancy = head(feet,1) # just one case
```

If there was more than one new case for which you want to calculate output values, you can create a new data frame with more than one case. For instance, here's how to make a data frame with 5 cases: nancy=head(feet,5)

Next, edit the values in the new data frame, nancy, so that they are what you want. An easy way to do this is with the edit command, which puts up a small spreadsheet in which new values can be entered. You invoke edit in the following way:

```
> nancy = edit(nancy)
```

This will cause the editing spreadsheet window to pop up.

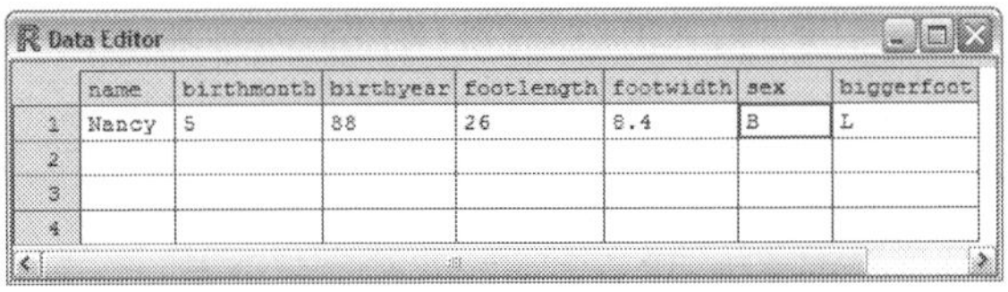

Data Editor

	name	birthmonth	birthyear	footlength	footwidth	sex	biggerfoot
1	Nancy	5	88	26	8.4	B	L
2							
3							
4							

Type in your changes and close the window. Once closed, the modified spreadsheet will be returned as the output of `edit`. Like any other operation in R, you need to save that output to an object if you want to use it later. That's why there is an assignment in the above R command: `nancy=edit(nancy)`. This will save the new values in the object called `nancy`. Without the assignment — just using `edit(nancy)` — will let you type in the spreadsheet, but the resulting modification will not be saved.

The figure shows the spreadsheet *after* changing the values to those appropriate for Nancy. You don't need to change all the variables, just the ones that are the inputs to your model. It doesn't matter what the value of the other variables are since the model doesn't use these. In particular, the value of the response variable doesn't need to be specified because this is not an *input* to the model; it will be the output from the model.

Once you have the edited data frame with the appropriate input values, you can feed this new data to the `predict` operator, which will do the arithmetic for you.

```
> predict(mod, newdata=nancy)
[1] 9.1553
```

It's your responsibility when editing data in this way to make sure that you are using proper labels for the categorical variables. The `edit` program is smart enough to keep track of all the of levels of each categorical variable, even if that level doesn't show up in the small subset being edited. But if you use a new level, `edit` will presume you want to add a new category, and give you a warning message to that effect. Here's the output that came out of the previous editing operation:

```
> nancy = edit(nancy)
Warning message:
added factor levels in 'name' in: edit.data.frame(nancy)
```

This message is pointing out that the child's name was changed to "Nancy," a new level for the categorical variable `name`. Since `name` isn't an input to the model, it doesn't matter what its value is.

But suppose rather than using G to indicate a girl, you had used lower-case g. The output of `edit` would be.

```
> nancy = edit(nancy)
Warning messages:
1: added factor levels in 'name' in: edit.data.frame(nancy)
2: added factor levels in 'sex' in: edit.data.frame(nancy)
```

Putting the improper level into the model will cause a problem:

```
> predict(mod, newdata=nancy)
Error ...
    factor 'sex' has new level(s) g
```

These error and warning messages should make it straightforward to track down the origin of any problem.

If you do make a mistake and add in a new categorical level, you won't be able to fix things by re-editing the spreadsheet because the new level will still be listed as a possibility. Instead, you need to start from scratch:

```
> nancy = head(feet,1) # just one case
> nancy = edit(nancy)
```

Keep in mind that if you want a list of the valid levels of any categorical variables, just use the `levels` operator:

```
> levels( feet$sex )
[1] "B" "G"
```

You won't need to include the quotes around the level when editing in the spreadsheet: `B` and `G` are just fine.

The `predict` operator won't work right unless you fit your models with the `data=` syntax, as is always done in this book. **Don't** use the `$` notation for your variables as in `lm(feet$width ~ feet$length + feet$sex)`

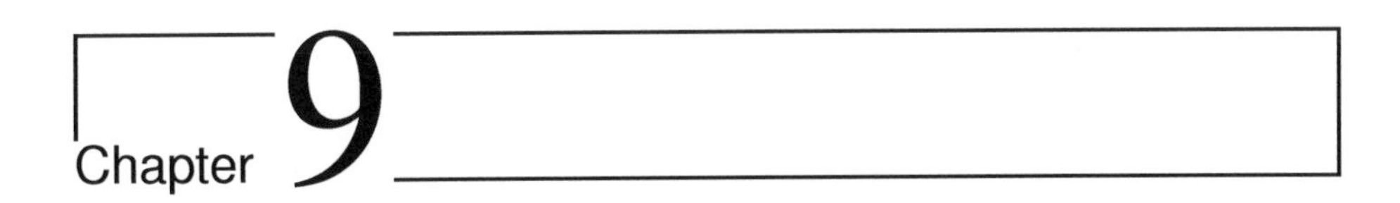

Model Vectors

By the aid of symbolism, we can make transitions in reasoning almost mechanically by the eye, which otherwise would require the higher faculties of the brain. ... Civilization advances by extending the number of important operations which we can perform without thinking about them.
— Alfred North Whitehead (1861-1947)

Many people are disappointed to trade the graphical display of models (as in Chapter 4) for model formulas and coefficients (as in Chapter 5). The formulas give an added ability to model complicated relationships involving multiple explanatory variables, but at the cost of losing touch. Without graphics, people can't apply their powerful human facility to visualize or make use of our cognitive strengths in recognizing spatial relationships.

The link between reasoning and our ability to perceive things in space is powerful and often unappreciated. It shows up, for instance, in the physical and spatial metaphors used to describe reasoning, learning, and understanding: turn it over, see it from a new perspective, get a grasp on it, take it apart, illuminate, pick up, take in, compare side-by-side, run through, sort out, put it together, have it at one's fingertips ....

This chapter introduces the tools and concepts needed to think about complicated, multivariate models using our powerful spatial intuition. The spatial/geometrical approach will make it easier to visualize and intuit the process of fitting, the sources of redundancy, and the meaning of correlation. The approach will also prepare you to apply statistical reasoning to your models: setting the explanations of our models in the context of what remains unexplained in the data.

The approach is founded on basic geometry: lengths, angles, directions, etc. But it is not the sort of scientific and statistical graphics you are used to: scatterplots of individual cases, histograms, etc. So be prepared to drop some of your preconceptions about the types of graphs that can display relationships.

9.1 Vectors

A **vector** is a mathematical idea that is deeply rooted in everyday physical experience. A vector is simply something that specifies a **length** and a **direction**. It lives in a space of some **dimension**.

You can draw vectors on a piece of paper. The paper is the two-dimensional space that the vectors inhabit. The convention is to draw a vector as an arrow, as in Figure 9.1. Note that each vector has a length that you can measure with a ruler. Each vector points in some direction.

Figure 9.1 shows vectors in different spaces: a one-dimensional space, a two-dimensional space, and a three-dimensional space. On a piece of paper, only vectors in a 2-dimensional space can be drawn in a natural way. But vectors can exist in three dimensions: point a pencil — it has a length and a direction. Drawing such vectors on paper requires tricks of perspective and adding a triple-coordinate axis in order to orient you.

You can also think of vectors in one-dimension, as if they were directions on a railroad track. Drawing such vectors requires laying down a track along the paper — a thin line — to emphasize that the vectors live only in a small part of the paper.

The mathematical notion of vectors is abstract: they don't have to be drawn or embodied as a pencil. One of the aspects of abstraction concerns position. Obviously, each of the vectors drawn in Figure 9.1 has a position where it has been drawn. But position is not an attribute of a mathematical vector; the only quantities that matter are length and direction.

If two vectors have the same length and direction, they are the same vector. Within each panel of Figure 9.1 there are two vectors with the same label. These two vectors are exactly the same — they have the same length and direction even though they have been drawn in different places.

You can position a mathematical vector wherever it is convenient for your purposes. For example, you can ask what is the **angle** between two vectors. To answer this, pick up the vectors — maintaining their direction — and position them tail to tail, like this: Then you can measure the angle with a protractor.

You can draw vectors as arrows, but a good way to think about them is in terms of movement: as steps. A vector is an instruction to take a step. Suppose one vector indicates a movement 1 meter to the north-east, another vector a movement 3 meters to the south. Such instructions make sense even without needing to specify where you are right now: position doesn't enter into it. There are two simple mathematical operations on such movements — scaling and addition — as shown in Figure 9.2.

Scale a vector. Make a change to a vector's length to make it longer or shorter, but keep the direction the same. Suppose the vector says, "move one meter to the north-east." Scaling the vector by a factor of 3 means to make it 3 times longer: "move three meters to the north-east." Scaling the vector by a factor of 0.5 means to make it half as long: "move $\frac{1}{2}$ meter to the north-east." A negative scale means to step backwards, so a scale of -2 means

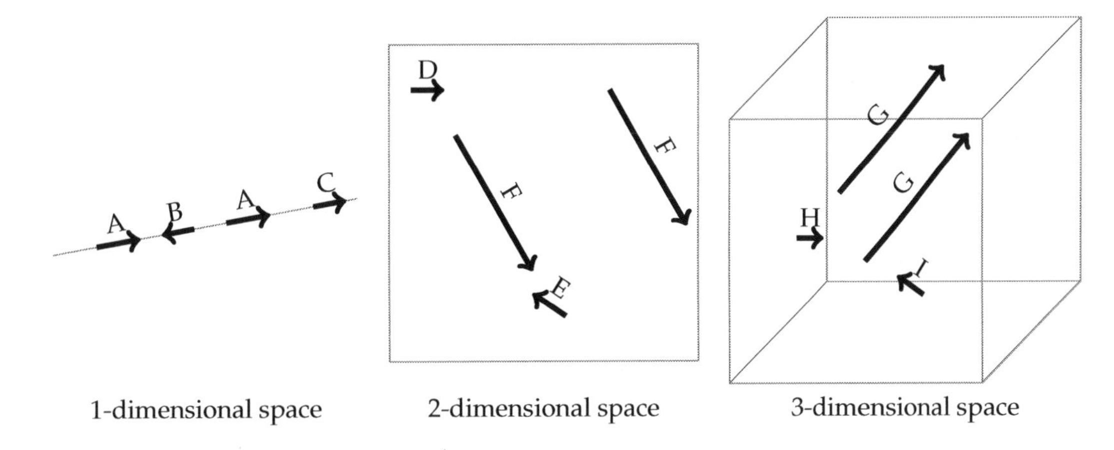

Figure 9.1: Some vectors in different spaces.

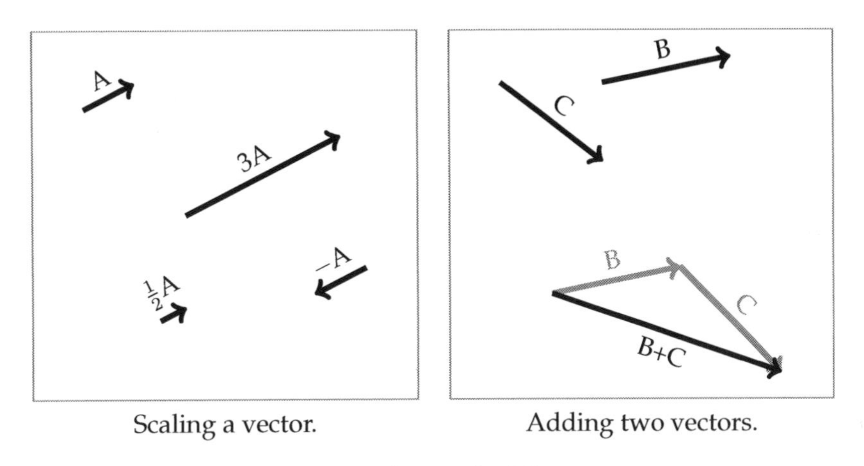

Figure 9.2: Scaling and adding vectors.

to move 2 meters backwards while facing north-east, effectively moving 2 meters to the south-west.

Add two vectors. Pick a place to start. Take one step according to the length and direction of the first vector. Then, from the point you ended up after the step, take another step according to the length and direction of the second vector. The overall distance and direction from the place you started to the place you ended up is the vector that results from adding up the two vectors.

Mathematics extends the notion of movement to high dimensions. Talking about 1- or 2- or 3-dimensional objects makes intuitive sense. An arrow drawn on a piece of paper inhabits a 2-dimensional space; a pencil lives in a three-dimensional space. But what is a 4-dimensional space, or a space that is 100-dimensional? The key to understanding is to consider length and direction abstractly without tying them to a physical embodiment like an arrow.

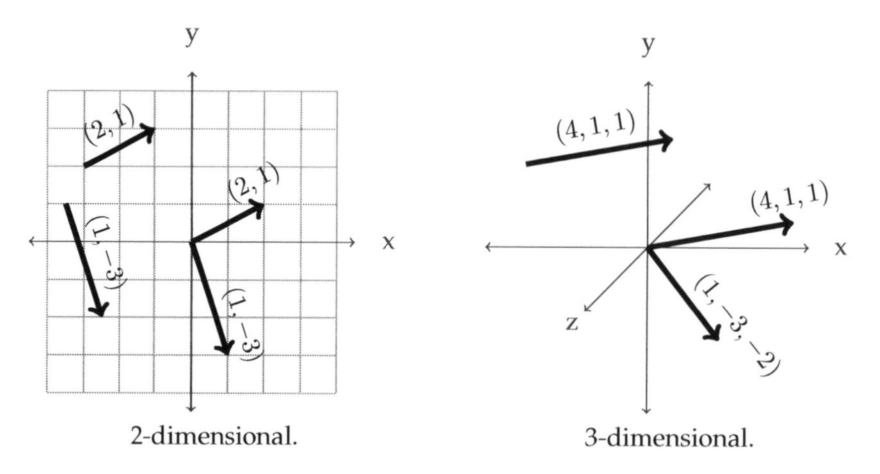

Figure 9.3: Vectors as coordinates.

For vectors in 2-dimensional space, you can specify a length and direction with coordinates on the Cartesian plane. (See Figure 9.3.) Most people are used to using the Cartesian plane to plot out position, but it works just as well for direction and length.

Figure 9.3 shows some vectors specified as coordinates. For example, $(2, 1)$ means move two units to the right and 1 unit up, altogether a length of 2.236 units directed to the east-north-east.[1] Where you end up — the position of the arrow head — depends on where you started. The vector, however, isn't the position, it's the step.

You can use the same idea of Cartesian coordinates to think about 3-dimensional vectors. Three coordinates are required — one along each axis. In geography, the coordinates system might be the east-west axis, the north-south axis, and the up-down axis. In a room, a convenient coordinate axis can be given by the three lines emerging from the corner: (1) the line where the two walls meet, (2) the line where one wall and the floor meet, and (3) the line where the other wall and the floor meet. What's key is that three numbers are needed to specify a coordinate in 3-dimensional pace.

The generalization to 4- and higher-dimensional spaces is straightforward. A 4-dimensional vector is represented by 4 numbers, a 5-dimensional vector by 5 numbers, and a 100-dimensional vector by 100 numbers. Although you can't see a 100-dimensional vector in the same way that you can see a pencil, it works perfectly well to think about it using your 2- and 3-dimensional capabilities.

9.2 Vectors as Collections of Numbers

The direct connection of vectors and statistical models comes from the representation of a vector as a collection of numbers, just as a quantitative variable or an

[1] 2.236 comes from the Pythagorean theorem: the movements to the right and up are the legs of a right triangle and the vector described is the hypothenuse of that triangle with length $\sqrt{2^2 + 1^2}$.

indicator variable is a collection of numbers. In this direct sense, vectors merely organize the calculations involved in fitting models. The indirect, geometrical connection with models is richer and will help in interpreting and understanding models.

As a collection of numbers, a vector is written conventionally as a column of numbers, for instance $\begin{bmatrix} 5 \\ 9 \\ 2 \\ 3 \end{bmatrix}$ is a vector in 4-dimensional space.

The two vector operations of scaling and vector addition have a simple arithmetic form.

To perform **vector addition**, add the corresponding terms of each vector. For instance

$$\begin{bmatrix} 5 \\ 9 \\ 2 \\ 3 \end{bmatrix} + \begin{bmatrix} 6 \\ 8 \\ 103 \\ -7 \end{bmatrix} \text{ gives } \begin{bmatrix} 11 \\ 17 \\ 105 \\ -4 \end{bmatrix}.$$

Of course, the two vectors being added have to be the same dimension, that is, they have to have the same number of components, 4 in this case.

Scaling a vector amounts to multiplication. For instance, to scale a vector to three times its length, multiply each component in the vector by 3:

$$3\begin{bmatrix} 1 \\ 0 \\ 1 \\ 1 \end{bmatrix} \text{ gives } \begin{bmatrix} 3 \\ 0 \\ 3 \\ 3 \end{bmatrix} \qquad 3\begin{bmatrix} 5 \\ 9 \\ 2 \\ 3 \end{bmatrix} \text{ gives } \begin{bmatrix} 15 \\ 27 \\ 6 \\ 9 \end{bmatrix}.$$

Notice that both vector addition and vector scaling produce an output that is a vector of the same dimension as the input vector or vectors.

A **linear combination** of vectors involves scaling the vectors and adding up the result. For instance, consider the two vectors $\begin{bmatrix} 1 \\ 0 \\ 1 \\ 1 \end{bmatrix}$ and $\begin{bmatrix} 2 \\ 4 \\ -1 \\ 8 \end{bmatrix}$. Here is a linear combination of them:

$$3\begin{bmatrix} 1 \\ 0 \\ 1 \\ 1 \end{bmatrix} - 6\begin{bmatrix} 2 \\ 4 \\ -1 \\ 8 \end{bmatrix} \text{ gives } \begin{bmatrix} -9 \\ -24 \\ 9 \\ -45 \end{bmatrix}.$$

Such combinations are at the heart of statistical modeling.

9.3 Model Vectors

Recall the distinction between variables and model terms that you've already seen. A modeler chooses the explanatory variables of interest and then needs to specify which terms involving these variables are to be included in a model design.

There is another step in the process that happens automatically: the translation from model terms to **model vectors**. This step doesn't involve any decision making or choice by the modeler, but mastering it is important in understanding and interpreting models.

In the translation each model term becomes one or more model vectors according to these rules:

1. The **intercept term** becomes a vector of all ones.

2. A **quantitative variable**, or a transformation of a quantitative variable, is already a collection of numbers, so it is already a vector.

3. A **categorical variable** becomes a set of vectors, the indicator vectors for each level of the variable.

4. An **interaction** between two terms becomes a vector by multiplying together the vectors from each of the terms. If one of the terms involves a categorical variable and is therefore a set of vectors, multiply each member of the set by the vector or vectors from the other terms.

Example 9.1: Variables, Model Terms, and Model Vectors Consider this very small subsample from the Current Population Survey wage data.

wage	educ	sex	married	age	sector
12.00	12	M	Married	32	manuf
8.00	12	F	Married	33	service
16.26	12	M	Single	32	service
13.65	16	M	Married	33	prof
8.50	17	M	Single	26	clerical

The categorical variable sex has two levels: F and M. It therefore is a set of two model vectors, the indicators for F and M respectively:

$$\overset{\text{sexF}}{\begin{bmatrix}0\\1\\0\\0\\0\end{bmatrix}} \text{ and } \overset{\text{sexM}}{\begin{bmatrix}1\\0\\1\\1\\1\end{bmatrix}}.$$

The little label over each vector reminds you where the vector comes from. The label is not a part of the vector — the vector is just the collection of numbers.

The categorical variable married has two vectors, one for each of its levels Married and Single. The indicator vectors are:

$$\overset{\text{Married}}{\begin{bmatrix}1\\1\\0\\1\\0\end{bmatrix}} \text{ and } \overset{\text{Single}}{\begin{bmatrix}0\\0\\1\\0\\1\end{bmatrix}}$$

The variable sector has eight levels — clerical, construction, manufacturing, professional, service, management, sales, and "other" — but only four of them show up in this small subset. The eight indicator vectors are:

$$\overset{\text{clerical}}{\begin{bmatrix}0\\0\\0\\0\\1\end{bmatrix}} \overset{\text{constr}}{\begin{bmatrix}0\\0\\0\\0\\0\end{bmatrix}} \overset{\text{manuf}}{\begin{bmatrix}1\\0\\0\\0\\0\end{bmatrix}} \overset{\text{prof}}{\begin{bmatrix}0\\0\\0\\1\\0\end{bmatrix}} \overset{\text{service}}{\begin{bmatrix}0\\1\\1\\0\\0\end{bmatrix}} \overset{\text{manag}}{\begin{bmatrix}0\\0\\0\\0\\0\end{bmatrix}} \overset{\text{sales}}{\begin{bmatrix}0\\0\\0\\0\\0\end{bmatrix}} \text{ and } \overset{\text{other}}{\begin{bmatrix}0\\0\\0\\0\\0\end{bmatrix}}.$$

Notice that some of these vectors are all zeros. That's because none of the five cases in this subsample happened to take on the levels for construction, management, sales, or "other."

The vectors from the quantitative variables wage, educ, age are simply reiterations of the values of those variables.

$$\overset{\text{wage}}{\begin{bmatrix} 12.00 \\ 8.00 \\ 16.26 \\ 13.65 \\ 8.50 \end{bmatrix}} \quad \overset{\text{educ}}{\begin{bmatrix} 12 \\ 12 \\ 12 \\ 16 \\ 17 \end{bmatrix}} \quad \overset{\text{age}}{\begin{bmatrix} 32 \\ 33 \\ 32 \\ 33 \\ 26 \end{bmatrix}}$$

An interaction between two terms is the pairwise product of the vectors from those terms. Here, for example, is the interaction between the two quantitative terms, education and age:

$$\overset{\text{educ:age}}{\begin{bmatrix} 384 \\ 396 \\ 384 \\ 528 \\ 442 \end{bmatrix}} \text{ is } \overset{\text{educ}}{\begin{bmatrix} 12 \\ 12 \\ 12 \\ 16 \\ 17 \end{bmatrix}} \times \overset{\text{age}}{\begin{bmatrix} 32 \\ 33 \\ 32 \\ 33 \\ 26 \end{bmatrix}}$$

The interaction between a categorical variable and a quantitative variable involves multiplying each of the vectors from the categorical variable by the vector from the quantitative variable. Here is the interaction of age and married:

$$\overset{\text{age:Married}}{\begin{bmatrix} 32 \\ 33 \\ 0 \\ 33 \\ 0 \end{bmatrix}} = \overset{\text{age}}{\begin{bmatrix} 32 \\ 33 \\ 32 \\ 33 \\ 26 \end{bmatrix}} \times \overset{\text{Married}}{\begin{bmatrix} 1 \\ 1 \\ 0 \\ 1 \\ 0 \end{bmatrix}} \text{ and } \overset{\text{age:Single}}{\begin{bmatrix} 0 \\ 0 \\ 32 \\ 0 \\ 26 \end{bmatrix}} = \overset{\text{age}}{\begin{bmatrix} 32 \\ 33 \\ 32 \\ 33 \\ 26 \end{bmatrix}} \times \overset{\text{Single}}{\begin{bmatrix} 0 \\ 0 \\ 1 \\ 0 \\ 1 \end{bmatrix}}$$

Interactions between two categorical variables involve all the combinations of vectors from the variables. So, the interaction of sex and married produces four vectors, since each sex and married has two. The four vectors stand for Female Married, Female Single, Male Married, and Male Single:

$$\overset{\text{Female Married}}{\begin{bmatrix} 0 \\ 1 \\ 0 \\ 0 \\ 0 \end{bmatrix}} \quad \overset{\text{Female Single}}{\begin{bmatrix} 0 \\ 0 \\ 0 \\ 0 \\ 0 \end{bmatrix}} \quad \overset{\text{Male Married}}{\begin{bmatrix} 1 \\ 0 \\ 0 \\ 1 \\ 0 \end{bmatrix}} \quad \overset{\text{Male Single}}{\begin{bmatrix} 0 \\ 0 \\ 1 \\ 0 \\ 1 \end{bmatrix}}$$

As it happens, there are no single females in the small subsample, so the second vector is all zeros.

The interaction of sector with its eight levels and sex with its two levels, produces a set of 16 vectors, more than it's appropriate to print here.

Finally, don't forget the intercept vector, which consists of a 1 for every case. For the five cases in the small data set, the intercept vector is

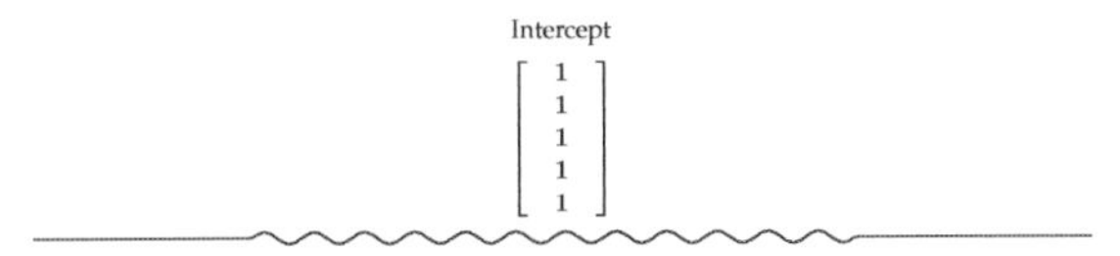

9.4 Model Vectors in the Linear Formula

The model formula for the linear model has a very simple structure in terms of model vectors: each coefficient multiplies a single vector — there's one coefficient for each vector. In constructing a linear model, you need to provide a list of model terms: the model design. The computer then translates this list of terms into a set of model vectors. (In mathematics, a set of vectors, all with the same dimension, is called a **matrix**.) The software then finds the coefficient for each vector.

To illustrate, consider the following model design fitted to a small subset of the Current Population Survey wage data:

$$\texttt{wage} \sim 1 + \texttt{age} + \texttt{sex}$$

There are four vectors associated with these three model terms: the intercept vector, the vector for age, and the two vectors associated with the two levels of sex:

$$\overset{\text{Intercept}}{\begin{bmatrix} 1 \\ 1 \\ 1 \\ 1 \\ 1 \end{bmatrix}}, \overset{\text{age}}{\begin{bmatrix} 32 \\ 33 \\ 32 \\ 33 \\ 26 \end{bmatrix}}, \overset{\text{sexM}}{\begin{bmatrix} 1 \\ 0 \\ 1 \\ 1 \\ 1 \end{bmatrix}}, \text{ and } \overset{\text{sexF}}{\begin{bmatrix} 0 \\ 1 \\ 0 \\ 0 \\ 0 \end{bmatrix}}$$

To describe the model formula, the software only needs to tell the coefficient on each vector. A typical report from statistical software looks like this:

```
              Estimate   Std. Error   t value   p value
(Intercept)     -7.21         9.04      -0.80     0.509
age              0.64         0.37       1.76     0.220
sexM            -3.71         3.30      -1.12     0.378
```

The actual coefficients are in the column labelled "Estimate." The other columns provide information about the reliability and the strength of evidence for each coefficient provided by the data. (The interpretation of these columns will be introduced in later chapters.)

There is one coefficient for each vector. As usual, one vector from a categorical term is dropped as being redundant. In this case, the sexF vector has been dropped.

The model values result from multiplying each vector by its coefficient, e.g.,

$$-7.21 \overset{\text{Intercept}}{\begin{bmatrix} 1 \\ 1 \\ 1 \\ 1 \\ 1 \end{bmatrix}} + 0.64 \overset{\text{age}}{\begin{bmatrix} 32 \\ 33 \\ 32 \\ 33 \\ 26 \end{bmatrix}} - 3.71 \overset{\text{sexM}}{\begin{bmatrix} 1 \\ 0 \\ 1 \\ 1 \\ 1 \end{bmatrix}} \text{ giving } \overset{\text{Model Values}}{\begin{bmatrix} 13.43 \\ 10.37 \\ 13.43 \\ 14.08 \\ 9.56 \end{bmatrix}}.$$

In other words, the model values are a **linear combination** of the model vectors.

9.5 Model Vectors and Redundancy

As a rule, there is one model coefficient for each model vector. The exception comes when one or more of the model vectors are **redundant**. In that case, the coefficients on the redundant vectors are dropped.

The most common situation where redundancy comes into play is when a model includes a categorical explanatory variable. It's always the case that one of the indicator vectors from the categorical variable will be dropped as redundant.

A model vector is redundant when that vector can be written as a linear combination of other model vectors. For categorical variables, it's easy to see why this happens. Recall the indicator vectors for the sex variable:

$$\overset{\text{sexM}}{\begin{bmatrix} 0 \\ 1 \\ 0 \\ 0 \\ 0 \end{bmatrix}} \text{ and } \overset{\text{sexF}}{\begin{bmatrix} 1 \\ 0 \\ 1 \\ 1 \\ 1 \end{bmatrix}}.$$

Consider the linear combination which comes from multiplying each of these vectors by 1 and then adding them up. Since the indicator vectors have 1s in complementary places, adding up all the indicators produces a vector of all ones.

$$1 \overset{\text{sexM}}{\begin{bmatrix} 0 \\ 1 \\ 0 \\ 0 \\ 0 \end{bmatrix}} + 1 \overset{\text{sexF}}{\begin{bmatrix} 1 \\ 0 \\ 1 \\ 1 \\ 1 \end{bmatrix}} = \overset{\text{Intercept}}{\begin{bmatrix} 1 \\ 1 \\ 1 \\ 1 \\ 1 \end{bmatrix}}.$$

The vector of all ones is, of course, the intercept vector. Thus, there is always redundancy between the set of indicator vectors for a categorical variable and the intercept vector.

9.6 Geometry by Arithmetic

It's fine to talk about 4- and 5-dimensional vectors, but how do you measure their geometrical properties, for instance their lengths or the angle between two vectors. After all, you can't sneak a ruler or a protractor into some hypothetical 5-dimensional space.

The trick is to perform the measurement by manipulating the numbers that represent the vector. Consider the vector labeled $(2,1)$ in Figure 9.3. The length of this ordinary, two-dimensional vector can be measured with a ruler. Or, you can use the Pythagorean theorem to calculate the length: $\sqrt{2^2+1^2}$. The formula for the length is very simple: square each of the coordinates, add up the squares, and take the square-root of the sum.

For dimensions higher than 2, simply *define* the length of a vector in the same way: the square root of the sum of squares of the coordinates.

An operation called the **dot product** simplifies the notation. The dot product takes two vectors as inputs, and produces a single number as an output. That number is found by multiplying the corresponding coordinates of the two vectors and adding up the results. For example

$$\begin{bmatrix} 3 \\ 1 \\ 0 \end{bmatrix} \cdot \begin{bmatrix} 4 \\ -3 \\ 2 \end{bmatrix} = 12 - 3 + 0 = 9.$$

You write the dot product between vectors A and B as A · B.

The length of a vector A is just the square root of A dotted with itself. Or, using mathematical notation where $\| A \|$ stands for the length of A,

$$\| A \| = \sqrt{A \cdot A}$$

Example 9.2: Vector Length Find the length of $\begin{bmatrix} 1 \\ -4 \\ 2 \\ 3 \end{bmatrix}$.

$$\sqrt{\begin{bmatrix} 1 \\ -4 \\ 2 \\ 3 \end{bmatrix} \cdot \begin{bmatrix} 1 \\ -4 \\ 2 \\ 3 \end{bmatrix}} = \sqrt{1+16+4+9} = \sqrt{30} = 5.477.$$

Angles can also be computed with dot products. The formula for the angle θ between vectors A and B is

$$\cos(\theta) = \frac{\mathrm{A} \cdot \mathrm{B}}{\parallel \mathrm{A} \parallel \parallel \mathrm{B} \parallel}.$$

You hardly ever need to use this formula, but it important that you know that angles can be found by arithmetic operations.

Example 9.3: Angles Between Vectors Find the angle between $\begin{bmatrix} 1 \\ -4 \\ 2 \\ 3 \end{bmatrix}$ and $\begin{bmatrix} 0 \\ 3 \\ -2 \\ 1 \end{bmatrix}$.

This involves three simple calculations:

- $\mathrm{A} \cdot \mathrm{B} = \begin{bmatrix} 1 \\ -4 \\ 2 \\ 3 \end{bmatrix} \cdot \begin{bmatrix} 0 \\ 3 \\ -2 \\ 1 \end{bmatrix} = -13$
- $\parallel \mathrm{A} \parallel = \sqrt{\begin{bmatrix} 1 \\ -4 \\ 2 \\ 3 \end{bmatrix} \cdot \begin{bmatrix} 1 \\ -4 \\ 2 \\ 3 \end{bmatrix}} = \sqrt{30}$
- $\parallel \mathrm{B} \parallel = \sqrt{\begin{bmatrix} 0 \\ 3 \\ -2 \\ 1 \end{bmatrix} \cdot \begin{bmatrix} 0 \\ 3 \\ -2 \\ 1 \end{bmatrix}} = \sqrt{14}$

Putting these together into the overall formula gives

$$\cos(\theta) = \frac{-13}{\sqrt{30}\sqrt{14}} = -0.6343.$$

Finding θ itself involves inverting the cosine. It turns out that $\cos(129.37°) = -0.6343$, so $\theta = 129.37°$ (or, equivalently, 2.258 radians).

What's remarkable here is that angles and lengths can be found entirely through arithmetic. This allows you to generalize the notions of angles and lengths to vectors in high-dimensional spaces.

Angles and lengths play an important role in the interpretation of statistical models. But it's inconvenient to have to take square-roots or to invert the cosine function. For this reason, statistics is often written in terms of **square lengths** and **cosines of angles**. The shorthand name for a square length in statistics is a **sum of squares**, reflecting that the square-length of a vector is calculated by squaring each of the components and adding them up. There is no standard shorthand name for the cosine of an angle, but in the setting where the cosine of an angle most frequently appears in statistics, it is the **correlation coefficient**.

9.7 Computational Technique

In your ordinary statistical work, you will not often compute explicitly with vectors in the ways talked about in this chapter. The geometrical calculations are contained implicitly within other operators such as `lm`. Still, in order to work with the geometrical concepts in the exercises, you need to know how to calculate the geometrical quantities numerically.

To illustrate, I'll use the kids' feet dataset :

```
> feet = ISMdata("kidsfeet.csv")
```

There are $n = 39$ cases in this data frame, so the vectors are 39-dimensional. Of course it seems impossible to visualize a 39-dimensional space, but the calculations are the same for 39-dimensions as they would be for 2-dimensional space or 2000-dimensional space.

9.7.1 Length and Dimension

There is no "dimension" operator in R. Instead, the `length` operator can be used. It simply counts how many numbers there are in a vector:

```
> length(feet$length)
[1] 39
```

The use of the name "length" for this operator can be confusing, since the geometrical length is different from the count of components returned by the `length` operator.

To find the geometrical length of a vector, use the Pythagorean formula: the square root of the sum of squares.

```
> sqrt( sum( feet$length^2 ))
[1] 154.6
```

Sometimes it's more convenient to talk about the "square-length" of a vector, which is just the same as the above, but without the square root:

```
> sum( feet$length^2 )
[1] 23904
```

Remember that the values are squared *before* they are summed. It's easy to make a mistake and put the square *after* the summation but this would give a wrong result.

Keep in mind that nothing about these computation requires the use of the $ sign. It's only because the vectors used in the examples happen to be variables in a data frame that the examples use $. If you had a vector with a different name, just use that.

```
> myvec = c(1,2,3,4)
> sqrt(sum( myvec^2 ))
[1] 5.477
```

9.7.2 Dot products and angles

A fundamental computation on two vectors is the dot product: the sum of the pairwise products of the components of the two vectors.

```
> sum( feet$length * feet$width )
[1] 8687
```

The angle between two vectors is important enough that the book software defines an operator:

```
> angle( feet$length, feet$width )
[1] 0.04602
```

The angle is reported in **radians**, since most other operators (such as `cos`) take their inputs in radians. People, however, are usually more comfortable with degrees. You can convert radians to degrees by multiplying by $180/\pi$:

```
> angle( feet$length, feet$width )*180/pi
[1] 2.637
```

For convenience, the `angle` operator can be instructed to output values in degrees.

```
> angle( feet$length, feet$width, degrees=TRUE )
[1] 2.637
```

Make sure not to pass this to another operator such as `cos`:

```
> cos(angle( feet$length, feet$width ))
[1] 0.999    # Correct
> cos(angle( feet$length, feet$width, degrees=TRUE ))
[1] -0.8751  # WRONG, WRONG, WRONG
```

9.7.3 Scaling and linear combinations

Scaling and addition of vectors is done in the usual way. For example, here is a linear combination of 3 times the `length` vector and -2 times the `width` vector:

```
> foo = 3*feet$length - 2*feet$width
```

Quantitative variables can be computed on directly. It's different for categorical variables. A categorical variable does not correspond to a single vector. Instead, there is one indicator vector for each level of the variable. To construct the indicator vector, use the == comparison operator for the particular level whose vector you want. For example, here is the indicator vector for `sex`**B**:

```
> boyvector = (feet$sex == "B")
> boyvector
 [1]  TRUE  TRUE  TRUE  TRUE  TRUE  TRUE  TRUE FALSE
 ... and so on ...
[33] FALSE  TRUE  TRUE FALSE FALSE FALSE FALSE
```

Even though these are printed as logical values, when you do arithmetic on them the value TRUE is treated as 1 and FALSE as 0:

```
> 3*boyvector
 [1] 3 3 3 3 3 3 3 0 0 3 3 3 3 3 0 0 0 0 0 0 3 3 0
[24] 0 0 3 0 3 3 3 0 0 0 3 3 0 0 0 0
```

Chapter 10

Statistical Geometry

I made straight for the ship, roused up the men
to get aboard and cast off at the stern.
They scrambled to their places by the rowlocks
and all in line dipped oars in the grey sea.
But soon an off-shore breeze blew to our liking —
a canvas-bellying breeze, a lusty shipmate
sent by the singing nymph with sunbright hair.
So we made fast the braces, and we rested,
letting the wind and steersman work the ship.
— The Odyssey, Book XII. Translation by Robert Fitzgerald.

This chapter introduces a way to think about model fitting without the arithmetic involved in minimizing sums of squares of residuals. Surprisingly, one can do statistical calculations with just a ruler and protractor: by drawing straight lines, measuring lengths and measuring angles. The intention, however, is not to replace the computer, which is capable of much more precise and rapid calculations. Rather, the point is to help develop a concise picture of the operations that the computer is implementing so that you can understand and anticipate the output that will be provided by the computer.

Underlying the geometry are two main metaphors. The first is that fitting a model corresponds to making a journey. The second metaphor has to do with the terrain on which the journey is to be made: spaces in which model vectors specify directions and destinations.

In the epic poems of the ancient Greeks, long voyages were often by sea. A ship's course was set by the wind and the winds themselves, favorable or unfavorable, were directed by the gods. In Homer's *Odyssey*, the hero Odysseus is being kept from his home by an opposing wind from hostile Poseidon. Athena, the gray-eyed daughter of Zeus, arranges a favorable wind that carries Odysseus's raft toward his destination. But "toward" is not "to." Odysseus follows the wind as far as advantageous, but then needs to take matters into his own hands

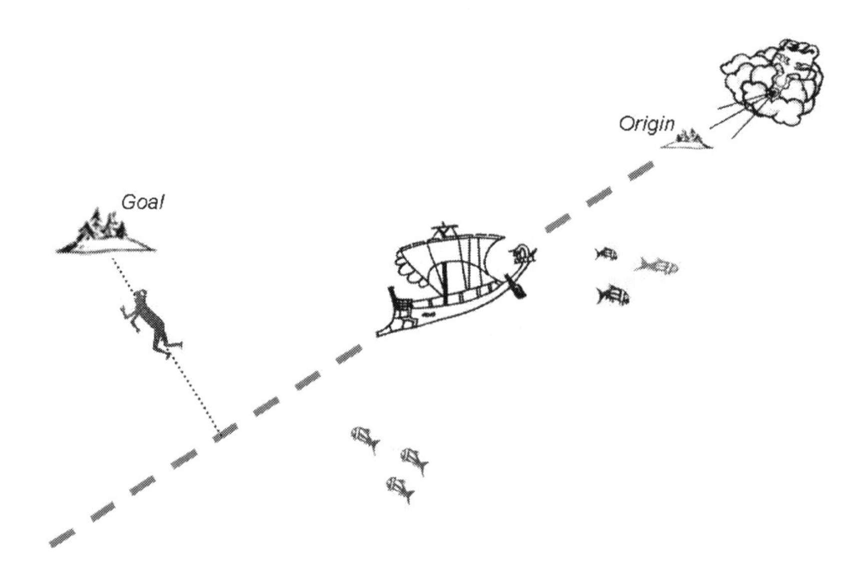

Figure 10.1: An ancient voyager could be carried downwind, but must complete his trip by swimming.

and swim. A sensible voyager tries to make the swimming leg of the trip short. Clever Odysseus jumps ship at the point where the ship passes as close as possible to the goal and then swims perpendicularly to the ship's course, directly toward the goal.

Fitting a statistical model is analogous. The modeler's explanatory variables are the winds, blowing where they will. The response variable is the goal. But the modeler does not cross the sea. Instead a more abstract mathematical space is being traversed: variable space.

10.1 Case Space and Variable Space

The scatter plot is a familiar graphical format for displaying data. Chapter 4 shows many of them. In a scatter plot, each case in the data frame is plotted as a single point. The whole data sample produces a scattering of points reflecting the case-to-case variation.

This chapter works with a different sort of plot, one that shows model vectors. When drawing model vectors, you are turning the scatter plot inside out. Instead of a point being a case and a variable being an axis, each variable is a vector and each case is an axis.

Both graphical configurations are useful. It's important not to confuse one for the other. To avoid confusion, it helps to have some nomenclature to refer to

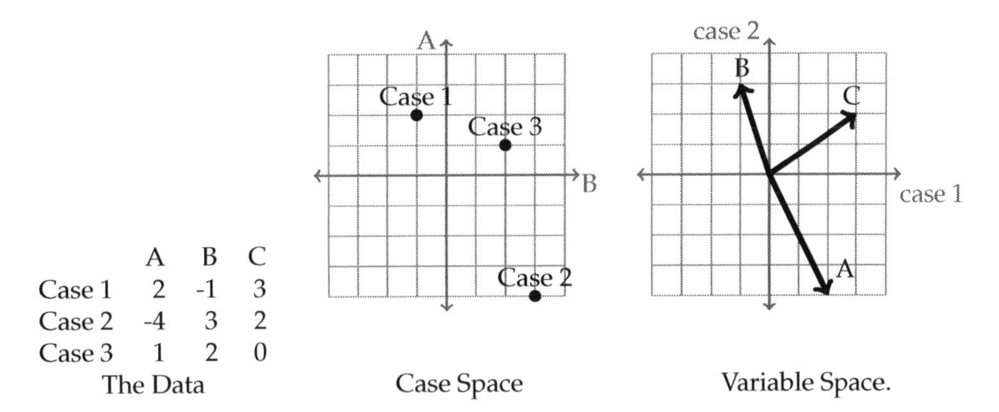

	A	B	C
Case 1	2	-1	3
Case 2	-4	3	2
Case 3	1	2	0

Figure 10.2: A small data set shown in both case- and variable-space formats. Case space can show only two variables at a time. Variable-space can show only two cases at a time.

the two different configurations:

Case Space. The space in which the usual scatter plot is made. Each point is a single case. Each axis is a variable. It's called "case space" because the axes define a space for plotting the cases.

Variable Space. The space in which model vectors are plotted. Each vector is a variable. Each case is a coordinate axis. It's called "variable space" because the coordinate axes define the space that the variables are drawn in.

Figure 10.2 shows a very small dataset as it would be plotted in both case space and variable space.

A variable-space plot drawn on paper can show only $n = 2$ cases, not enough data to be interesting statistically. The value of variable space is not in displaying data but in diagramming the geometrical relationships that underlie statistical modeling. Simple concepts that one can visualize in two or three dimensions — lengths, angles, and so on — give a framework for understanding the operations of statistical modeling.

The geometrical relationships that people capture visually can also be revealed by appropriate arithmetic computations — vector addition, vector scaling, dot products. The computer programs that perform model fitting are actually doing the geometry, but using numbers rather than pictures. This enables the computer to work with large samples, much larger than the $n = 2$ cases that can be graphed on paper.

10.2 Subspaces and Geometrical Operations

A **subspace** is a part of the entire space. But it's not just any arbitrary part; it has a very specific definition that is related to vectors. Each vector has a direction

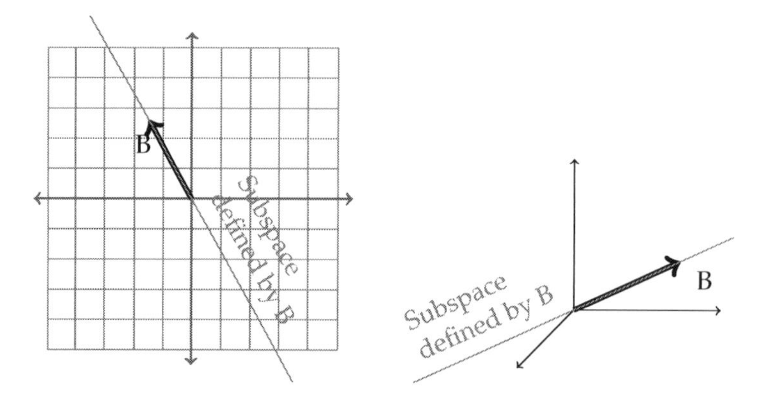

Figure 10.3: A vector and its subspace in $n = 2$ (left) and $n = 3$ (right) dimensional spaces.

and a length. The subspace defined by a vector is all the points in space that can be reached by moving from the origin in the direction of that vector. That is, the subspace is all the points that can be reached by scaling a vector. This is illustrated for $n = 2$ and $n = 3$ in Figure 10.3. The subspace extends out indefinitely far.

When dealing with more than one vector, each vector has its own subspace, as shown in Figure 10.4. Different vectors define different subspaces, although the subspaces from two aligned vectors will be identical.

Collections of two or more vectors also define subspaces: all the points that can be reached by moving from the origin first in the direction of one vector, then turning and moving in the direction of the other (and so on if there are more than two vectors). That is, the subspace defined by a set of vectors is all the points that can be reached by linear combinations of the vectors. A collection of two vectors defines a subspace that is a plane — unless the two vectors happen to be exactly aligned, in which case the subspace is just a line.

There are three geometrical operations that you will need for fitting models to data: projection, finding coefficients, and centering.

The operation of **projecting** a vector A onto a subspace defined by B finds the single point in the subspace of B that is as close as possible to A. As notation, write the projection of A onto B as $A_{\|B}$. Figure 10.5 shows an example. The parallel sign ($\|$) in the notation is intended as a reminder that the projection of A onto B is on the B subspace. So, the vector $A_{\|B}$ is always aligned with B.

Projection of A onto B corresponds to fitting the model A $\sim$ B. In fitting, as described in Chapter 6, you are looking for the coefficient c in the model formula $\underset{\text{Values}}{\text{Model}}$ A $= c$B, that is, multiply vector B by the number c. This is just vector scaling, so the quantity c lies somewhere on the subspace defined by B. When you fit the model, you are finding the point on that subspace that is the best possible match for A, that is, the point in the subspace that is as close as

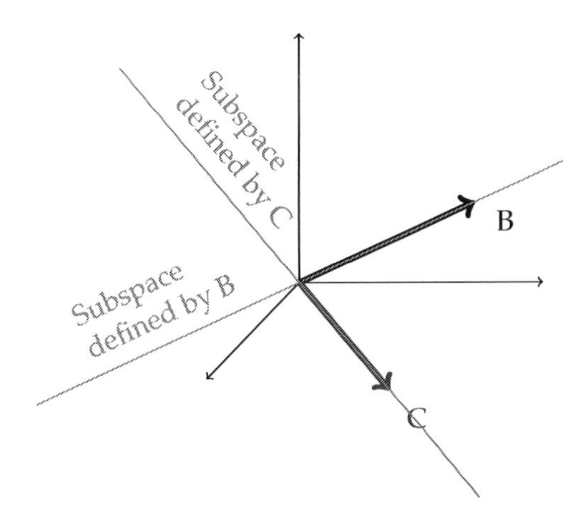

Figure 10.4: Each vector defines its own subspace.

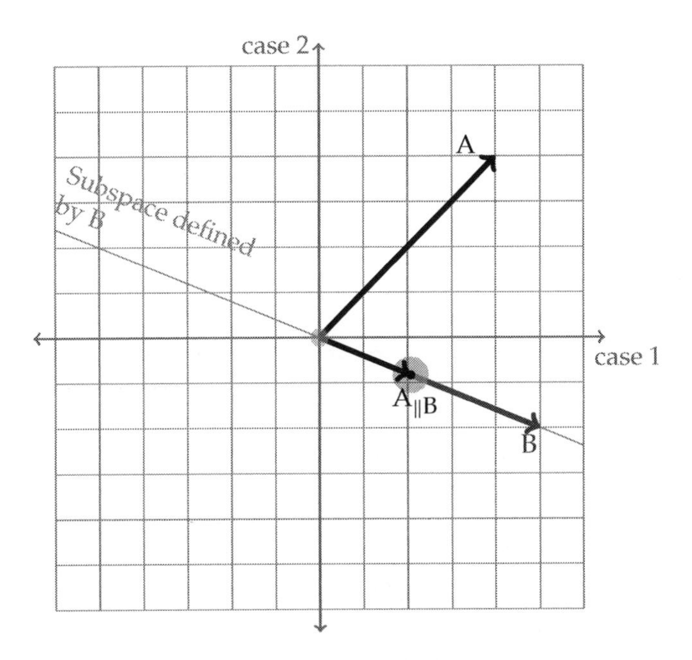

Figure 10.5: Projection of vector A onto vector B means finding the point in the subspace of B that is as close as possible to A. This point is labelled $A_{\|B}$, meaning "the part of A aligned with B."

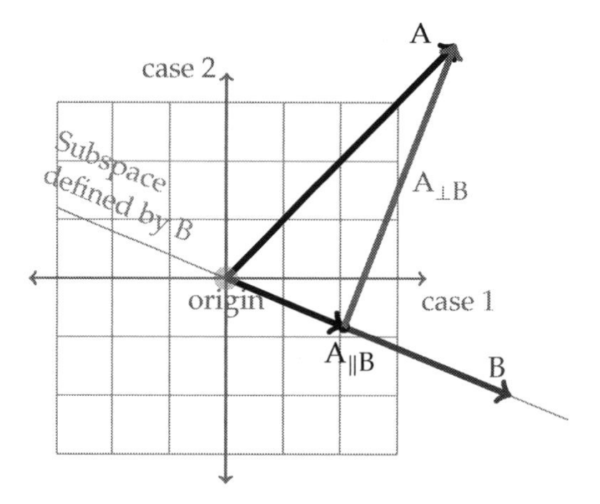

Figure 10.6: The residual of A $\sim$ B is $A_{\perp B}$. It connects the fitted model values $A_{||B}$ to the response variable A.

possible to B. So, $A_{||B} = c$B.

Compare Figure 10.1 to Figure 10.5. The explanatory model term (B in the figure) sets the direction of the wind. Odysseus starts at the origin and seeks to get to A. But the wind is not blowing in exactly the right way to accomplish this; it's blowing toward B. The model subspace is the set of all points that the ship can reach directly. The voyager jumps ship when it is as close as possible to the goal A. The remaining part of the journey, the part that needs to be swum, is the residual vector.

What's the point of describing things this way? After all, you already have a computer that will calculate exact values for coefficients, not just the vague approximations you can get from a graph. But the geometrical perspective makes it clear that there is no magic involved in fitting; just a simple idea of getting as close as possible to a goal point.

Once you have found the projected point $A_{||B}$, you can translate this into a coefficient. The coefficient is the multiplier on B that extends the vector to the point $A_{||B}$. For the example shown in the above drawing, the coefficient is about 0.4, since 0.4 steps in the direction of B will reach $A_{||B}$.

There is a vector between the fitted point, $A_{||B}$, and the goal vector A. This vector is the **residual**, denoted $A_{\perp B}$. Calculating the residual is just a matter of finding the vector that connects $A_{||B}$ with A. In the language of vector addition, $A_{\perp B}$ = A - $A_{||B}$, as shown in Figure 10.6.

The perpendicular symbol $\perp$ is used in the notation $A_{\perp B}$ because the residual is always perpendicular to the explanatory model vector. Why perpendicular? Because the length of the residual tells how far $A_{||B}$ is from A and the fitting process makes that distance as small as possible.

These geometrical operations are all that is needed to fit models. This chapter emphasizes fitting models with a single explanatory vector. The next chapter moves on to fitting with more than one explanatory vectors.

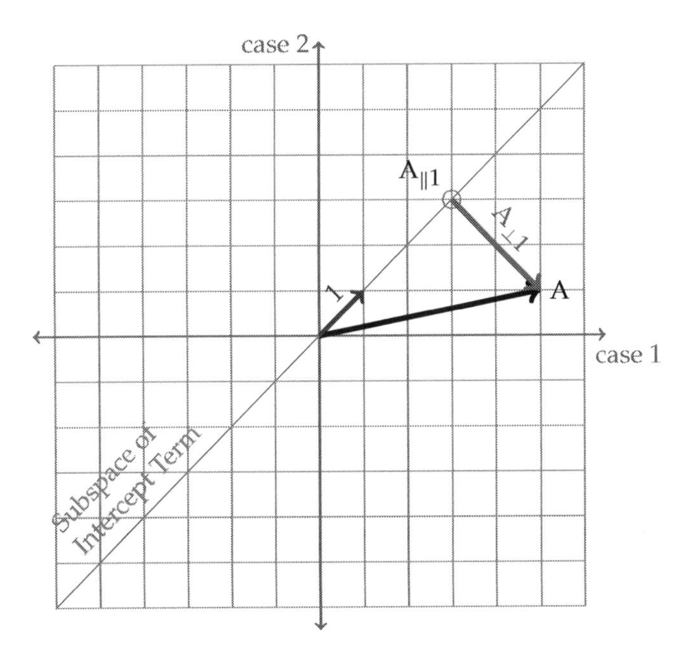

Figure 10.7: The mean of A is the coefficient of projecting A onto the intercept vector 1. The standard deviation reflects the length of $A_{\perp 1}$.

Example 10.1: Finding the Mean Geometrically

As described in Chapter 5, finding the mean of a variable A is equivalent to fitting the model $A \sim 1$. In this model, you can read off the mean of A as the coefficient on the intercept term 1.

To illustrate, let A be the vector $\begin{bmatrix} 5 \\ 1 \end{bmatrix}$. Obviously, the mean of A is 3. The standard deviation turns out to be 2.83.

Fitting the model to the data gives a model formula: Model Values $A = 3 \begin{bmatrix} 1 \\ 1 \end{bmatrix}$. The fitted model values are therefore $A_{\|1} = \begin{bmatrix} 3 \\ 3 \end{bmatrix}$ and the residuals are

$$A_{\perp 1} = A - A_{\|1} = \begin{bmatrix} 5 \\ 1 \end{bmatrix} - \begin{bmatrix} 3 \\ 3 \end{bmatrix} = \begin{bmatrix} 2 \\ -2 \end{bmatrix}.$$

Figure 10.7 shows the calculation of the mean graphically rather than arithmetically. Notice that the fitted model values are $\begin{bmatrix} 3 \\ 3 \end{bmatrix}$ and the coefficient is 3 — the mean of A. It's also worth pointing out that the standard deviation of A is the length of the residual vector $A_{\perp 1}$.

There's no good reason to use geometry to compute a mean or standard deviation: the arithmetic formulas do the job nicely. But the geometry serves to emphasize that the mean and standard deviation are two complementary aspects of a variable. The mean corresponds to the fitted model vector $A_{\|1}$ and the standard deviation corresponds to the length of the residual vector $A_{\perp 1}$. As always, the fitted plus the residual exactly equals the response variable.

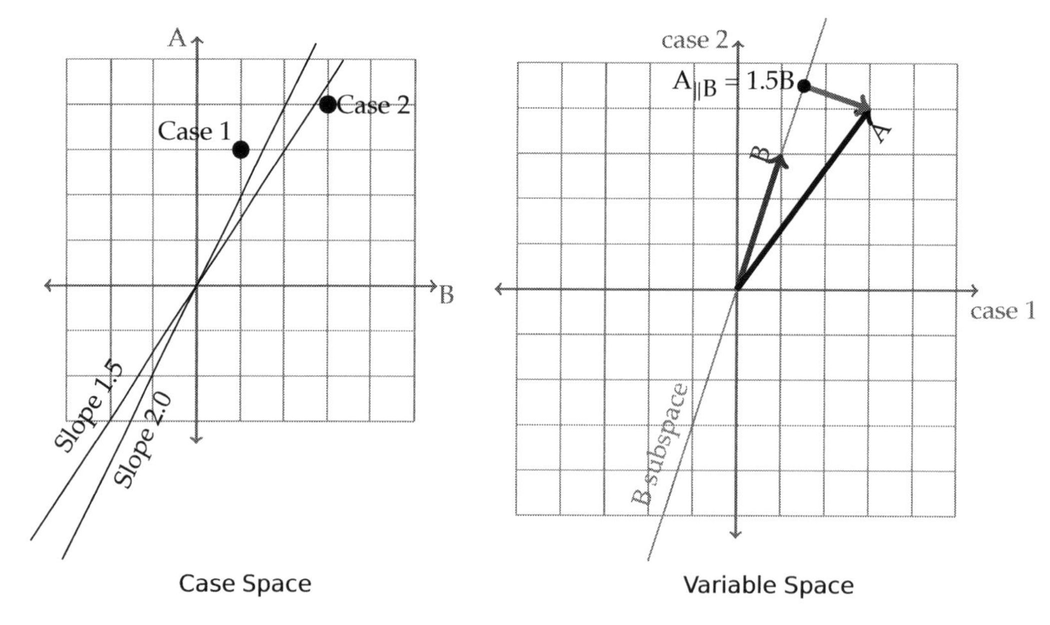

Figure 10.8: The model A ~ B. Left: In case space the model is a straight line through the origin. Right: In variable space, the fit is a projection.

Example 10.2: Fitting a Line through the Origin

Recall that the model that describes a straight-line relationship between A and B is A ~ 1 + B. This has two model terms. The techniques of this chapter can only fit a model with one term — the next chapter will expand things to multiple terms. To give another illustration of fitting, consider the unusual model A ~ B. Lacking an intercept, this model is a straight line that is forced to go through the origin. The only coefficient is the slope.

Figure 10.8 shows an example of the situation in both case and variable space. There are only $n = 2$ cases. Ordinarily, the best fitting line would connect the two points. But since the model A ~ B is a straight line going through the origin, the best fitting line doesn't touch either point. In the figure, two plausible candidates are shown for the best straight-line fit. It's hard to see which is the better one; you have to measure the vertical deviations of the points from the line, square them, and add them up.

In variable space, the situation is more abstract, but clearer. The best-fitting model values are represented by a single point, $A_{||B}$, which is the point on the B subspace that is closest to A. The model coefficient is the multiplying scale on B that reaches $A_{||B}$. With a ruler, you can see that it's 1.5, so that $A_{||B}$ = 1.5 B.

Both the case-space and the variable-space representations are useful. Case space shows the data and the model form clearly; variable space shows clearly why the best fitting coefficient is what it is.

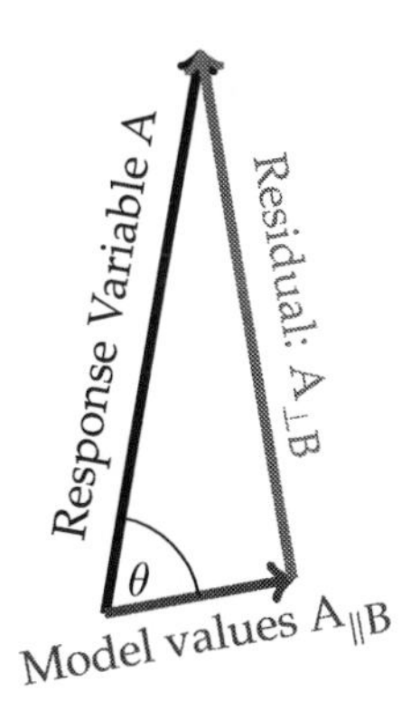

Figure 10.9: The model triangle for A $\sim$ B relates the response values A to the fitted model values $A_{\|B}$ and the residuals $A_{\perp B}$. It is always a right triangle, since the residual vector is always perpendicular to the model subspace. θ labels the angle between the hypotenuse and the adjacent side, that is, between A and $A_{\|B}$.

10.3 The Model Triangle

Consider three of the quantities that are involved in a fitting a model such as A $\sim$ B: the response values A, the fitted model values $A_{\|B}$, and the residuals $A_{\perp B}$. In terms of the model formula, each of the quantities is a set of numbers. Geometrically, it is a vector.

The most fundamental relationships regarding fitting stem from a couple of observations that relate the three vectors:

1. The three vectors A, $A_{\|B}$, and $A_{\perp B}$ form a triangle called the **model triangle**.

2. The residual vector $A_{\perp B}$ is always **perpendicular** to the fitted model vector $A_{\|B}$. Thus, the model triangle is a right triangle with the response values A being the hypotenuse and the fitted model values and the residuals being the legs, as shown in Figure 10.9.

That three vectors always form a triangle results from the way the residuals are defined: the difference between the response values and the fitted model values

$$\text{response values} = \text{fitted model values} + \text{residuals.}$$

The model triangle shows two ways to get to the goal point. One is to walk the vector of the response values: the direct route. The other way is to walk the vector of the fitted model values, then turn 90 degrees and walk the vector of the residuals.

Since the model triangle is a right triangle, the Pythagorean theorem relates the lengths of the three sides: the square-length of the response values equals the sum of the square-length of the fitted model values plus the square-length of the residuals.

The square-length of any vector is the sum of squares of the column of numbers that specifies the vector. So, the partitioning is:

$$\begin{array}{rl} & \text{sum of squares of fitted model values } A_{\parallel B} \\ + & \text{sum of squares of residuals } A_{\perp B} \\ \hline = & \text{sum of squares of response values } A \end{array}$$

This relationship will be fundamental in the interpretation of statistical models and underlies the use of sums of squares and of variances in characterizing statistical models.

10.4 Simple Statistics: Geometrically

Using variable space to look at fitting the model $A \sim 1$ shows that the mean and the standard deviation are two legs of the same model triangle. The mean corresponds to the length of the fitted model values, the standard deviation to the length of the residual vector. Beyond this, there is a more general connection between the geometry of fitting and simple descriptive statistics such as variance and correlation.

10.4.1 Variation is Length

in variable space

Variation can be quantified in several ways: standard deviations, variances, sums of squares, coverage intervals. With model vectors, you can *see* the variation directly in the vector itself.

Variation is the length of the vector.

Long vectors represent quantities which have a lot of variation from case to case. Short vectors are quantities with little variation.

Actually, the statement above is not quite right, but it turns out that it is an excellent way to think about things. A more accurate statement is that the length of a vector, once you have subtracted out the mean of the components of the vector, is proportional to the standard deviation of the numbers listed in the vector.

This business of subtracting out the mean is a detail that is important in doing computations, but the main use for vectors is drawing schematic diagrams to understand the relationships in the data. So, it will work very well for you to forget the detail when interpreting schematic vector diagrams.

Recall that the standard deviation of a variable A is the typical size of the residual from the simple model $A \sim 1$. Of course, the model values from this model are just the mean of A, so the residuals are just A − mean(A).

Perhaps it's clear that the standard deviation is an appropriate measure of the typical size of a residual. To find the standard deviation, take the sum of squares, divide by $n-1$, and take the square root. The variance is even simpler: Take the sum of squares and divide by $n-1$. Even simpler: just take the sum of squares — it's proportional to the variance. But recall that the sum of squares is the square-length of a vector. When you look at the length of a vector, you're looking at the square-root of the square-length — that's proportional to the standard deviation.

Example 10.3: Vector Lengths and Standard Deviations Consider these 5-dimensional vectors (each of which has already had the mean subtracted away)

$$A = \begin{bmatrix} 3 \\ 1 \\ -4 \\ 2 \\ -2 \end{bmatrix} \quad B = \begin{bmatrix} -2 \\ -3 \\ 4 \\ 5 \\ -4 \end{bmatrix} \quad C = \begin{bmatrix} 21 \\ -32 \\ 19 \\ -28 \\ 20 \end{bmatrix} \quad D = \begin{bmatrix} -217 \\ 302 \\ 126 \\ -156 \\ -55 \end{bmatrix} \quad E = \begin{bmatrix} 1000 \\ -3100 \\ 2400 \\ -1200 \\ 900 \end{bmatrix}$$

For each vector you can calculate the length — just the square root of the sum of squares — and the standard deviation. Here they are:

Vector	Length	Std. Dev.
A	5.8	2.9
B	8.4	4.2
C	54.8	27.4
D	426.0	213.0
E	4315.0	2157.5

For each vector, the length is exactly twice the standard deviation. That is, length is proportional to standard deviation. The constant of proportionality is 2 for these 5-dimensional vectors. In general, for an n-dimensional vector, the constant of proportionality will be $\sqrt{n-1}$.

For computations, this constant of proportionality is important. But leave it to the computers. When you're looking at a vector diagram, you are comparing only vectors with the same dimension. It's the same constant of proportionality for all the vectors. So just consider the length as a stand-in for standard deviation, or the square length as a stand-in for variance. Or, if you always want to be exactly right, think about the sum of squares which is always the same thing as the square length.

10.4.2 Correlation is Alignment

I'll start with a statement and then justify it later. (Refer to Figure 10.10.)

> **The correlation coefficient r between two variables A and B is related to the angle θ between the vector A and the vector B. The relationship is $r = \cos(\theta)$.**

This means that a perfect correlation, $r = 1$ corresponds to exact alignment of A and B: the angle $\theta = 0$. No correlation at all, $r = 0$ corresponds to A and B

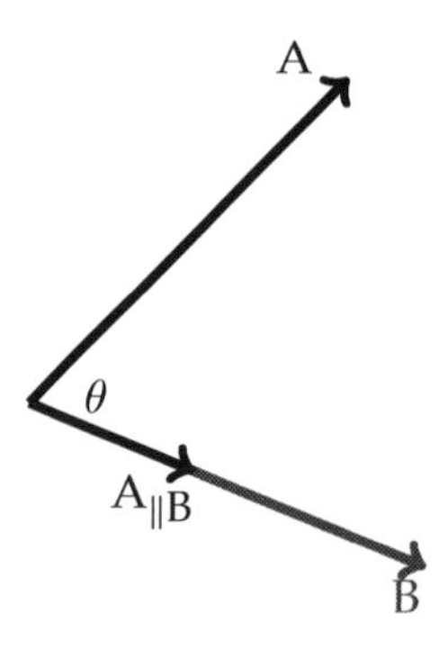

Figure 10.10: The correlation coefficient between A and B compares the length of the vector $A_{\|B}$ to the length of the vector A. Trigonometrically, this ratio is $\cos(\theta)$.

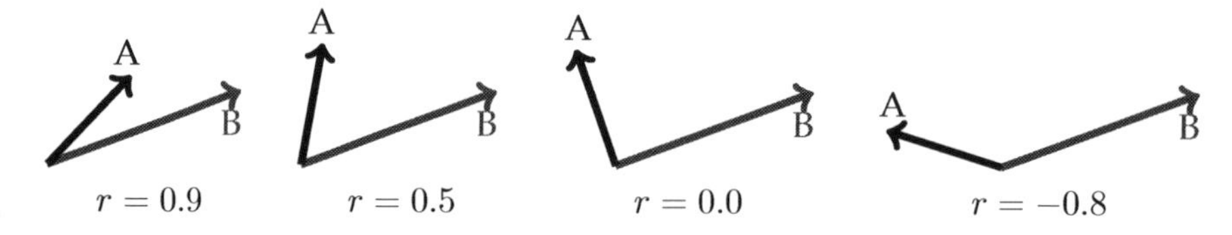

Figure 10.11: The correlation coefficient corresponds to the angle between vectors.

being at right angles: $\theta = 90°$. A negative correlation, $r < 0$, corresponds to A and B pointing in opposite directions. Figure 10.11 shows some examples.

Vectors that are perpendicular are said to be **orthogonal**.

For the statement to be precisely true, A and B must be vectors where the mean is zero. That is, the mean must have been subtracted from each vector before looking at the angle. But, as in the previous section, you can regard this as a computational detail and assume that the drawing has already taken the mean into account.

Recall from Chapter 7 that the correlation coefficient r and its square r^2 describe in simple terms the relationship between two variables. If two variables A and B are related in some manner, then knowing the value of B would gives some information about the value of A. The correlation coefficient measures how much of the variation in A is explained by B using the simple model A ~ 1+B.

Correlation has a simple geometric interpretation using the model triangle. Consider the vectors A and B shown in Figure 10.10. The r^2 compares the variance in the fitted model values $A_{\|B}$ with the variance in the response variable A; r^2 is the ratio of those two quantities. Now recall from the previous section that the variance is proportional to the square length of the vector. So, r^2 is the ratio of the square length of the vector marked $A_{\|B}$ to the square length of the vector A.

Or, if you don't like square lengths, think about the correlation coefficient r: it's the ratio of the length of the vector $A_{\|B}$ to the length of the vector A.

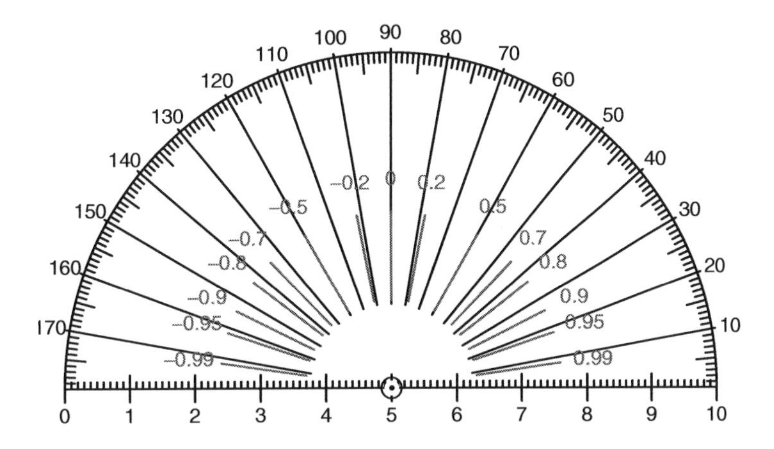

Figure 10.12: Measure correlation with a protractor! Each value of the correlation coefficient r corresponds to an angle between vectors. This protractor is marked both in degrees and in the corresponding values of r. Notice that moderate values of r correspond to angles that are close to 90°.

Recall the model triangle and that $A_{\|B}$ is the leg of a right triangle of which A is the hypotenuse. In trigonometry, the sines, cosines and tangents of angles are defined in terms of the ratio of lengths of the sides of right triangles. As the ratio of the adjacent side to the hypotenuse, the coefficient r is the cosine of the angle between A and B, marked θ in Figure 10.10.

Since a correlation comes from an angle, you can measure it with a protractor. Figure 10.12 shows a protractor with angles marked both in degrees and as a correlation coefficient.

10.5 Drawing Vector Diagrams

When drawing schematic diagrams of variables using vectors, it's helpful to get the standard deviations of the variables and the correlations between the variables right. For example, consider the discussion of SAT scores and fraction of students taking the exam in Chapter 5. (See page 104.) The diagram of the variables sat and frac, should reflect that the standard deviation of sat and frac are 26.76 and 74.82 respectively and the correlation between the two variables is -0.887. This correlation corresponds to an angle θ of 152.5 degrees. (You can read the angle off of the protractor in Figure 10.12 if you like.) So, in drawing a vector diagram of this situation, draw frac as about three times as long as sat and have the vectors pointing in roughly opposite directions, like this:

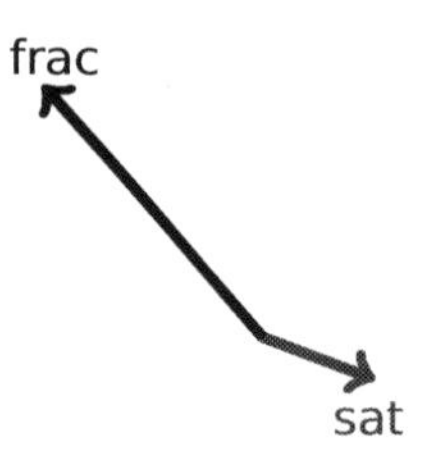

You may object. What's the point in trying to draw vector diagrams that show correlations and standard deviations when the real quantities are vectors in high-dimensional space, not the two-dimensional space. Indeed, the SAT data have one case for each of the 50 states in the US, so the vectors are really 50-dimensional. Certainly a 2-dimensional piece of paper can't do justice to 50-dimensional vectors.

Actually, it can. Whatever the dimension of the space, any two vectors always live in a two-dimensional subspace. That is, if you could magically draw out the vectors in the 50-dimensional space and then could sneak a two-dimensional piece of paper into that space, you could always find a way to lay the paper flat onto the vectors. To see why this is, make a V-sign with your index and ring fingers: two vectors in three-dimensional space. Whichever way that V is oriented, you can always place a two-dimensional piece of paper so that it is aligned with both fingers.

DATA FILE
`galton.csv`

As another example, consider the relationship in Galton's height data between the mother's height and the child's height. The standard deviations of these variables are 2.31 and 3.58 inches respectively, the correlation between them is 0.20. (Presumably the standard deviation of the mothers is smaller than of the children because all the mothers are female. The children are of both sexes, adding another source of variation in height.) Similarly, in the relationship between mothers and fathers, the correlation coefficient is only 0.074. So, the vectors will look like this:

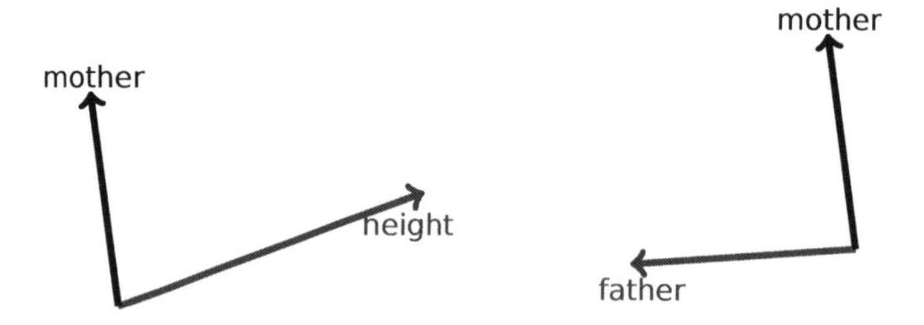

Notice that mother and father are practically uncorrelated — the vectors are almost exactly at right angles. If tall women tended to marry tall men, there should be a correlation. Evidently, they don't.

In contrast, there is a discernible correlation between the child's height and the mother's height, as would be expected for a heritable trait such as height.

It's tempting to draw all three vectors, height, mother, and father on the same diagram. That way you can use both mother and father to explain height. But when drawing all three vectors on one picture, keep in mind that the three vectors do not lie on a plane; they are really in 898-dimensional space. Still, just as any two vectors lie on a plane, any three vectors lie in a three-dimensional space. So you can think about the relationships among them using your three-dimensional intuition. That's the subject of the next chapter.

10.6 Computational Technique

The point of studying the geometry of vectors is to illuminate statistical methods. You won't ordinarily make geometrical computations in the course of statistical work — the computer will do that for you, behind the scenes. However, to develop an understanding of the concepts, it can be useful to do some geometrical calculations directly.

10.6.1 Projections

The basic program is still `lm` — the software for fitting linear models. Interpreted geometrically, it is performing projections. It returns the coefficients, the fitted model values, and the residuals.

For example, to project the vector $A = \begin{bmatrix} 3 \\ 4 \end{bmatrix}$ onto $B = \begin{bmatrix} 1 \\ 3 \end{bmatrix}$, the computer commands look like this:

```
A = c(3,4)
B = c(1,3)
mod = lm( A ~ B - 1 )
```

There's no need to include a `data` argument to `lm`, since A and B are not being taken from a data frame.

As aways, the coefficients, fitted model values, and residual vector can be found using the `coef`, `fitted`, and `resid` operators:

```
> mod = lm( A ~ B - 1 )
> coef(mod)
  B
1.5
> fitted(mod)
  1   2
1.5 4.5
> resid(mod)
   1    2
 1.5 -0.5
```

Keep in mind that `lm` will include an intercept term by default, unless you explicitly instruct it not to by including `-1` in the model description. So, the formula `A~B` really means `A~1+B` and will produce two coefficients:

```
> mod2 = lm(A ~ B)
> coef(mod)
```

```
  B
1.5
> coef(mod2)
(Intercept)           B
        2.5         0.5
```

If you want the model $A \sim 1$, the model formula needs to be `A~1`

```
> mod3 = lm(A~1)
> coef(mod3)
(Intercept)
        3.5
```

10.6.2 Subtracting Out the Mean

Subtracting out the mean of a variable can be done easily:

```
> A = A - mean(A)
> B = B - mean(B)
```

The assignment has overwritten the values of A and B. Alternatively, the assignment could be made to a new object.

10.6.3 Correlation and Angles

The correlation coefficient, r, between two quantitative variables can be calculated with the `cor` operator. Here is the correlation between hourly wages and the worker's age in the Current Population Survey wage data.

```
> cps = ISMdata("cps.csv")
> cor( cps$wage, cps$age)
[1] 0.1770
```

When drawing vector diagrams to represent real data, it can be helpful to translate this to an angle. Keeping in mind that the correlation coefficient is the cosine of the angle, the angle is therefore the arc-cosine of the correlation coefficient:

```
> acos(0.1770)
[1] 1.393
```

This angle is reported in radians. To convert it to degrees, multiply the radian measure by $180/\pi$:

```
> acos(0.1770)*180/pi
[1] 79.8
```

The angle $79.8°$ is very close to $90°$ suggesting that the vectors should be drawn almost perpendicular to one another.

You could also calculate the angle between the vectors themselves:

```
> angle( cps$wage, cps$age, degrees=TRUE)
[1] 31.26
```

This is quite different than the angle based on the correlation coefficient. The reason is that the correlation coefficient corresponds to the angle *after* the mean has been subtracted from each vector.

A coefficient of determination, R^2, can also be converted to an angle. Remember to take the square root *before* computing the arc-cosine.

```
> mod = lm(wage ~ age, data=cps)
> var( fitted(mod)) / var(cps$wage)
[1] 0.03132
> acos( sqrt(0.03132))*180/pi
[1] 79.8
```

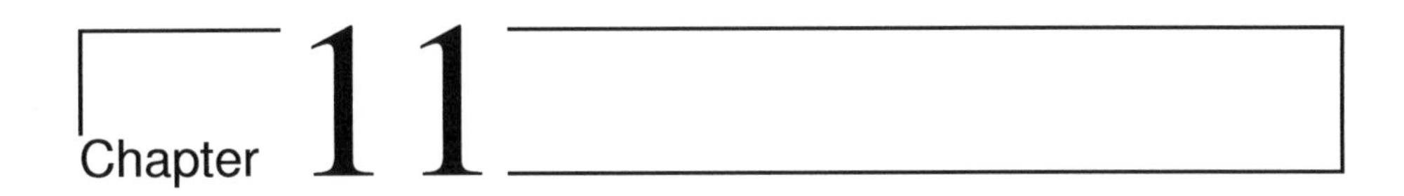

Geometry with Multiple Vectors

Philosophy is written in this grand book - the universe - which stands continuously open to our gaze. But the book cannot be understood unless one first learns to comprehend the language and interpret the characters in which it is written. It is written in the language of mathematics, and its characters are triangles, circles, and other geometrical figures, without which it is humanly impossible to understand a single word of it; without these one is wandering about in a dark labyrinth. — Galileo Galilei (1564-1642)

Consider the data on SAT scores introduced in Chapter 5 and modeled with two explanatory variables: the fraction of students taking the test and the level of expenditures on education. This model involves three vectors: sat, expend, and frac.

Since any two vectors lie in a plane, you can easily draw a realistic representation of any two model vectors. For instance, Figure 11.1 shows the relationship between frac and expend. The angle between the vectors is based on the correlation between the two variables: $r = 0.59$ so $\theta = 53.7$ degrees.

The variance of frac is much larger than the variance of expend; they have completely different units. Since variation is shown by the length of the vectors, the lengths of frac and expend in the figure are very different. To mark the relative directions clearly, the figure shows the subspaces defined by expend and by frac.

Where to draw the sat vector? In general, three vectors don't lie on a plane, so it would be misleading to draw sat in the plane. But, seeing as this book is printed on paper, there isn't much choice — the drawing has to be done on the page! A key insight is given by a model formula derived by fitting the model

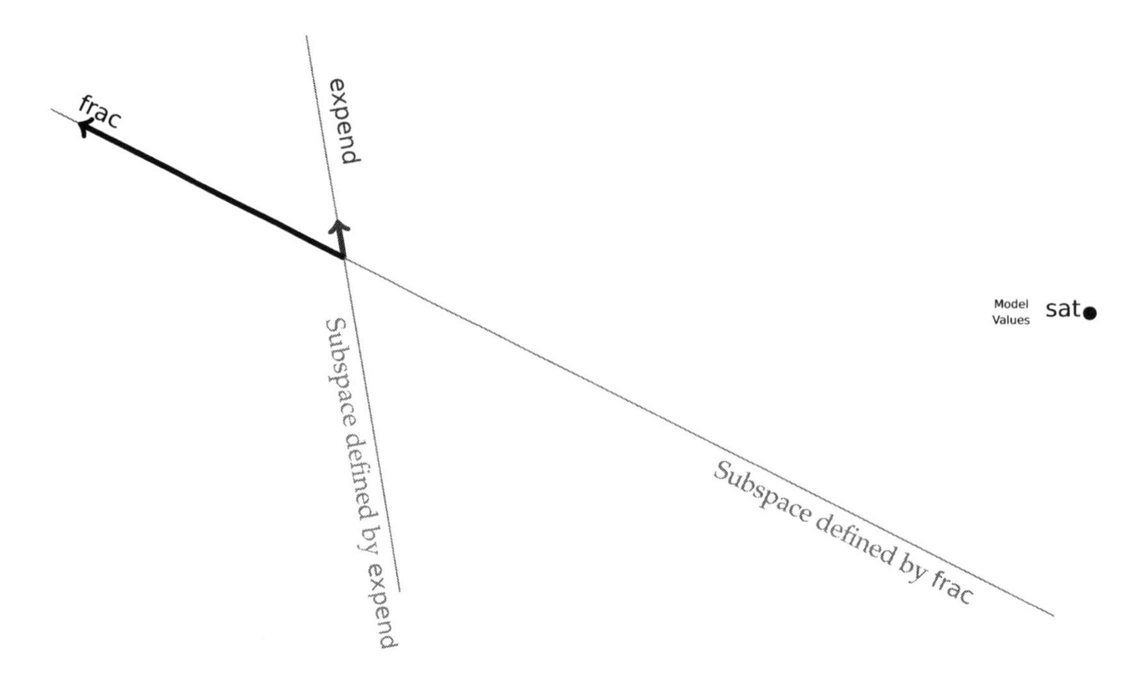

Figure 11.1: The fitted model values of sat from the model sat $\sim$ frac + expend lie in the plane defined by frac and expend.

sat $\sim 1 +$ frac + expend:

$$\text{Model Values sat} = 994 - 2.85\ \text{frac} + 12.3\ \text{expend}.$$

The model values of sat are a linear combination of frac and expend, so they can be plotted in the plane defined by frac and sat. (The intercept is ignored in these diagrams. See Aside 11.1.)

The model values of sat are the dot in the far right of the figure. You can verify that this is drawn in the right place by following the directions in the model formula: walk 2.85 frac steps backwards and then take 12.3 expend steps. You will arrive at the Model Values sat point.

DATA FILE
sat.csv

Where is the sat point itself? It's out of the plane, hovering in the air above the Model Values sat point. The vector that connects the hovering sat point and the Model Values sat point in the plane is the residual vector. The residual is perpendicular to the plane.

The general situation for a model A $\sim$ B+C is shown in Figure 11.2. The two explanatory vectors B and C define a plane. The vector A will not in general be in that plane, but the fitted model values $A_{\|(B,C)}$ will be on the plane. Indeed, any linear combination of B and C is a point in the (B,C)-plane.

The idea of a model triangle still holds. Since the point $A_{\|(B,C)}$ is the closest point in the (B,C) plane to A, the residual vector $A_{\perp(B,C)}$ is perpendicular to the plane. Vector A is the hypotenuse of the triangle. The two vectors $A_{\|(B,C)}$ and

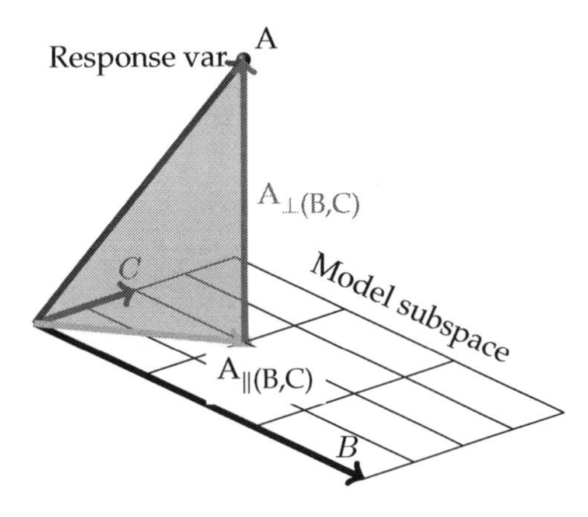

Figure 11.2: Projecting A onto the subspace defined by a set of two model vectors, B and C. The model triangle is shaded.

$A_{\perp(B,C)}$ are the legs and meet at a right angle.

No matter how many model vectors there are, the residual will be orthogonal to every one of them.

Take a look again at Figure 11.1. The actual sat is floating in the air above the Model Values sat point. By finding the length of the residual vector, you can figure out how far above the plane the sat vector is. It turns out to be about 4 cm when you scale things to the picture.

Hold your fingertip 4 cm directly above the Model Values sat point in Figure 11.1. Now find the point on the subspace defined by expend that is as close as possible to your finger. You have just fitted the model sat $\sim 1 +$ expend. If you did it right, you found a point very near the first "d" in "Subspace defined by expend."

Now do the same thing, but rather than starting from the point sat in the air, start from the Model Values sat point on the plane and project onto the same subspace, the one defined by expend. You should get exactly the same result, ending up near the "d."

The point of all this is that you can get the same result in two different ways: project a vector (such as sat) directly onto a subspace, or project it onto a plane containing the subspace and then project that point onto the subspace. The reason has to do with the residual being perpendicular to the plane.

This means that in thinking about modeling a response variable with two explanatory variables, you only need to draw a plane. Rather than drawing the response variable itself as a vector, you can draw it as the projection down onto the plane defined by the two explanatory variables.

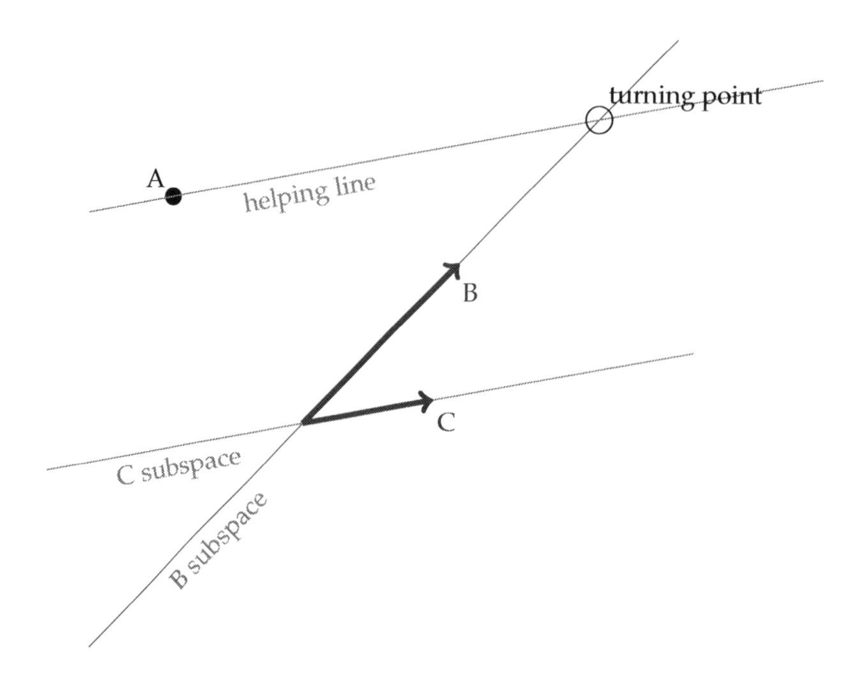

Figure 11.3: Finding coefficients with two explanatory vectors.

11.1 Finding Coefficients

To fit a model involving multiple explanatory terms, first project the response variable onto the subspace defined by those terms. So, for the model A $\sim$ B+C, the first step is to project A onto the plane spanned by B and C. Figure 11.3 shows an example where the projection has already been done.

The next step is to find the coefficients on B and C. In thinking about this geometrically, imagine that you are finding how far to walk along B so that you can reach a point where you can turn in the direction of C. After turning, head in the C direction to reach the goal A.

The hard part is finding the turning point. One way to figure this out is to draw a helping line: a pathway parallel to C but passing through A. The place where this pathway crosses the B subspace tells where you will make the turn. In the figure, the turning point is marked with a circle. Finding the coefficient on B is a matter of scaling vector B to get from the origin to the turning point. This is a little less than 2 in the figure. With a ruler, by measuring the total path length along the B subspace to the turning point and dividing by the length of B, you would get 1.9. The coefficient on C is the number of steps of length C to take along the helping line in order to get from the turning point to A. This is a little more than 3 steps, in the negative direction — opposite to the way the C arrow points. With a ruler, measuring the path length along the helping line and dividing by the length of C, you would get a more precise number, -3.3, with the negative sign indicating the direction relative to C.

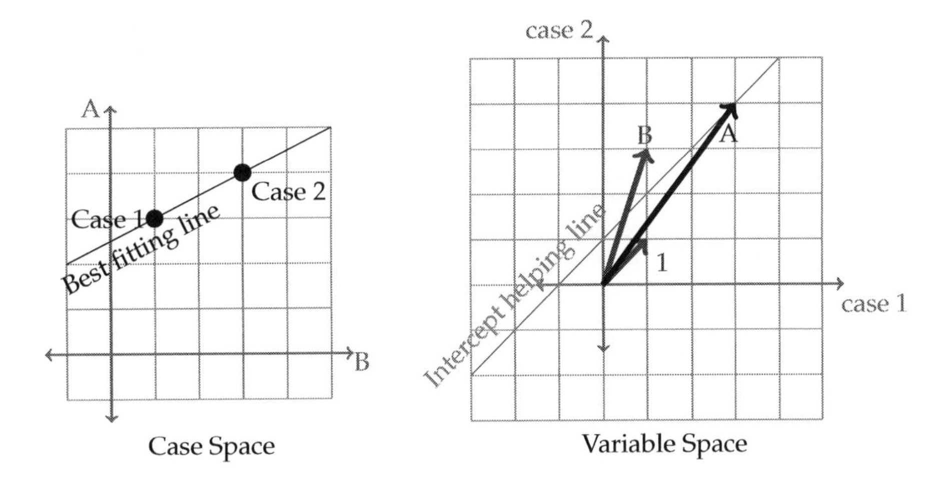

Figure 11.4: The model A ~ 1+B is a straight line in case space. Finding the intercept and slope — the coefficients on 1 and B — can be done in variable space.

Example 11.1: Fitting a Line Consider fitting the simple model A ~ 1+B, which is a straight line in case space. In the example in Section 10.2 in the last chapter, the intercept term 1 was missing so that only candidate lines going through the origin were admissible. But here, with the intercept term included, any line is possible. Figure 11.4 shows the situation with two cases plotted both in case space and in variable space. Since there are only two cases, the best fitting line goes exactly through the cases in case space; fitting is a trivial process.

In variable space, fitting is a matter of finding the coefficients on B and 1 to bring the fitted model values as close as possible to the response vector A. Drawing a helping line parallel to the intercept makes it easy to see that the turning point is half-way up B, so the coefficient on B is 0.5. To find the coefficient on 1 — the intercept — walk along the helping line to A. This involves 2.5 steps.

Because the fit is perfect, there is no residual vector. More precisely, the residual vector has length 0; the walk along B and 1 arrives exactly at A.

11.2 The Coefficient of Determination

The **coefficient of determination** R^2 is used to quantify how much of the variation in the response variable is accounted for by a model.

In Chapter 7, R^2 was defined as the ratio of the variance in the fitted model values to the variance in the response variable.

This has a simple interpretation in terms of the model triangle. The situation is exactly analogous to the interpretation of the simple correlation coefficient in

Aside. 11.1 Subtracting out each vector's mean

The intercept vector is missing in Figure 11.1. To fit the model, it's essential to include an intercept term. But in drawing the picture the intercept has been left out. One way to think about this is that the frac and expend vectors have been arranged as if the intercept vector were pointing directly out of the page. That is, the model subspace is three dimensional, involving the subspace defined by the triplet of vectors frac, expend, and the intercept. To draw it on the page, this three-dimensional subspace has been turned so that the intercept vector points directly toward you, out of the page.

This means that the lengths of frac and expend in the diagram, and the angle between them in the diagram, are not the lengths and angle of the vectors themselves, but the lengths and angle as seen in perspective, with the intercept vector pointing toward you.

This perspective has a close tie to the standard deviation and correlation coefficient as numerical quantities. Recall that the correlation coefficient r between vectors A and B is associated with the angle θ between A and B. Similarly, the standard deviation of a variable is associated with the length of that variable as a vector. Both of these statements are true *only when the variable had mean zero*, that is, when the mean is subtracted from the variable.

Geometrically, subtracting out the mean is equivalent to displaying the vectors in the intercept-toward-you perspective.

The rest of this chapter assumes that vectors have already had their mean value subtracted away. This means that the intercept vector $\left[\begin{smallmatrix}1\\ \vdots\\ 1\end{smallmatrix}\right]$ becomes $\left[\begin{smallmatrix}0\\ \vdots\\ 0\end{smallmatrix}\right]$ — nothing: a vector of zero length.

This makes the vector picture easier to interpret. It doesn't actually change any of the coefficients, so long as the mean is subtracted from *every* model vector as well as the response.

Chapter 9. Like the correlation coefficient r, the coefficient of determination R^2 is related to the angle θ between the response variable (the hypotenuse) and the fitted model values: $R^2 = \cos^2(\theta)$.

When R^2 is close to zero, then the model accounts for little of the variance in the response variable. Geometrically, this means that θ is close to 90 degrees, or, in terms of the model triangle, that the residual is much longer than the fitted model values. When R^2 is close to one, θ is close to zero and therefore the residual is much smaller than the fitted model values.

11.3 Collinearity

When two or more explanatory variables are correlated, you will get different model coefficients depending on which of the vectors you include in a model. The SAT example shows this. As shown in Chapter 5, the coefficient on expend differs strongly depending on whether or not frac is included as an explanatory variable.

Remember that in drawing Figure 11.1, the position of the model values of sat — that is the projection of sat onto the plane defined by frac and expend —

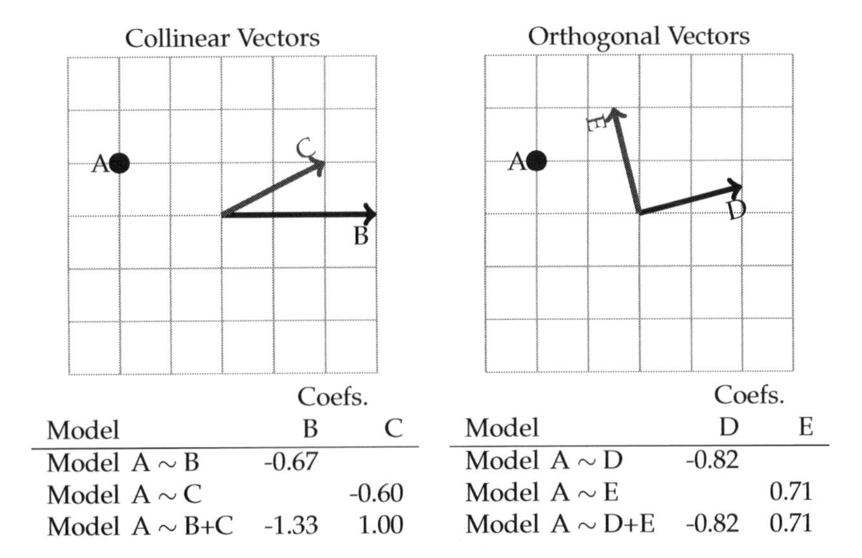

Model	Coefs. B	C
Model A ~ B	-0.67	
Model A ~ C		-0.60
Model A ~ B+C	-1.33	1.00

Model	Coefs. D	E
Model A ~ D	-0.82	
Model A ~ E		0.71
Model A ~ D+E	-0.82	0.71

Figure 11.5: With collinear explanatory variables (left), model coefficients depend on which variables are included in the model. When explanatory variables are orthogonal, there is no such dependence.

was given by the model formula: a linear combination -2.85 frac $+ 12.3$ expend.

Now look at what happens in Figure 11.1 when you project sat onto the subspace defined by expend, ignoring frac. The coefficient of this projection is about -21, although the expend vector is so small that it can be hard to see. (It's the same coefficient as in the model sat ~ 1+expend on page 104.) So, depending on whether frac is included in the model or not, the coefficient on expend will be either -21 or $+12.3$. Simpson's paradox.

The vectors frac and expend are entangled in their explanation of sat because frac and expend are correlated. Geometrically, this means that frac and expend are aligned, even if only partially. The word often used to describe the alignment is **collinearity**.

To be collinear, vectors do not need to be exactly aligned: any angle other than 90 degrees means that there is some collinearity. The effects of collinearity will be stronger the more closely aligned the vectors are. The more closely explanatory vectors are aligned, the stronger will be the change in a coefficient when one of the vectors is added or dropped from a model.

Variables that are orthogonal — that is, at an angle of 90 degrees — are uncorrelated simply because $r = \cos(\theta) = 0$ when $\theta = 90$ degrees. With orthogonal explanatory variables, the coefficients do not depend on which explanatory variables are included in the model. (See Figure 11.5.)

Collinearity is a fact of life; explanatory variables are often closely related. For example, in studying poverty, income is related to education, health, and race or ethnicity, among other things. Depending on which of these variables you include in a model, you can get very different results.

Example 11.2: Home Ownership Rates

The US Census Bureau collects data on race and home ownership.[18] The table shows the fraction of households that own their house, versus the race of the householder:

Home Ownership (%)			
	Year		
Race/Ethnicity	1996	2000	2004
Black	44.1	47.2	49.1
White, non-Hispanic	71.7	73.8	76.0
Hispanic	42.8	46.3	48.1

There's a pretty clear relationship here: Black and Hispanic households have much lower home ownership rates than White households.

There's also a relationship between income and race/ethnicity:[18]

Percentage of Households below $25000 Annual Income		
	Year	
Race/Ethnicity	1995	2006
Black	45.1	39.6
White	26.8	23.3
Hispanic	43.3	31.8

A smaller fraction of White households than Black or Hispanic households have an income below $25,000 per year.

Since race/ethnicity and income are correlated, what will happen when both race/ethnicity and income are included in a model of home ownership. Both explanatory variables are correlated, so it's certain that the coefficients of the model `ownership ~ race+income` will be different than the coefficients from `ownership ~ race`. They might even be reversed in sign — Simpson's Paradox — and show that once you account for income, Blacks and Hispanics have higher home ownership rates than Whites. Or, perhaps income will make the relationship of ownership with race even stronger.

Are you waiting now for the answer? Which one is it? Unfortunately, there isn't an answer for you. Getting an answer requires having data which record on a case-by-case basis the three variables ownership, race/ethnicity, and income. (You might want additional variables, too.) But such a data set is hard to come by: the Census Bureau and other organizations try to preserve confidentiality by publishing summary tables rather than case-by-case data.

What's easy is to find is tables like the above ones, that break down one variable (e.g., ownership rates) by another variable (e.g., race/ethnicity). Such summary tables — cross-tabulations — are very widely used in policy discussions. But they can involve only two variables and must therefore be regarded as simplistic and very likely misleading.

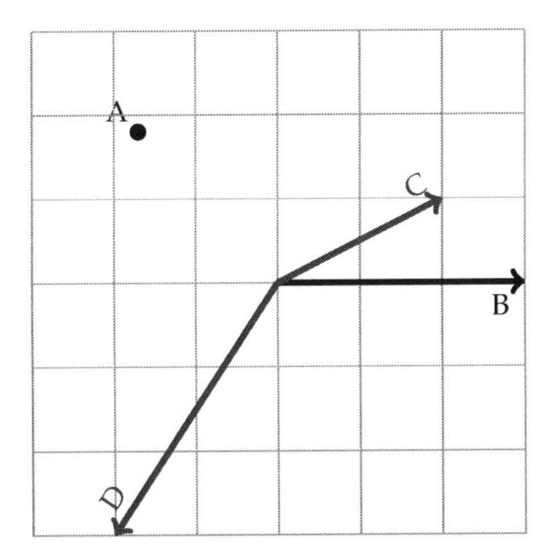

Figure 11.6: With three vectors B, C, and D in a plane, one of the vectors is redundant. This means that there are many linear combination of B, C, and D that can reach a point such as A.

11.4 Redundancy

Consider the three vectors in Figure 11.6. All three vectors, B, C and D, are in the plane. With a linear combination of any two of the vectors, you can reach anywhere in the plane. The third vector is redundant.

In a set of vectors like B, C, and D, one of the vectors is redundant when that vector is a linear combination of the other vectors in the set. In mathematical terminology, the vectors in the set are said to be **linearly dependent**.

A set of vectors always defines a subspace. The dimension of the subspace — 1 for a line, 2 for a plane, and so on — is called the **degrees of freedom** of the subspace. The degrees of freedom cannot be any larger than the number of vectors that define the subspace, but it will be less if there is redundancy among the vectors.

11.5 Multi-Collinearity

Collinearity is not just a relationship between two vectors, it can also describe larger sets of vectors, in which case it is referred to as **multi-collinearity**. Collinearity and multi-collinearity can be a problem, partly because they create problems interpreting models. More important, as you will see in Chapter 14, collinearity and multi-collinearity can cause models to become unreliable.

Consider the simple models of wages presented in Chapters 4 and 5, these involved two variables: age and sex. But why only these two? Why not other potentially relevant factors, for instance education and years of experience?

Once you learn how to build models with multiple variables, it's tempting to add in all the variables that you have. With today's computers, it's just as easy

	Model	Coefficient on age	educ	exper
Model 1:	wage ~ 1+age	0.08		
Model 2:	wage ~ 1+educ		0.75	
Model 3:	wage ~ 1+age + educ	0.11	0.82	
	Adding exper to the models ...			
Model 4:	wage ~ 1+age + exper	0.92		-0.82
Model 5:	wage ~ 1+educ + exper		0.93	-0.82
Model 6:	wage ~ 1+age + educ + exper	-0.02	0.95	0.13

Table 11.1: Adding a collinear explanatory variable can strongly affect model coefficients. exper is collinear with age and strongly multi-collinear with age and educ.

to build the model wage ~ sex + age + educ + exper as it is to build a simpler model. Sometimes this is the right thing to do. But one of the issues that weighs against this is collinearity and multi-collinearity.

In the CPS data, age and educ are not strongly correlated. The correlation coefficient is $r = -0.15$, corresponding to an angle of 99 degrees. The slight negativity of r, and correspondingly the slightly obtuse angle (bigger than 90 degrees), suggests that older people have slightly less education, consistent with the spread of high-school and college education over the decades.

Since age and educ are close to orthogonal, it's easy to interpret the models that include them — the coefficients on each of them don't depend strongly on whether the other is included in the model. You can see this in the first three lines of Table 11.1. Those lines of the table compare three different models involving age and educ as explanatory variables. All three of the models give consistent results: a year's additional education is associated with about 80 cents per hour increase in wage, a year's additional age is associated with about 10 cents per hour increase.

In contrast, age and exper are very strongly correlated; the older you are, chances are the more work experience you will have. Indeed, this is seen in the data. The correlation coefficient between age and exper is very high: $r = 0.978$, corresponding to an angle of 12 degrees. When building a model with both age and exper (e.g., Models 4, 5, and 6 in Table 11.1), the coefficient on age is very different from what it is in a model that doesn't include exper (e.g., Models 1, 2, and 3).

As might be expected, exper and educ are not strongly correlated: $r = -0.35$, or an angle of 111 degrees. This makes something in Table 11.1 a surprise. Compare Models 4 and 6. The only difference in the models is the addition of the explanatory variable educ, which is more or less uncorrelated with each of the other explanatory variables. So why are the coefficients in Model 6 on age and exper so utterly different than they were in Model 4?

The reason is multi-collinearity. Experience is strongly correlated with age — recall that the angle between exper and age is 12 degrees. But experience is even more strongly correlated with the *pair* of variables age and educ; the angle between exper and the subspace defined by the pair age and educ is less than 1 degree.

How do you measure multi-collinearity? An easy way is to look at the R^2 for a model that examines one of the explanatory variables modeled by the other explanatory variables. For instance, the model exper ~ age + educ produces an R^2 of 0.9998: almost perfect, indicating very strong multi-collinearity.

This strong multi-collinearity isn't a a surprise to someone who looks closely at the data. In the CPS data, exper was estimated assuming that people started school at age 6, then went to school, and then went to work. So knowing age and educ gives exper. It's only because of a slight discrepancy that R^2 was not exactly 1.00. If it had been, the software would have recognized the multi-collinearity as a redundancy, because one vector could have been written exactly as a linear combination of the other two.

When vectors are redundant, there is an infinity of sets of different coefficients that are equally consistent with the data, as described in Section 6.3; the software chooses one of the possible sets. But when there is collinearity rather than exact redundancy, the sets of coefficients are not exactly equally consistent but they are very close. The software will choose the very best one, but its preference for this one over the others likely reflects random variability. When the collinearity is extreme, as with age, educ, and exper, the coefficients may be unacceptably influenced by the random variability. That is, building a model with strongly collinear explanatory vectors can give unacceptably unreliable results. In order to identify such situations, you will need additional tools, including ways to measure the precision of coefficients. Chapter 14 gives more information about this.

Aside. 11.2 Multi-Collinearity and Interaction Terms

An important source of multi-collinearity is the use of interaction terms between two variables. Recall that an interaction term is generated by the product of the two variables. (See Chapter 9.) Often, that product is strongly correlated with the two variables, even when the variables themselves are not strongly correlated.

For example, in the CPS data, the variables educ and age are weakly correlated: an angle of 99 degrees. The interaction term age:educ is moderately correlated with each of the variables: an angle of 40 degrees to age and of 60 degrees to educ. Yet, age:educ is almost completely determined by a linear combination of age and educ; the R^2 for the model age:educ ~ age + educ is 0.9652, corresponding to an angle of only 11 degrees between age:educ and the subspace spanned by age and educ.

11.6 Higher Dimensions

It takes three dimensions to show the relationships in the model `sat ~ frac + expend`, although the trick of projecting `sat` onto the planar subspace defined by `frac` and `expend` lets the picture be drawn on paper. (The whole space for the SAT data is 50-dimensional, since there are 50 cases in the data frame.)

What happens with more than two explanatory vectors? For example, there are four explanatory vectors involved in the model `sat ~ frac + expend + salary + ratio`. Those four variables define a 4-dimensional subspace. Even worse, when a categorical explanatory variable gets translated into model vectors, there is one vector for each level of the variable. So a single variable with 8 levels will correspond geometrically to an 8-dimensional subspace. How is that to be drawn? Hopeless!

It's important to keep in mind the purpose of the vector drawings. They aren't intended to replace the calculations for model fitting; the computer will do that for you. One purpose of the drawings is to show you how explanatory variables relate to one another and how including them or not affects the coefficients found on other variables. Another purpose, which you'll see later, involves understanding how chance and randomness plays a role in models. Both of these purposes can be served by drawing the diagrams in 2- and 3-dimensions, but treating the explanatory vectors as representing an *entire subspace* due to sets of variables or model terms.

A three-dimensional diagram is adequate to show schematically the relationship between the variables of interest (drawn as one vector in the diagram), the covariates (drawn as another vector), and the residuals (a third vector). Since the residual vector is always perpendicular to the model space, a two-dimensional picture can suffice.

11.7 Computational Technique

No new software is needed for projections onto subspaces defined by multiple explanatory vectors: `lm` does it.

Keep in mind that investigating multi-collinearity among expanatory model terms involves constructing models where one of the model vectors is treated as the response and the others as the explanatory variables.

Optional: Readers with a strong background in linear algebra might prefer to study multi-collinearity with some of the tools used in that field. The `model.matrix` operator extracts from a model the matrix of explanatory model vectors — this is just the set of explanatory model vectors organized in matrix form. This matrix can be studied with tools such as singular value decomposition (`svd`) and QR decomposition (`qr`).

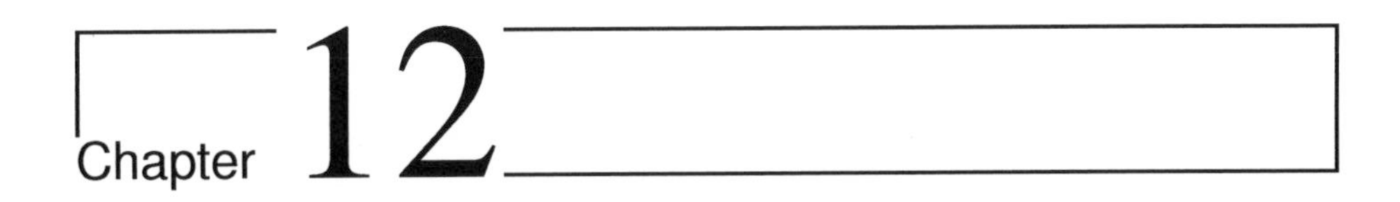

Chapter 12

Modeling Randomness

> *The race is not to the swift, nor the battle to the strong, neither yet bread to the wise, nor yet riches to men of understanding, nor yet favor to men of skill; but time and chance happeneth to them all.* — Ecclesiastes

Until now, emphasis has been on the deterministic description of variation: how explanatory variables can account for the variation in the response. Little attention has been paid to the residual other than to minimize it in the fitting process.

It's time now to take the residual more seriously. It has its own story to tell. By listening carefully, the modeler gains insight into even the deterministic part of the model. Keep in mind the definition of statistics offered in Chapter 1:

> Statistics is the explanation of variability **in the context of what remains unexplained.**

The next two chapters develop concepts and techniques for dealing with "what remains unexplained." In later chapters these concepts will be used when interpreting the deterministic part of models.

12.1 Describing Pure Randomness

Consider an **event** whose **outcome** is completely random, for instance, the flip of a coin. How to describe such an event? Even though the outcome is random, there is still structure to it. With a coin, for instance, the outcome must be "heads" or "tails" — it can't be "rain." So, at least part of the description should say what are the possible outcomes: H or T for a coin flip. This is called the **outcome set**. (The outcome set is conventionally called the **sample space** of the event. This terminology can be confusing since it has little to do with the sort of sampling encountered in the collection of data nor with the sorts of spaces that vectors live in.)

For a coin flip, one imagines that the two outcomes are equally likely. This is usually specified as a **probability**, a number between zero and one. Zero means "impossible." One means "certain."

A **probability model** assigns a probability to each member of the outcome set. For a coin flip, the accepted probability model is 0.5 for H and 0.5 for T — each outcome is equally likely.

What makes a coin flip purely random is *not* that the probability model assigns equal probabilities to each outcome. If coins worked differently, an appropriate probability model might be 0.6 for H and 0.4 for T. The reason the flip is purely random is that the probability model contains all the information; there are no explanatory variables that account for the outcomes in any way.

Example 12.1: Rolling a Die The outcome set is the possibilities 1, 2, 3, 4, 5, and 6. The accepted probability model is to assign probability $\frac{1}{6}$ to each of the outcomes. They are all equally likely.

Now suppose that the die is "loaded." This is done by drilling into the dots to place weights in them. In such a situation, the heavier sides are more likely to face down. Since 6 is the heaviest side, the most likely outcome would be a 1. (Opposite sides of a die are arranged to add to seven, so 1 is opposite 6, 2 opposite 5, and 3 opposite 4.) Similarly, 5 is considerably heavier than 2, so a 2 is more likely than a 5. An appropriate probability model is this:

Outcome	1	2	3	4	5	6
Probability	0.28	0.22	0.18	0.16	0.10	0.06

One view of probabilities is that they describe how often outcomes occur. For example, if you conduct 100 trials of a coin flip, you should expect to get something like 50 heads. According to the **frequentist** view of probability, you should base a probability model of a coin on the relative proportion of times that heads or tails comes up in a very large number of trials.

Another view of probabilities is that they encode the modeler's assumptions and beliefs. This view gives everyone a license to talk about things in terms of probabilities, even those things for which there is only one possible trial, for instance current events in the world. To a **subjectivist**, it can be meaningful to think about current international events and conclude, "there's a one-quarter chance that this dispute will turn into a war." Or, "the probability that there will be an economic recession next year is only 5 percent."

Example 12.2: The Chance of Rain. Tomorrow's weather forecast calls for a 10% chance of rain. Even though this forecast doesn't tell you what the outcome will be, it's useful; it contains information. For instance, you might use the forecast in making a decision not to cancel your picnic plans.

But what sort of probability is this 10%? The *frequentist* interpretation is that it refers to seeing how many days it rains out of a large number of trials of identical days. Can you create identical days? There's only one trial — and that's tomorrow. The *subjectivist* interpretation is that 10% is the forecaster's way of giving you some information, based on his or her expertise, the data available, etc.

Saying that a probability is subjective does not mean that anything goes. Some probability models of an event are better than others. There is a reason why you look to the weather forecast on the news rather than gazing at your tea leaves. There are in fact rules for doing calculations with probabilities: a "probability calculus." Subjective probabilities are useful for encoding beliefs, but the probability calculus should be used to work through the consequences of these beliefs. The **Bayesian** philosophy of probability emphasizes the methods of probability calculus that are useful for exploring the consequences of beliefs.

Example 12.3: Flipping two coins. You flip two coins and count how many heads you get. The outcome space is 0 heads, 1 head, and 2 heads. What should the probability model be? Here are two possible models:

Model 1

# of Heads	0	1	2
Probability	$\frac{1}{3}$	$\frac{1}{3}$	$\frac{1}{3}$

Model 2

# of Heads	0	1	2
Probability	$\frac{1}{4}$	$\frac{1}{2}$	$\frac{1}{4}$

If you are unfamiliar with probability calculations, you might decide to adopt Model 1. However heartfelt your opinion, though, Model 1 is not as good as Model 2. Why? Given the assumptions that a head and a tail are equally likely outcomes of a single coin flip, and that there is no connection between successive coin flips — that they are **independent** of each other — the probability calculus leads to the $\frac{1}{4}, \frac{1}{2}, \frac{1}{4}$ probability model.

How can you assess whether a probability model is a good one, or which of two probability models are better? One way is to compare observations to the predictions made by the model. To illustrate, suppose you actually perform 100 trials of the coin flips in Example 12.1 and record your observations. Each of the models also gives a prediction of the expected number of outcomes of each type:

	# of Heads		
	0	1	2
Observed	28	45	27
Model 1	33.3	33.3	33.3
Model 2	25	50	25

The discrepancy between the model predictions and the observed counts tells something about how good the model is. A strict analogy to linear modeling would suggest looking at the residual: the difference between the model values and the observed values. For example, you might look at the familiar sum of squares of the residuals. For Model 2 this is $(28-25)^2 + (45-50)^2 + (27-25)^2$. More than 100-years ago, it was worked out that for probability models this is not the most informative way to calculate the size of the residuals. It's better to adjust each of the terms by the model value, that is, to calculate $\frac{(28-25)^2}{25} + \frac{(45-50)^2}{50} + \frac{(27-25)^2}{25}$. The details of the measure are not important right now, just that there is a way to quantify how well a probability model matches the observations.

12.2 Settings for Probability Models

A statistical model describes one or more variables and their relationships. In contrast, a probability model describes the outcomes of a random event, sometimes called a **random variable**. The **setting** of a random variable refers to how the event is configured. Some examples of settings:

- The number of radio-active particles are given off by a substance in a minute.
- A particular student's score on a standardized test consisting of 100 questions.
- The number of blood cells observed in a small area of a microscope slide.
- The number of people from a random sample of 1000 voters who support the incumbent candidate.

In constructing a probability model, the modeler picks a form for the model that is appropriate for the setting. By using expert knowledge (e.g., how radioactivity works) and the probability calculus, probability models can be derived for each of these settings.

It turns out that there is a reasonably small set of standard probability models that apply to a wide range of settings. You don't always need to derive probability models, you can use the ones that have already been derived. This simplifies things dramatically, since you can accomplish a lot merely by learning which of the models to apply in any given setting.

Each of the models has one or more **parameters** that can be used to adjust the models to the particular details of a setting. In this sense, the parameters are akin to the model coefficients that are set when fitting models to data.

12.3 Models of Counts

12.3.1 The Binomial Model

Setting. The standard example of a **binomial** model is a series of coin flips. A coin is flipped n times and the outcome of the event is a count: how many heads. The outcome space is the set of counts $0, 1, \ldots, n$.

To put this in more general terms, the setting for the binomial model is an event that consists of n trials. Each of those trials is itself a purely random event where the outcome set has two possible values. For a coin flip, these would be heads or tails. For a medical diagnosis, the outcomes might be "cancer" or "not cancer." For a political poll, the outcomes might be "support" or "don't support." All of the individual trials are identical and independent of the others.

The name binomial — "bi" as in two, "nom" as in name — reflects these two possible outcomes. Generically, they can be called "success" and "failure." The outcome of the overall event is a count of the number of successes out of the n trials.

Parameters. The binomial model has two parameters: the number of trials n and the probability p of success in each trial.

Example 12.4: Multiple Coin Flips Flip 10 coins and count the number of heads. The number of trials is 10 and the probability of "success" is 0.5. Here is the probability model:

Outcome	0	1	2	3	4	5
Probability	0.001	0.010	0.044	0.117	0.205	0.246
Outcome		6	7	8	9	10
Probability		0.205	0.117	0.044	0.01	0.001

Examples of Binomial Models

$n = 4, p = 0.30$

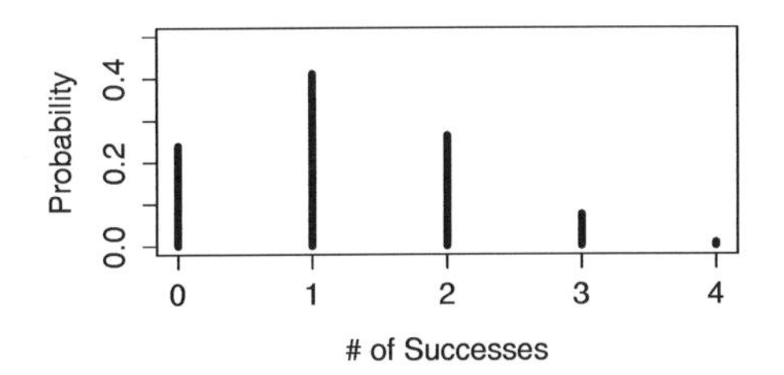

$n = 100, p = 0.30$

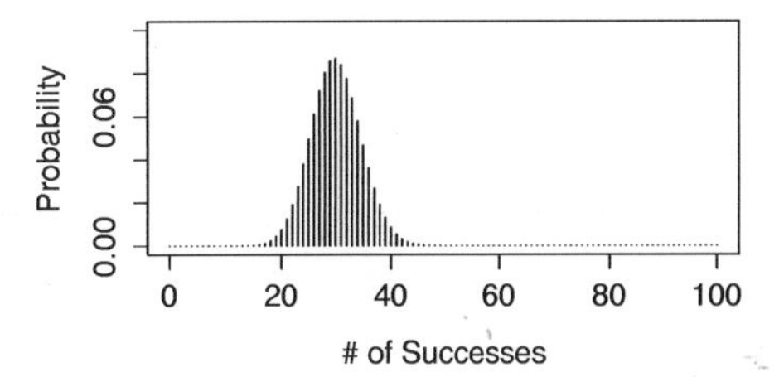

Example 12.5: Houses for Sale Count the number of houses on your street with a for-sale sign. Here n is the number of houses on your street. p is the probability that any randomly selected house is for sale. Unlike a coin flip, you likely don't know p from first principles. Still, the count is appropriately modeled by the binomial model. But until you know p, or assume a value for it, you can't calculate the probabilities.

12.3.2 The Poisson Model

Setting. Like the binomial, the **Poisson** model also involves a count. It applies when the things you are counting happen at a typical rate. For example, suppose you are counting cars on a busy street on which the city government claims the typical traffic level is 1000 cars per hour. You count cars for 15 minutes to check whether the city's claimed rate is right. According to the rate, you expect to see 250 cars in the 15 minutes. But the actual number that pass by is random and will likely be somewhat different from 250. The Poisson model assigns a probability to each possible count.

Parameters. The **Poisson** model has only one parameter: the average rate at which the events occur.

Example 12.6: The Rate of Highway Accidents In your state last year there were 300 fatal highway accidents: a rate of 300 per year. If this rate continued, how many accidents will there be today?

Since you are interested in a period of one day, divide the annual rate by 365 to give a rate of $300/365 = 0.8219$ accidents per day. According to the Poisson model, the probabilities are:

Outcome	0	1	2	3	4	5	...
Probability	0.440	0.361	0.148	0.041	0.008	0.001	...

People often confuse the binomial setting with the Poisson setting. Notice that in a Poisson setting, there is not a fixed number of trials, just a typical rate at which events occur. The count of events has no firm upper limit. In contrast, in a binomial setting, the count can never be higher than the number of trials, n.

Examples of Poisson Models

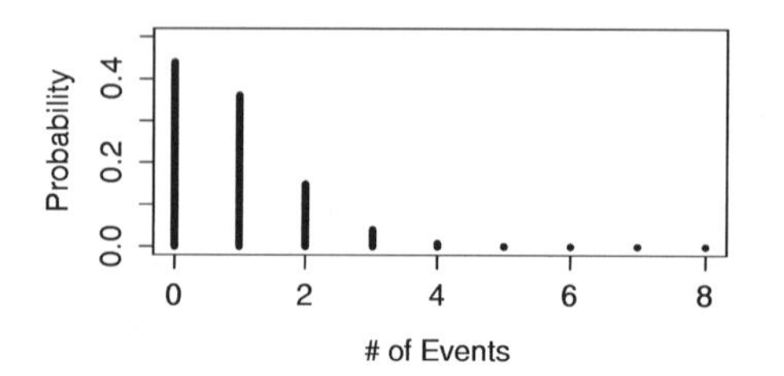

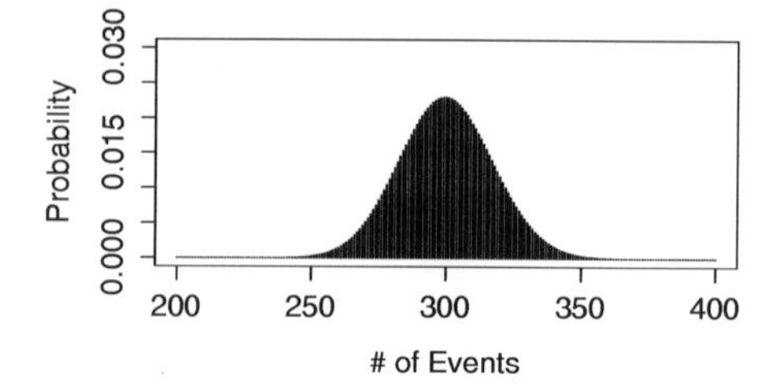

12.4 Common Probability Calculations

A probability model assigns a probability to each member of the outcome set. Usually, though, you are not interested in these probabilities directly. Instead, you want other information that is based on the probabilities:

Percentiles. Students who take standardized tests know that the test score can be converted to a percentile: the fraction of test takers who got that score or less. Similarly, each possible outcome of a random variable falls at a certain percentile. It often makes sense to refer to the outcomes by their percentiles rather than the value itself. For example, the 90th percentile is the value of the outcome such that 90 percent of the time a random outcome will be at that value or smaller.

Percentiles are used, for instance, in finding what range of outcomes is likely, as in calculating a coverage interval for the outcome.

Example 12.7: A Political Poll You conduct 500 interviews with likely voters about their support for the incumbent candidate. You believe that 55% of voters support the incumbent. What is the likely range in the results of your survey?

This is a binomial setting, with parameters $n = 500$ and $p = 0.55$. It's reasonable to interpret "likely range" to be a 95% coverage interval. The boundaries of this interval are the 2.5- and 97.5-percentiles, or, 253 and 297 respectively.

Quantiles or **Inverse percentiles.** Here, instead of asking what is the outcome at a given percentile, you invert the question and ask what is the percentile at a given outcome. This sort of question is often asked to calculate the probability that an outcome exceeds some limit.

Example 12.8: A Normal Year? Last year there were 300 highway accidents in your state. This year saw an increase to 321. Is this a likely outcome if the underlying rate hadn't changed from 300?

This is a Poisson setting, with the rate parameter at 300 per year. According to the Poisson model, the probability of seeing 321 or more is 0.119: not so small.

Calculation of percentiles and inverse percentiles is almost always done with software. The "computational technique" section gives specific instructions for doing this.

12.5 Continuous Probability Models

The binomial and Poisson models apply to settings where the outcome is a count. There are other settings where the outcome is a number that is not necessarily a counting number.

12.5.1 The Uniform Model

Setting. A random number is generated that is equally likely to be anywhere in some range. An example is a spinner that is equally likely to come to rest at any orientation between 0 and 360 degrees.

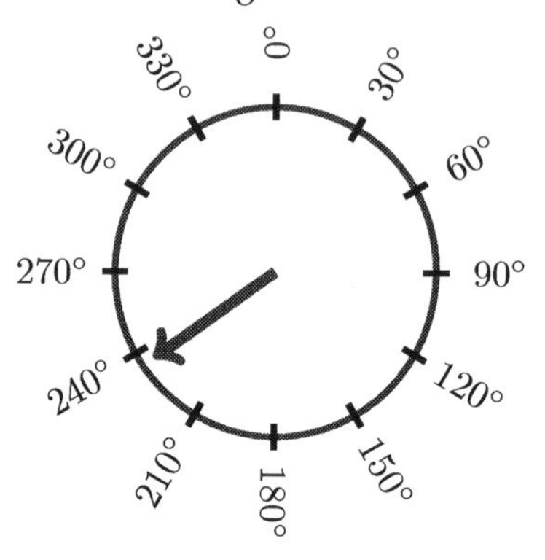

Parameters. Two parameters: the upper and the lower end of the range.

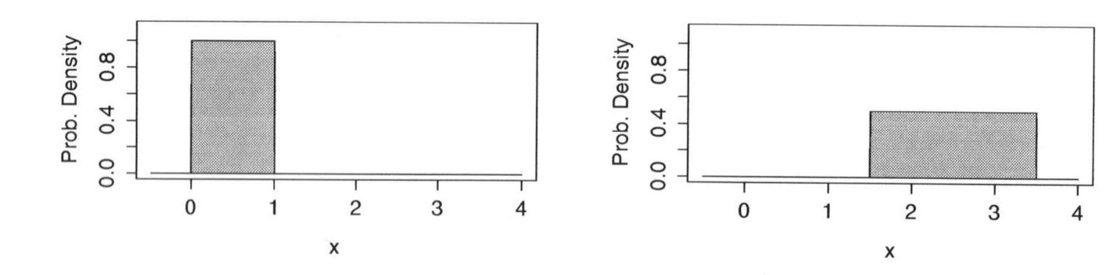

Example 12.9: Equally Likely A uniform random number is generated in the range 0 to 1. What is the probability that it will be smaller than 0.05? Answer: 0.05. (Simple as it seems, this particular example will be relevant in studying p-values in a later chapter.)

12.5.2 The Normal Model

Setting. The normal model turns out to be a good approximation for many purposes, thus the name "normal." When the outcome is the result of adding up **lots** of independent numbers, the normal model often applies well. In statistical modeling, model coefficients are calculated with dot products, involving lots of addition. Thus, model coefficients are often well described by a normal model.

The normal model is so important that Section 12.6 is devoted to it exclusively.

Parameters. Two. The **mean** which describes the most likely outcome, and the **standard deviation** which describes the typical spread around the mean.

Examples of Normal Models

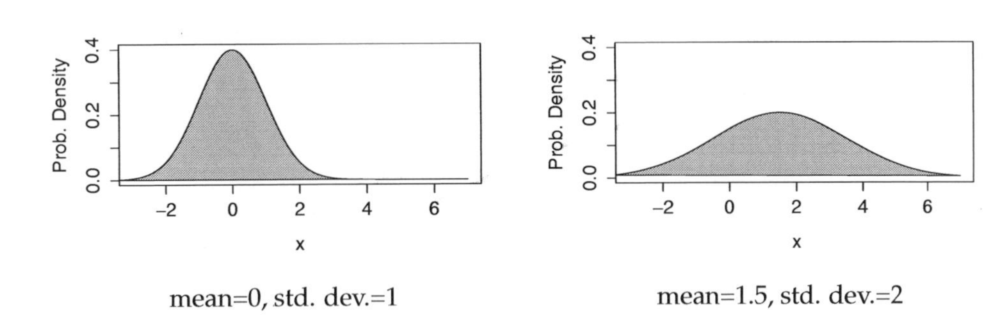

mean=0, std. dev.=1 mean=1.5, std. dev.=2

Example 12.10: IQ Test Scores Scores on a standardized IQ test are arranged to have a mean of 100 points and a standard deviation of 15 points. What's a typical score from such a test? Answer: According to the normal model, the 95% coverage interval is 70.6 to 129.4 points.

12.5.3 The Log-normal Model

Setting. The normal distribution is widely used as a general approximation in a large number of settings. However, the normal distribution is symmetrical and fails to give a good approximation when a distribution is strongly skewed. The log-normal distribution, is often used when the skew is important. Mathematically, the log-normal model is appropriate when taking the logarithm of the outcome would produce a bell-shaped distribution.

Parameters. In the normal distribution, the two parameters are the mean and standard deviation. Similarly, in the log-normal distribution, the two parameters are the log of the mean and the log of the standard deviation.

Example 12.11: High Income In most societies, incomes have a skew distribution; very high incomes are more common than you might expect. In the United States, for instance, data on middle-income families in 2005 suggests a mean of about \$50,000 per year and a standard deviation of about \$20,000 per year (taking "middle-income" to mean the central 90% of families). If incomes were distributed in a bell-shaped, normal manner, this would imply that the 99th percentile of income would be \$100,000 per year. But this understates matters. Rather than being normal, the distribution of incomes is better approximated with a log-normal distribution.[19] Using a log-normal distribution with parameters set to match the observations for middle-income familes gives a 99th percentile of \$143,000.

Examples of Log-normal Models

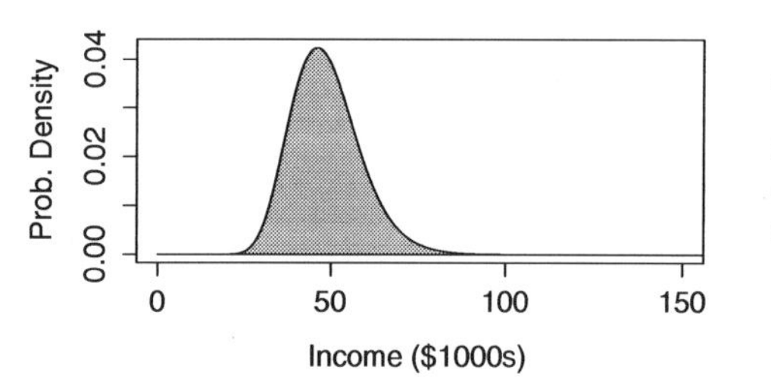

logmean = 3.89, logsd=0.20

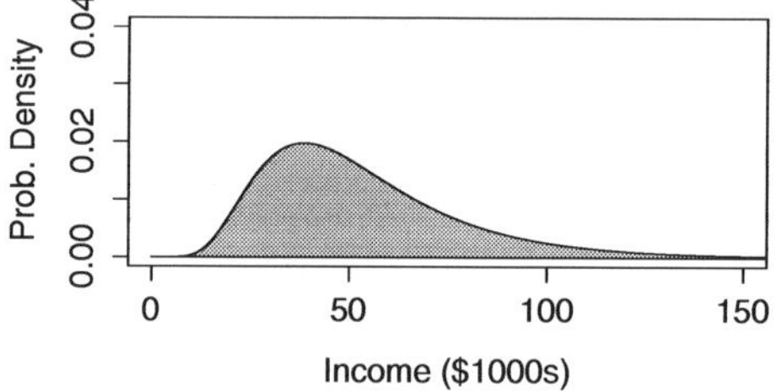

logmean = 3.89, logsd=0.47

12.5.4 The Exponential Model

Setting. The exponential model is closely related to the Poisson. Both models describing a setting where events are happening at random but at a certain rate. The Poisson describes *how many* events happen in a given period. In contrast, the exponential model describes the *spacing* between events.

Parameters. Just like the Poisson, the exponential model has one parameter, the rate.

Example 12.12: Times between Earthquakes In the last 202 years, there have been six large earthquakes in the Himalaya mountains. The last one was in 2005. This suggests a typical rate of $\frac{6}{202} = 0.03$ earthquakes per year — about one ever 33 years. If earthquakes happen at random at this rate, what is the probability that there will be another large earthquake within 10 years of the last, that is, by 2015? According to the exponential model, this is 0.259. That's a surprisingly large probability if you expected that earthquakes won't happen again until a suitable interval passes. There can also be surprisingly large gaps between earthquakes. For instance, the probability that another large earthquake won't occur until 2105 is 0.05 according to the exponential model.

Examples of Exponential Models

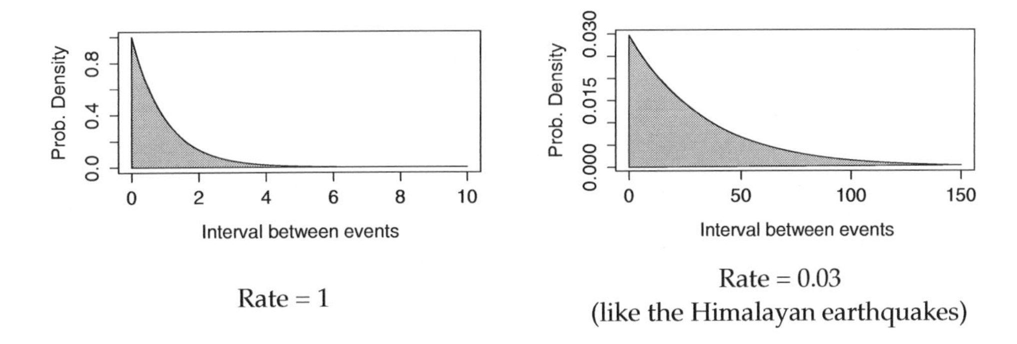

Rate = 1

Rate = 0.03
(like the Himalayan earthquakes)

12.6 The Normal Distribution

Many quantities have a bell-shaped distribution so the normal model makes a good approximation in many settings. The parameters of the normal model are the mean and standard deviation. It is important to realize that a mean and standard deviation can be calculated for other models as well.

To see this, imagine a large number of trials generated according to a probability model. For the models discussed in this chapter, the outcomes are always numbers. So it makes sense to take the mean or the standard deviation of the trial outcomes.

When applied to probability models, the mean describes the center or typical value of the random outcome, just as the mean of a variable describes center of the distribution of that variable. The standard deviation describes how the outcomes are spread around the center.

The calculation of percentiles, coverage intervals, and inverse percentiles is so common, that a shorthand way of doing it has been developed for normal models. At the heart of this method is the **z-score**.

The z-score is a way of referring a particular value of interest to a probability model or a distribution of values. The z-score measures how far a value is from the center of the distribution. The units of the measurement are in terms of the standard deviation of the distribution. That is, if the value is x in a distribution with mean m and standard deviation s, the z-score is:

$$z = \frac{x - m}{s}.$$

Example 12.13: IQ Percentiles from Scores. In a standard IQ test, scores are arranged to have a mean of 100 and a standard deviation of 15. A person gets an 80 on the exam. What is their z-score? Answer: $z = \frac{80-100}{15} = -1.33$.

In some calculations, you know the z-score and need to figure out the corresponding value. This can be done by solving the z-score equation for x, or

$$x = m + sz.$$

Example 12.14: IQ Scores from Percentiles. In the IQ test, what is the value that has a z-score of 2? Answer: $x = 100 + 15 \times 2 = 130$.

Here are the rules for calculating percentiles and inverse percentiles from a normal model:

95% Coverage Interval is bounded by the two points with $z = -2$ and 2.

Percentiles The percentile of a value is indicated by its z-score.

z-score	-3	-2	-1	0	1	2	3
Percentile	0.1	2.3	15.9	50	84.1	97.7	99.9

Values with z-scores greater than 3 or smaller than -3 are very uncommon, at least according to the normal model.

Example 12.15: Bad Luck at Roulette? You have been playing a roulette game all night, betting on red, your lucky color. Altogether you have played $n = 100$ games and you know the probability of a random roll coming up red is $16/34$. You've won only 42 games — you're losing serious money. Are you unlucky? Has red abandoned you?

If you have a calculator, you can find out approximately how unlucky you are. The setting is binomial, with $n = 100$ trials and $p = 16/34$. The mean and standard deviation of a binomial random variable are np and $\sqrt{np(1-p)}$ respectively; these formulas are ones that all skilled gamblers ought to know. The z-score of your performance is:

$$z = \frac{42 - 100\frac{16}{34}}{\sqrt{100\frac{16}{34}(1 - \frac{16}{34})}} = \frac{42 - 47.06}{5} = -1$$

The score of $z = -1$ translates to the 16th percentile — your performance isn't good, but it's not too surprising either.

When you get back to your hotel room, you can use your computer to do the exact calculation based on the binomial model. The probability of winning 42 games or fewer is 0.181.

12.7 Computational Technique

One of the most surprising outcomes of the revolution in computing technology has been the discovery of diverse uses for randomness in the analysis of data and in science generally. Most young people have little trouble with the idea of a computer generating random data; they see it in computer games and simulations. Older people, raised with the idea that computers do mathematical operations and that such operations are completely deterministic, sometimes find computer-generated randomness suspect. Indeed, conventional algebraic notation ($+$, $-$, $\surd$, cos, and so on) has no notation for "generate at random."

One of the simplest operators for generating random events is `resample`. This takes two arguments: the first is a set of items to choose from at random, the second is how many events to generate. Each item is equally likely to be choosen. For example, here is a simulation of a coin flip:

```
> coin = c("H","T")
> resample(coin, 5)
[1] "H" "T" "T" "H" "T"
> resample(coin, 5)
[1] "T" "H" "T" "H" "H"
```

The first command creates an object holding the possible outcome of each event, called `coin`. The next command generated five events, each event being a random choice of the outcomes in `coin`.

Another example is rolling dice. First, construct a set of the possible outcomes: the numbers 1, 2, 3, 4, 5, 6.

```
> die = seq(1,6)
> die
[1] 1 2 3 4 5 6
```

Then generate random events. Here is a roll of two dice.

```
> resample(die,2)
[1] 4 5
```

The `resample` operator is also useful for selecting cases at random from a data frame. This use will be the basis for statistical methods introduced in later chapters.

12.7.1 Random Draws from Probability Models

Although `resample` is useful for random sampling, it can work only with finite sets of possible outcomes such as H/T or 1/2/3/4/5/6 or the cases in a data frame. By default in `resample`, the underlying probability model is **equiprobability** — each possible outcome is equally likely. You can specify another probability model by using the `prob=` argument to `resample`. For instance, to flip coins that are very likely to come up heads:

```
> resample( coin, 10, prob=c(.9,.1))
 [1] "H" "H" "T" "H" "H" "H" "H" "H" "H" "H"
```

R provides other operators that allow draws to be made from outcome sets that are infinite.

For example, the `rnorm` operator makes random draws from a normal probability distribution. The required argument tells how many draws to make. Optional, named arguments let you specify the mean and standard deviation of the particular normal distribution that you want. To illustrate, here is a set of 15 random numbers from a normal distribution with mean 1000 and standard deviation 75:

```
> samps = rnorm(15, mean=1000, sd=75)
> samps
 [1]  949.1 1007.7  885.8  956.2 1006.8  985.0 1019.3
 [8]  824.2 1132.5  881.6  905.7  976.9  926.1 1104.9
[15] 1081.1
```

In this example, the output was assigned to an object `samps` to facilitate some additional computations to the values. For instance, here is the mean and standard deviation of the sample:

```
> mean(samps)
[1] 976.2
> sd(samps)
[1] 86.47
```

Don't be surprised that the mean and standard deviation of the sample don't match exactly the parameters that were set with the arguments `mean=1000, sd= 75`. The sample was drawn at random and so the sample statistics are going to vary from one sample to the next. Part of the statistical methodology to be studied in later chapters has to do with determining how close the statistics calculated from a sample are likely to be to the parameters of the underlying population.

Often you will generate very large samples. In these situations you usually don't want to display all the samples, just do calculations with them. The practical limits of "large" depend on the computer you are using and how much time you are willing to spend on a calculation. For an operator like `rnorm` and the others to be introduced in this chapter, it's feasible to generate samples of size 10,000 or 100,000 on an ordinary laptop computer.

```
> samps = rnorm(100000, mean=1000, sd=75)
> mean(samps)
[1] 999.8
> sd(samps)
[1] 74.65
```

Notice that the sample mean and standard deviation are quite close to the population parameters in this large sample. (Remember not to put commas in as punctuation in large numbers: it's `100000` not 100,000.)

The simulations that you will do in later chapters will be much more elaborate than the simple draws here. Even with today's computers, you will want to use only a few hundred trials.

12.7.2 Standard Probability Models

R provides a large set of operators like `rnorm` for different probability models. All of these operators work in the same way:

- Each has a required first argument that gives the number of draws to make.
- Each has an optional set of parameters that specify the particular probability distribution you want.

All the operators start with the letter `r` — standing for "random" — followed by the name of the probability model:

Family	R name	Parameters	
Normal	`rnorm`	`mean,sd`	continuous
Uniform	`runif`	`min,max`	continuous
Binomial	`rbinom`	`size,prob`	discrete
Poisson	`rpois`	Average rate (written `lambda`)	discrete
Exponential	`rexp`	Same rate as in poisson but the parameter is called `rate`.	continuous
Lognormal	`rlnorm`	Mean and sd of the natural logarithm. `meanlog`, `sdlog`	continuous
χ^2	`rchisq`	Degrees of freedom (`df`)	continuous
t	`rt`	Degrees of freedom (`df`)	continuous
F	`rf`	Degrees of freedom in the numerator and in the denominator (`df1`, `df2`)	continuous

To use these operators, you first must choose a particular probability model based on the setting that applies in your situation. This setting will usually indicate what the population parameters should be. Some examples:

- You are in charge of a hiring committee that is going to interview three candidates selected from a population of job applicants that is 63% female. How many of the interviewees will be female? Modeling this as random selection from the applicant pool, a binomial model is appropriate. The `size` of each trial is 3, the probability of being female is 63%:

```
> samps = rbinom(40, size=3, prob=0.63)
> samps
 [1] 2 2 1 1 1 3 3 3 2 2 3 3 1 2 1 2 3 2 1 2 2 3 3
[28] 2 1 2 1 3 2 1 0 3 2 2 2 3 0 1 1 2
```

There are 40 trials here, since the first argument was set to 40. Remember, each of the trials is a simulation of one hiring event. In the first simulated event, two of the interviewees were female; in the third only one was female. Typically, you will be summarizing all the simulations, for example to see how likely each possible outcome is.

```
> table(samps)
samps
 0  1  2  3
 2 11 16 11
```

- You want to simulate the number of customers who come into a store over the course of an hour. The average rate is 15 per hour. To simulate a situation where customers arrive randomly, the poisson model is appropriate:

```
> rpois(25, lambda=15)
 [1] 18 11 11 12 13 19 11 17 11 22 11 15 19  8 16
[20] 19 11 21 19  9 21 15 12 25 11
```

- You want to generate a simulation of the interval between earthquakes and in Example 12.5.4. To simulate the random intervals with a typical rate of 0.03 earthquakes per year, you would use

```
> rexp( 15, rate=0.03 )
 [1]  0.4709 15.6561 21.3271 13.7895 89.0088 87.8006
 [7]  1.3473 14.7994 49.0427 49.8065 20.4635  9.3172
[13] 23.3749 51.1697 23.3353
```

  Notice the huge variation in the intervals, from less than half a year to almost 90 years between earthquakes.

12.7.3 Coverage Intervals

You will often need to compute coverage intervals in order to describe the range of likely outcomes from a random process. R provides a series of operators for this purpose; a separate operator for each named probability model. The operators all begin with q, standing for **quantiles**. In all cases, the first argument is the set of quantiles you want to calculate for the particular probability model. The optional named arguments are the parameters.

Remember that to find a 95% coverage interval you need the 0.025 and 0.975 quantiles. For a 99% interval, you need the 0.005 and 0.995 quantiles.

To illustrate, here are 95% coverage intervals for a few probability models.

- A normal distribution with mean 0 and standard deviation 1:

```
> qnorm( c(0.025, 0.975), mean=0, sd=1)
[1] -1.96  1.96
```

- The hiring committee situation modelled by a binomial distribution with size=3 and prob=0.63:

```
> qbinom( c(0.025, 0.975), size=3, prob=0.63)
[1] 0 3
```

Perhaps you are surprised to see that the coverage interval includes all the possible outcomes. That's because the number of cases in each trial ($n = 3$) is quite small.

- The number of customers entering a store during an hour as modelled by a poisson distribution with an average rate of 15 per hour.

```
> qpois( c(0.025, 0.975), lambda=15)
[1]  8 23
```

- The interval between earthquakes modelled by an exponential distribution with a typical rate of 0.03 earthquakes per year:

```
> qexp( c(.025, .975), rate=0.03)
[1]   0.844 122.963
```

You can also use the `q` operators to find the value that would be at a particular percentile. For example, the exponential model with `rate=0.03` gives 25th percentile of the interval between earthquakes as:

```
> qexp( c(.25), rate=0.03)
[1] 9.59
```

A quarter of the time, the interval between earthquakes will be 9.59 years or less.

It's entirely feasible to calculate percentiles and coverage intervals by combining the random-number generators with `quantile`. For example, here is the 95% coverage interval from a normal distribution with mean 0 and standard deviation 1:

```
> samps = rnorm(10000, mean=0, sd=1)
> qdata( c(.025, .975), samps )
  2.5%  97.5%
-1.961  1.939
```

The disadvantage of this approach is that it is a simulation and the results will vary randomly. By making the sample size large enough — here it is $n = 10000$ — you can reduce the random variation. Using the `q` operators uses mathematical analysis to give you what is effectively an infinite sample size. For this reason, it's advisable to use the `q` operators when you can. However, for many of the techniques to be introduced in later chapters you will have to generate a random sample and then apply `quantile` to approximate the coverage intervals.

12.7.4 Percentiles

A percentile computation applies to situations where you have a measured value and you want to know where that value ranks relative to the entire set of possible outcomes. You have already seen percentiles computed from samples; they also apply to probability models.

It's easy to confuse percentiles with quantiles because they are so closely related. Mathematically, the percentile operators are the inverse of the quantile operators. To help you remember which is which, it's helpful to distinguish them based on the type of argument that you give to the operator:

Percentile The input argument is a measured value, something that could be the output of a single draw from the probability distribution. The output is always a number between 0 and 1 — a percentile.

Quantile The input is a percentile, a number between 0 and 1. The output is on the scale of the measured variable.

Example: You have just gotten your score, 670, on a professional school admissions test. According to the information published by the testing company, the scores are normally distributed with a mean of 600 and a standard deviation of 100. So, your ranking on the test, as indicated by a percentile, is:

```
> pnorm(670, mean=600, sd=100)
[1] 0.758
```

Your score is at about the 75th percentile.

Example: Unfortunately, the professional school that you want to go to accepts only students with scores in the top 15 percent. Your score, at 75.8%, isn't good enough. So, you will study some more and take practice tests until your score is good enough. How well will you need to score to reach the 85%?

```
> qnorm(0.85, mean=600, sd=100)
[1] 703.6
```

12.8 Simulations

Most systems of interest are not purely random: there are relationships between the variables involved that are partly deterministic. Models that generate outcomes from such systems are called **simulations**.

Simulations are useful for studying real systems or, more precisely, learning how to study real systems. By using a simulation, it's possible to generate data inexpensively and in large quantities. Although the simulation may not duplicate the real system it is meant to model, the simulation can still provide insight into important issues. How much data would be needed to study the real system? How well would a data analysis model capture the actual dynamics? These questions can be answered because, when using a simulation, it is known exactly what is happening in the system, so it's possible to compare the results of data analysis with the known structure of the simulation. Insofar as the simulation captures the important aspects of the real system, the results found by studying the simulation can provide lessons for studying data from the real system.

In this book, simulations will be used to explore how many cases are needed to get reliable results when fitting a model and to indicate the consequences of adding or omitting an explanatory variable from a model. Later, the simulations will be used to show how to design experiments and analyze the data from them.

The `ISM.Rdata` software contains programs to simulate several different systems. To illustrate, consider a simulated system called `heights` which presents one idea of how a person's height is determined by a combination of genes inherited from the mother and from the father as well as by the person's sex.

Running the simulation is easy. The operator `run.sim` takes two arguments: the name of the simulated system and the number of cases to generate:

```
> samps = run.sim(heights,5)
> samps
  mgenes fgenes kgenes Sex FHeight MHeight KHeight
1  0.761   3.58  2.171   M    59.4    78.2    70.6
2  2.162   2.74  2.453   M    65.5    71.1    72.9
3 -2.367   1.20 -0.582   M    63.4    73.2    67.5
4 -0.508   1.81  0.653   M    66.8    70.5    68.1
5 -0.947  -1.68 -1.314   F    63.2    69.5    67.0
```

It's entirely practical to generate thousands or tens of thousands of cases.

You can get information about the simulated system by typing the name of the system:

```
> heights
Causal Network with  7  vars:
mgenes, fgenes, kgenes, Sex,
MHeight, FHeight, KHeight
===============================================
mgenes is exogenous
fgenes is exogenous
kgenes <== mgenes & fgenes
Sex is exogenous
MHeight <== mgenes
FHeight <== fgenes
KHeight <== kgenes & Sex
```

This gives a listing of the variables involved in the system as well as showing which variables are connected to which. When a variable is described as **exogenous**, it means that the variable is set entirely at random. In the `heights` system, the level of the variables mgenes, fgenes, and Sex are exogenous. (mgenes represents the mother's contribution to the child's height, fgenes represents the father's contribution.) Variables that are non exogenous are determined, at least in part, by other variables in the system. For example, the mother's height MHeight is set by her genes mgenes. The kid's height KHeight is set by the kid's genes kgenes which are is in turn set by the mother's and father's genes. The kid's height is also set by the kid's Sex.

A more detailed description of the system is given by the equations that implement the simulation. These are open for inspection with the equations operator:

```
> equations(heights)
mgenes <== rnorm(nsamps, mean=0,sd=2)
fgenes <== rnorm(nsamps, mean=0,sd=2)
kgenes <== .5*mgenes + .5*fgenes
Sex <== sample(c('F','M'),nsamps,replace=TRUE)
MHeight <== mgenes + rnorm(nsamps,mean=69,sd=2.5)
FHeight <== fgenes + rnorm(nsamps,mean=65,sd=2.5)
KHeight <== kgenes + rnorm(nsamps,mean=69,sd=2.5) -
            4*(Sex=='F')
```

It takes a bit of skill to interpret these equations, and most of the time the system will be described in words. Still, you can see that in the heights system the mother's and father's gene values are simply random variables; that's why they are listed as being exogenous. The kid's genes are a half and half combination of the father's and mother's genes. You can also see that the mother's height is the sum of her genes and a random variable.

The main thing, however, is that you can generate data from the simulated system and then analyze it. Comparing your results to what is known to be true about the simulation lets you see how good your analysis is. For example, the simulation has been set up so that males (SexM) are 4 inches taller, on average, than females. But with only the 5 cases generated above, the result found from a fitted model is not particularly close:

```
> mod = lm( KHeight ~ Sex, data=samps)
> coef(mod)
(Intercept)         SexM
      67.04         2.72
```

This shouldn't be a surprise when looking at the data; there's only one female among the five cases.

Using more data gives a better result:

```
> samps = run.sim(heights,5000)
> mod = lm( KHeight ~ Sex, data=samps)
> coef(mod)
(Intercept)        SexM
      64.98        4.13
```

Other simulated systems will be introduced as needed.

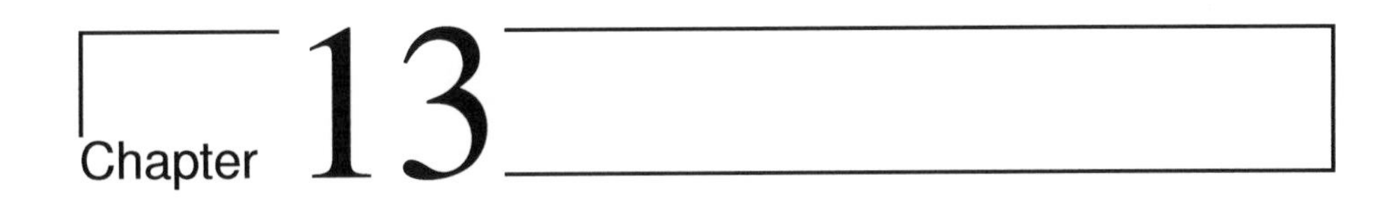

Geometry of Random Vectors

Oh, many a shaft at random sent
Finds mark the archer little meant!
And many a word at random spoken
May soothe, or wound, a heart that's broken! — Walter Scott (1771-1832)

In the next several chapters, models will be interpreted using ideas of randomness. Random simulations of residuals will be generated to explore the repeatability of model coefficients. Explanatory variables will be replaced by random variables to see whether the genuine variable is any better than junk.

This chapter introduces some of the geometry of randomness — for instance, what happens when model vectors are generated at random and used in fitting models. Some of it will contradict your intuition. For example, even though a random vector is equally likely to point in any direction, when you compare a random vector to another vector, they are very likely to meet at an angle near 90 degrees.

13.1 Random Angles

A statistician from the National Union of Transcendent Science (NUTS) approaches you. He claims to have mathematical proof that everything in the universe is linked. He discovered this when he decided to model the attendance count at the last two NUTS meetings. Taking this as his response variable A, he chose as an explanatory variable the last two stock market closing prices, B. In constructing the model A ~ B, he found a non-zero coefficient on B. But if there had been no relationship between NUTS attendance and the stock market, the coefficient would be expected to be zero.

Then he went on to try many other possible B: the distance to Mars and Jupiter at the time of the NUTS meetings; the altitudes of the two highest mountains in his home state; the number of times his dog barked during the NUTS

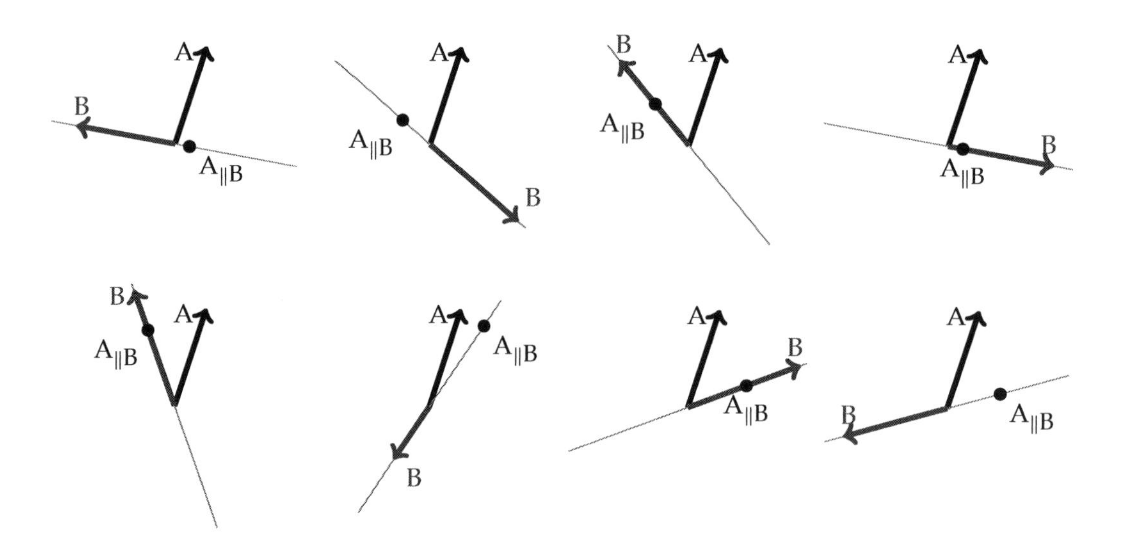

Figure 13.1: The model A ∼ B for several random B. The fitted model values point $A_{||B}$ is not usually at the origin, so the coefficient on B is not usually zero.

meetings. Hundreds of different B variables. In almost every case, the coefficient on B in A ∼ B was non-zero. "This proves that everything is linked!" he says enthusiastically.

You patiently try to explain things to the NUTS guy. The coefficient on B tells a lot about how A and B are aligned as vectors, but it doesn't necessarily tell very much about how the underlying variables are connected. What he proved is not that almost everything is linked, but that even random vectors can be aligned.

Figure 13.1 illustrates the situation. Whatever direction B points in, it's likely that the fitted model values $A_{||B}$ will lie at some distance along B and so the coefficient on B will be non-zero. The only situation in which the coefficient will be zero is when B is exactly perpendicular to A.

It's correct that, in the absence of a link, the expected value of the coefficient on B will be zero. But the expected value tells only part of the story about the distribution of the coefficient on B. The spread around the expected value is also important.

It's easier to think about things in terms of the angle θ between A and B. There's only one situation where the coefficient will be zero, when $\theta = 90°$. But there are many situations where θ is different from $90°$. So it's much more likely that the coefficient will be non-zero than it will be exactly zero.

Picking a random direction in 2-dimensional space amounts to picking a random point anywhere on a circle, as shown in Figure 13.2. Since the angles are evenly spaced, each angle is equally likely to be chosen. Thus, the distribution of angles between A and a random vector B is uniform.

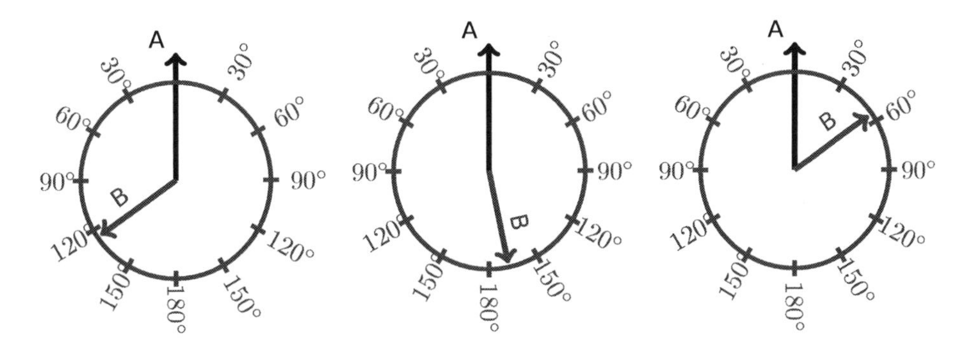

Figure 13.2: In $n = 2$ dimensions, picking a random direction corresponds to picking a random point on the circle. The angle between A and B is equally likely to be anything between 0 and $180°$.

This idea that all angles are equally likely matches well with most people's intuition. Unfortunately, that intuition is misleading when it comes to angles in higher-dimensional spaces.

Consider the situation in 3-dimensional space. Picking a random direction can be done by picking a random point on a sphere. Figure 13.3 shows a vector A piercing the pole of a sphere. The angle of a vector B going through some other point on the sphere can be read off from the circle of latitude on the sphere. For example, the vector B shown in the figure points to the circle of latitude marked $45°$, so it is at an angle of $45°$ to A.

The probability that a randomly chosen B will point to any particular circle of latitude is proportional to the circumference of that circle. From the figure, you can see that the $15°$ circle is short compared to the $30°$ circle, which is itself short compared to the $60°$ or $90°$ circles. That is, a randomly selected point on a sphere is much more likely to be near the equator (the $90°$ circle) than it is to be near either the north or south pole.

This pattern where angles near $90°$ are much more likely than other angles gets even stronger as the dimension of the space increases. Figure 13.4 shows the probability of angles between random vectors for a variety of dimensions n.

One way to summarize the distribution of angles is with the mean and standard deviation of the distribution. The mean angle is $90°$ for any n — on average, random vectors are perpendicular. This is why the NUTS guy was reasonable in thinking that, on average, the coefficient on B in his model A ~ B will be zero if there is no relationship between A and B. But "on average" isn't the same as "always." To know how far from zero a coefficient is likely to be when A is unrelated to B, you have to know the standard deviation of the distribution.

The standard deviation of the angle between random vectors depends on the dimension n of the space. You can see this in Figure 13.4: the distributions become narrower for larger n. The precise form of the standard deviation is particularly simple if you consider not the angle directly, but the cosine of the angle — the quantity that's relevant when dealing with model coefficients.

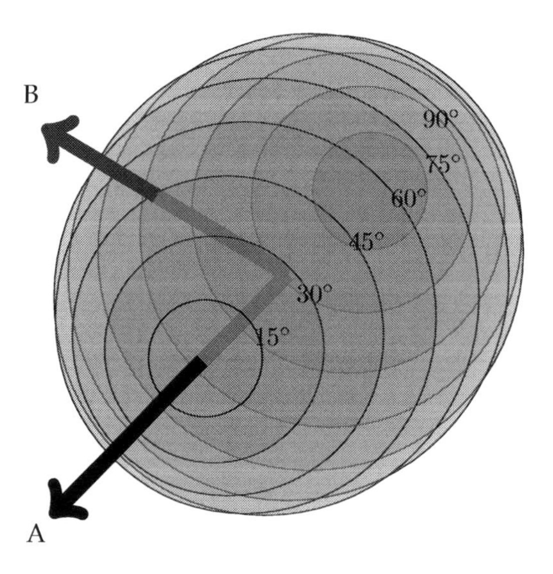

Figure 13.3: Picking a random direction in 3-dimensional space amounts to picking a random point on the sphere.

> For the cosine of angles between random vectors in n-dimensional space, the mean is 0 and the standard deviation is $\frac{1}{\sqrt{n}}$.

In practice, this means that random variables are always approximately perpendicular to one another, so long as n is big enough. This is the mathematical form of the intuition that two different random variables should be uncorrelated. They will be, so long as you have enough data: large enough n. That is, large n makes it hard to have accidental correlations of the sort the NUTS guy used in building his theory of universal linkage.

13.2 Random Models

Consider what will happen if you use one or more random vectors to model some response variable A. Why would you do this? Perhaps because you are skeptical that the explanatory variables you are actually using are any better than random.

The previous section indicated that random vectors tend to be perpendicular to other vectors, at least for large enough sample size n. That suggests that random vectors won't be particularly effective for modeling; walking along their subspaces won't bring you much closer to the target A.

On the other hand, the correlation r between the target A and a random vector, measured by the cosine of the angle between them, has a typical size $\frac{1}{\sqrt{n}}$. This means that the typical size of $r^2 = \frac{1}{n}$. Recall from Chapter 10 that an interpretation of r^2 is the fraction of the square distance to A that the model value carries you. So, a random vector typically gets you $\frac{1}{n}$ of the square distance to

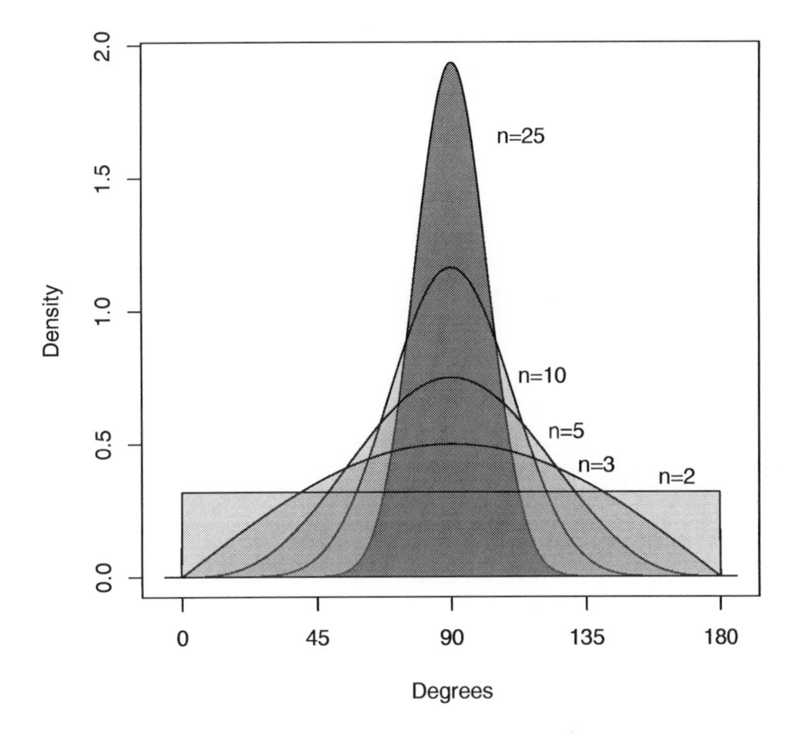

Figure 13.4: The distribution of angles between random vectors in n dimensional space.

the goal.

This pattern continues as more random vectors are added to a model. With two model vectors, $R^2 = \frac{2}{n}$, with three model vectors it's $R^2 = \frac{3}{n}$, and so on. And n random vectors will typically bring the fitted model values all the way to A, giving $R^2 = \frac{n}{n} = 1$.

This result will be important in later chapters. It provides a way to decide if a model vector is no better than what would be expected by luck.

13.3 Random Walks

Many of the processes encountered in nature and life involve accumulation, collecting incremental effects. Possibly the purest example of this is the irregular, jiggling motion of individual molecules in gas and liquid. These are far too small for the unaided eye to perceive directly. The first such observations with a microscope were made by the botanist Robert Brown (1773-1858). In 1827 he was examining pollen grains and later reported, "These motions were such as to satisfy me, after frequently repeated observations, that they arose neither from currents in the fluid nor from its gradual evaporation, but belonged to the particle itself." [20]

Figure 13.5 shows some paths of what came to be called **Brownian motion**.

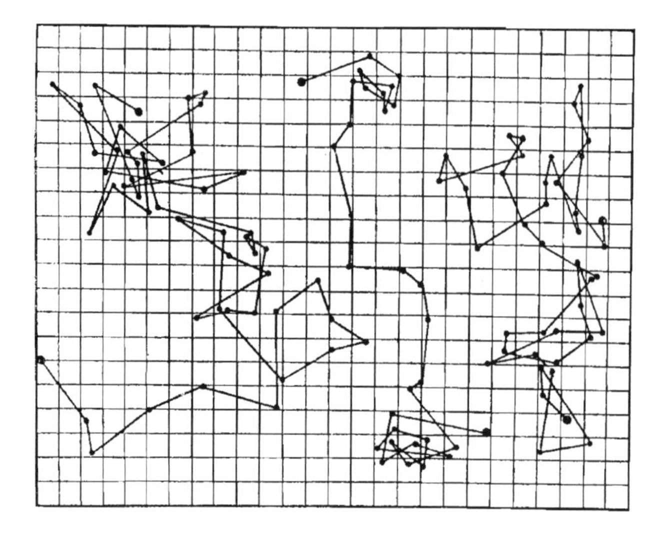

Figure 13.5: M. Jean Perrin's observations in 1909 of the motion of very small granules of mastic in fluid.[21]

Each dot is the position of a small particle at one instant in time. The line drawn between dots shows the overall step taken by the particle in getting to the next plotted position.

The position of the particle at each instant is an accumulation of the previous steps it has taken; each step adds on to the previous steps. Such accumulation is familiar. People get places by taking a series of steps. Each step is a small matter, but by accumulating many of them a person can move large distances.

At least, that is how things happen when walking in a purposeful, directed way. But Brown's pollen grains did not have a goal: they were not displaying a vital force. As has been understood since Einstein's 1905 paper on Brownian motion,[22] each segment of the motion is the result of a random collision with some other particle. Einstein worked out the theoretical properties of such random motion. A few years later, M. Jean Perrin confirmed Einstein's predictions experimentally, work for which he received the Nobel Prize.

The motion itself is a complicated tangle. In commenting on his drawing of the trajectory where the dots are observations of position every 30 seconds, Perrin noted, "[The drawings] only give a very feeble idea of the podigiously entangled character of the real trajectory. If the positions were indicated from second to second, each of these rectilinear segments would be replaced by a polygonal contour of 30 sides, relatively as complicated as the drawing here reproduced, and so on."[21]

Even if the motion itself is very complicated, the accumulation of it has a relatively simple form. Imagine a particle that takes n steps each of length s. The total length of the path will be sn. The net distance travelled — the distance between the start point and the end point — will also be sn if the motion is in a straight line. But if the motion is random, the average net distance will be only

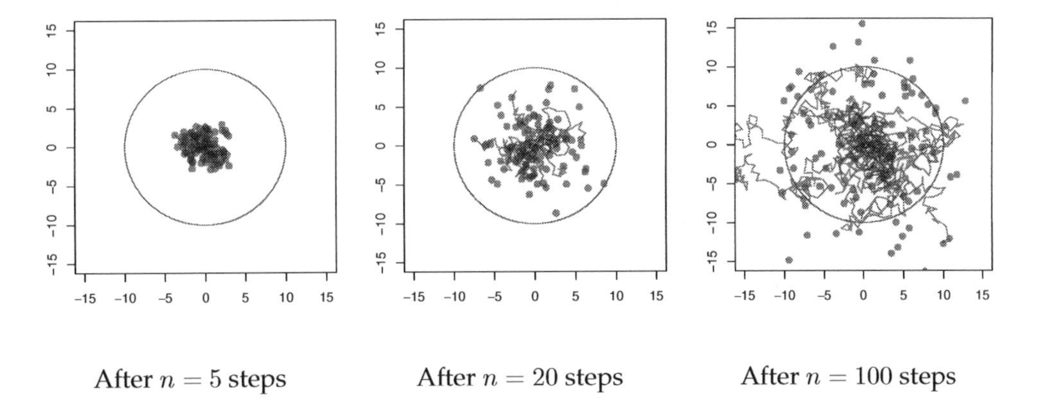

After $n = 5$ steps | After $n = 20$ steps | After $n = 100$ steps

Figure 13.6: 100 random walkers starting in a circle of radius 10 steps. For a typical walker to reach the circle will take $10^2 = 100$ steps.

$D = s\sqrt{n}$.

This is the *average* net distance. On any individual walk, the net distance is random. Figure 13.6 shows 100 random walks starting from the center of a circle. The circle has radius of $10s$, so if these were purposeful walks, each of the walkers would reach the circle after 10 steps. They don't. Even after 20 steps, only a couple of the walkers have reached the circle. It takes 100 steps for a typical walker to have reached the circle and even then a good fraction of the walkers have not yet made it.

One consequence of $D = s\sqrt{n}$ is that some familiar conventions don't make sense. For example, how to measure the step length? Take the overall distance after n steps and divide by n. For the purposeful walk, this gives $D/n = sn/n = s$. Makes sense. But for the random walk, the step length calculated in this way is $D/n = s\sqrt{n}/n = s/\sqrt{n}$. The apparent step length depends on how long one watches the particle.

Things are made much simpler with random walks by thinking of distances in terms of square lengths. The square distance of a random walk is, on average, $s^2 n$. So, the square length *per* step is $s^2 n/n = s^2$. In other words, the square length per step doesn't depend on n.

Example 13.1: The Stock Market Suppose the typical daily change in stock prices is up or down about 0.5% — that is, the market close changes from day to day by a random amount with a standard deviation of about half a percent. If this pattern continues for the approximately 250 trading days in a year, what is the typical *yearly* change in stock prices.

Answer: $0.5 \times \sqrt{250} = 0.5 \times 15.8 = 7.9$.

Aside. 13.1 Why Square-Lengths Add in Random Walks

The logic behind $D = s\sqrt{n}$ is easy to understand if you think about random vectors. Consider a walk of just two steps, each of length s. Here are some of the possibilities:

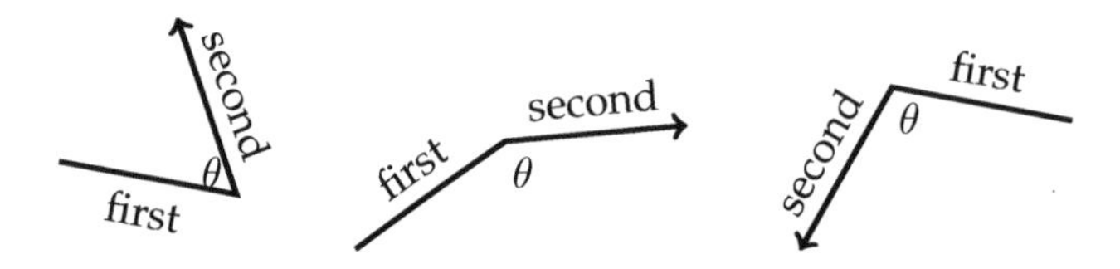

On average, the angle θ will be $90°$. So, on average, the walk looks like this

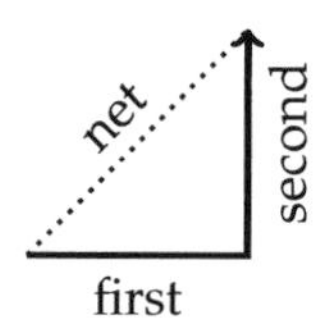

The net distance is the hypotenuse of the right triangle: $D^2 = s^2 + s^2$ which implies $D = \sqrt{2}s$. This same logic can be iterated. Add on a third step of length s to the first two steps of net distance $\sqrt{2}s$. The third step will, on average, be perpendicular to the net result of the first two. The net distance after the third step will be the hypotenuse of a right triangle with sides of length s and $\sqrt{2}s$. Adding these in the Pythagorean way gives a net distance after three steps of $\sqrt{3}s$.

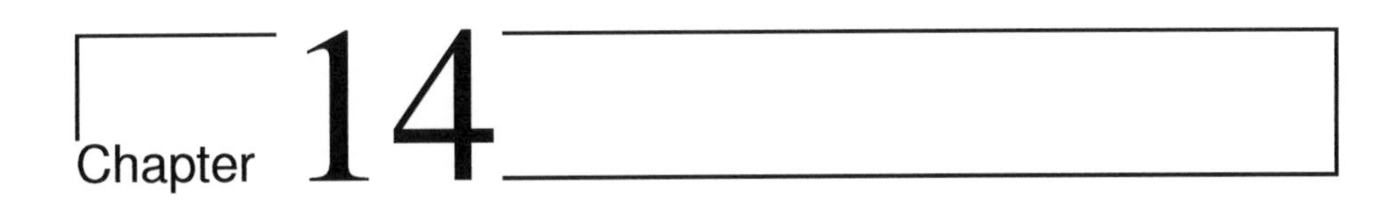

Chapter 14

Confidence in Models

To know one's ignorance is the best part of knowledge. — Lao-Tse

If you are a skilled modeler, you try to arrange things so that your model coefficients are random numbers.

That statement may sound silly, but before you jump to conclusions it's important to understand where the randomness comes from and why it's a good thing. Then you can use the tools in the previous two chapters to deal with the randomness and interpret it.

Recall the steps in building a statistical model:

1. Collect data. This is the hardest part, often involving great effort and expense.

2. Design your model, choosing the response variable, the explanatory variables, and the model terms. (It's sensible to have the design in mind *before* you collect your data, so you know what data are needed.)

3. Fit the model design to your data.

Step (3) is entirely deterministic. Given the model design and the data to which the model is to be fitted, fitting is an automatic process that will give the same results every time and on every computer. There is no randomness there. (There is some choice in choosing which to remove from a set of vectors containing redundancy, but that is a choice, not randomness. In any event, the fitted model values are not affected by this choice.)

Step (2) appears, at first glance, to leave space for chance. After all, when explanatory terms are collinear, as they often are, the fitted coefficients on any term can depend strongly on which other terms have been included in the model. The coefficients depend on the modeler's choices. But this means only that the coefficients reflect the beliefs of the modeler. If those beliefs aren't random, then the randomness of coefficients doesn't stem from the modeler.

It's step (1) that introduces the randomness. In collecting data, the sample cases are selected from a population. As described in Chapter 2, it's advantageous to make a random selection from the population; this helps to make the sample representative of the population.

Saying something is random means that it is uncertain, that if the process were repeated again the result might be different. When a sample is selected at random, the particular sample that is produced is just one of a set of possibilities. Once can imagine other possibilities that might have come about from the luck of the draw.

Insofar as the sample is random, the coefficients that come from fitting the model design to the sample are also random. The randomness of model coefficients means that the coefficients that come from a model design fitted to any particular data set are not likely to be an exact match to what you would get if the model design were fitted to the *entire population.* After all, the randomly selected sample is unlikely to be an exact match to the entire population.

It's helpful to know how close the results from the sample are to what would have been obtained if the sample had been the entire population. Ultimately, the only way to know this for sure is to create a sample that is the entire population. Usually this is impractical and often it is impossible.

But there is an approach that will gives insight, even if it does not gives certainty. To start, imagine that the sample were actually a census: a sample that contains the entire population. Repeating the study with a new sample would give exactly the same result because the new sample would be the same as the old one; it's the same population.

When the sample is not the entire population, repeating the study won't give the same result every time because the sample will include different members of the population. If the results vary wildly from one repeat to another, you have reason to think that the results are not a reliable indication of the population. If the results vary only a small amount from one repeat to another, then there's reason to think that you have closely approximated the results that you would have gotten if the sample had been the entire population. The repeatability of the process indicates how well the modeler knows the coefficients, or, in a word, the **precision** of the coefficients.

Knowing the precision of coefficients is key in drawing conclusions from them. Consider Galton's problem in studying whether height is a heritable trait. Had Galton known about modeling, he might have constructed a model like height ~ 1 + mother + father + sex. A relationship between the mother's height and her child's height should show up in the coefficient on mother. If there is a relationship, that coefficient should be non-zero.

Fitting the model to Galton's data, the coefficient is 0.32. Is this non-zero? Yes, for this particular set of data. But how might it have been different if Galton repeated his sampling, selecting a new set of cases from the population? How precise is the coefficient? Until this is known, it's mere bravado to say that a result of 0.32 means anything.

This chapter introduces methods that can be used to estimate the precision of coefficients from data. A standard format for presenting this estimation is the

confidence interval. For instance, from Galton's data, the estimated coefficient on mother is 0.32 ± 0.06, giving confidence that a different sample would also show a non-zero coefficient.

It's important to contrast precision with **accuracy**. Precision is about repeatability. Accuracy is about how the result matches the real world. Ultimately, accuracy is what the modeler wants. But the results of a model always depend on the choices the modeler makes, e.g., which explanatory variables to include, how to choose a sample, etc. The results can be accurate only if the modeler makes good choices. Knowing whether this is the case depends on knowing how the world really works, and this is what you are seeking to find out in the first place. Accuracy is elusive knowledge. Precision will have to suffice.

14.1 The Sampling Distribution

In principle, the way to see how much variation in model coefficients is introduced by the sampling process is to repeat the process of sampling and model fitting. The coefficients from each repetition can be collected; their distribution is called the **sampling distribution**. This is illustrated schematically in Figure 14.1, which shows just three random samples selected from the population. In reality, a much larger number of samples should be used, not just three.

To illustrate, consider the ten-mile race dataset. Thisis a census containing the running times for all 8636 registered participants in the Cherry Blossom Ten Mile race held in Washington, D.C. in April 2005.

The variable net records the start-line to finish-line time of each runner. There are also variables age and sex. Any model would do to illustrate the sampling distribution. Try

net ~ 1 + age + sex

Just for reference, here are the coefficients when the model is fit to the entire population:

	Intercept	age	sexM
Population	5540	16.9	-727

The coefficients indicate that runners who are one year older tend to take about 17 seconds longer to run the 10 miles.

Of course, if you knew the coefficients that fit the whole population, you would hardly need to collect a sample! But the purpose here is to demonstrate the effects of a random sampling process. The table below gives the coefficients from several sampling draws; each sample has $n = 100$ cases randomly selected from the population. That is, each sample simulates the situation where someone has randomly selected $n = 100$ cases.

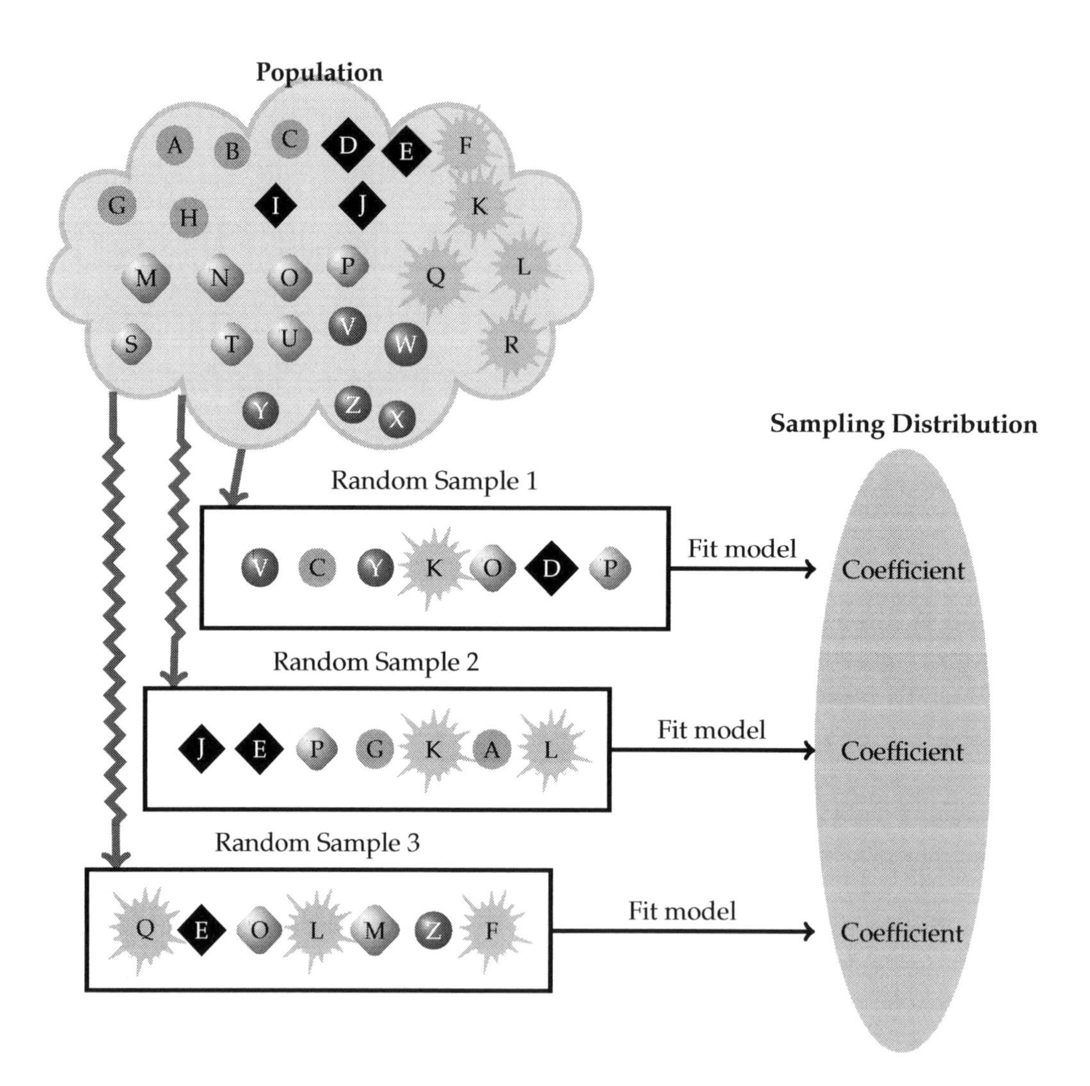

Figure 14.1: The sampling distribution reflects the variation in model coefficients from one random sample to another.

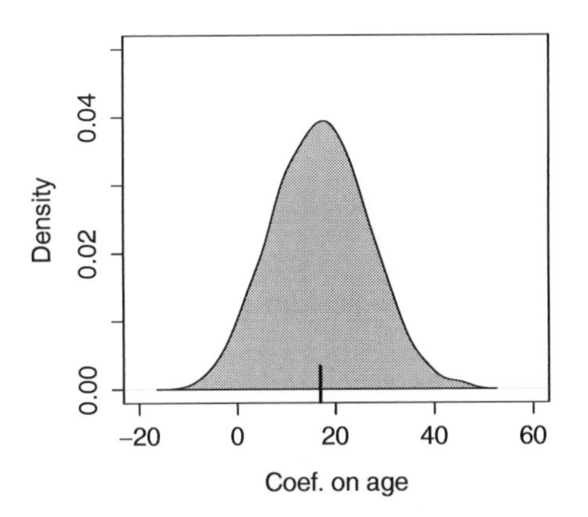

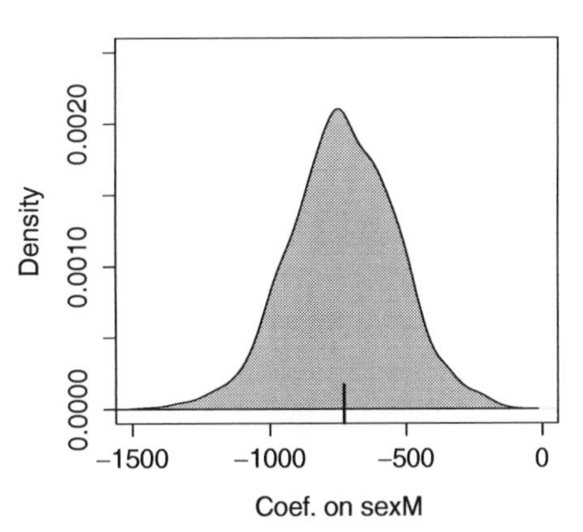

Figure 14.2: Sampling distributions for the age and sex coefficients for a sample size of $n = 100$. The bell-shaped distributions are centered on the population value (shown as a tick).

	Intercept	age	sexM
Sample 1	5570	12.6	-867
Sample 2	4630	31.2	-430
Sample 3	5980	2.1	-765
Sample 4	5850	5.0	-772
Sample 5	5420	16.5	-1010
Sample 6	5950	3.1	-583
Sample 7	4770	33.7	-779
Sample 8	5420	11.9	-551

Judging from these few samples, there is a lot of variability in the coefficients. For example, the age coefficient ranges from 2.1 to 33.7 in just these few samples. Figure 14.2 shows the distribution of coefficients for the age and sex coefficients for 1000 repeats of the sampling process. Each of the samples has 100 cases randomly selected from the population. These distributions, which reflect the randomness of the sampling process, are called **sampling distributions**. It's very common for sampling distributions to be bell-shaped and to be centered on the population value.

The width of the sampling distribution shows the reliability or repeatability of the coefficients. The two most common ways to summarize the width are the standard deviation and the 95% coverage interval.

An important item to add to your vocabularly is the "standard error."

> Sampling distributions, like other distributions, have standard deviations. The term **standard error** is used to refer to the standard deviation of a sampling distribution.

For the age coefficient, the standard error is about 9.5 seconds per year (keeping in mind the units of the data), and the standard error in **sexM** standard error is about 192 seconds. These standard errors were calculated by drawing 10,000 repeats of samples of size $n = 100$, and fitting the model to each of the 10000 samples. This is easy to do on the computer, but practically impossible in actual field work.

Every model coefficient has its own standard error which indicates the precision of the coefficient. The size of the standard error depends on several things:

- The quality of the data. The precision with which individual measurements of variables is made, or errors in those measurements, translates through to the standard errors on the model coefficients.
- The quality of the model. The size of standard errors is proportional to the size of the residuals, so a model that produces small residuals tends to have small standard errors. However, collinearity or multi-collinearity among the explanatory variables tends to inflate standard errors.
- The sample size n. Standard errors tend to get tighter the more data is used to fit the model. A simple relationship holds very widely; it's worth remember this rule:

> The standard error typically gets smaller as the sample size n increases, but slowly; it's proportional to $1/\sqrt{n}$. This means, for instance, that to make the standard error 10 times smaller, you need a dataset that is 100 times larger.

14.2 Standard Errors and the Regression Report

Finding the standard error of a model coefficient would be straightforward if one could follow the procedure above: repeatedly draw new samples from the population, fit the model to each new sample, and collect the resulting coefficients from each sample to produce the sampling distribution. This process is impractical, however, because of the expense of collecting new samples.

Fortunately, there can be enough information in the original sample to make a reasonable guess about the sampling distribution. How to make this guess is the subject of Section 14.6. For now, it suffices to say that the guess is based on an approximation to the sampling distribution called the **resampling distribution**.

A conventional report from software, often called a **regression report**, provides an estimate of the standard error. Here is the regression report from the model net $\sim$ 1 + age + sex fitted to a sample of size $n = 100$ from the running data:

	Estimate	Std. Error	t value	Pr(>|t|)
Intercept	5110	340.0	14.89	0.0000
age	29	9.1	3.21	0.0018
sexM	-860	188.0	-4.57	0.0000

The coefficient itself is in the column labelled "Estimate." The next column, labelled "Std. Error," gives the standard error of each coefficient. You can ignore the last two columns of the regression report for now. They present the information from the first two columns in a different format.

Example 14.1: The standard error of the mean

Even the simple model A ~ 1 involves a sampling distribution. The coefficient from this model is, of course, the sample mean of A and so the standard error of the coefficient is called the **standard error of the mean**.

To illustrate, consider the widespread process of calculating a student's "grade-point average." Where grades are given as letters, these are converted to numbers (i.e., A=4, B=3, and so on) and the numbers are averaged. Here is an example of part of the transcript, for the student whose ID is 31509.

sessionID	grade	sid	grpt	dept	level	sem	enroll	iid
C1959	B+	S31509	3.33	Q	300	FA2001	13	i323
C2213	A-	S31509	3.66	i	100	SP2002	27	i209
C2308	A	S31509	4.00	O	300	SP2002	18	i293
C2344	C+	S31509	2.33	C	100	FA2001	28	i140
C2493	S	S31509		n	300	FA2002	10	i500
C2562	A-	S31509	3.66	Q	300	FA2002	22	i327
C2585	A	S31509	4.00	O	200	FA2002	19	i310
C2737	A	S31509	4.00	q	200	SP2003	11	i364
C2764	S	S31509		j	100	SP2003	54	i447
C2851	A	S31509	4.00	O	300	SP2003	14	i308
C2928	B+	S31509	3.33	O	300	FA2003	22	i316
C3036	B+	S31509	3.33	q	300	FA2003	21	i363
C3443	A	S31509	4.00	O	200	FA2004	17	i300

(The course name and other identifying information such as the department and instructor have been coded for confidentiality.)

The student's grade-point average is just the mean of the last column: 3.60. It would be nice to know how precise this number is.

Calculating a standard error of the mean is easy. The regression report on the model `gradepoint` ~ 1 gives:

	Estimate	Std. Error	t value	Pr(>|t|)
Intercept	3.6036	0.1549	23.27	0.0000

There is actually a simple formula for the standard error of the mean:

$$\text{standard error} = \frac{\text{std. dev. of A}}{\sqrt{n}}.$$

This will give the same result as the regression report.

Formulas like this are useful when making back-of-the-envelope calculations, for instance when trying to interpret "statistics" given in newspaper articles. If you can make a reasonable guess about the standard deviation of the variable, you can often estimate the standard error from data given in the article. Journalists often give information about means and sample sizes, but rarely about standard errors. Reporter James Surowiecki writes, "As many studies have shown, people don't have an intuitive understanding of things like margins of error and random sampling; they prefer to focus on a single number, even if it's falsely precise, and so end up overemphasizing [that] number."[23]

Notice in the formula that the standard error depends on the sample size n: it is proportional to $1/\sqrt{n}$.

14.3 Confidence Intervals

The regression report gives the standard error explicitly, but it's common in many fields to report the precision of a coefficient in another way, as a **confidence interval**.

A confidence interval is a little report about a coefficient that is written like this: "15.8 ± 17.8 with 95% confidence." The report involves three components:

Point Estimate The center of the confidence interval: 15.8 here. Read this directly from the regression report.

Margin of Error The half-width of the confidence interval: 17.8 here. This is two times the standard error from the regression report.

Confidence Level The percentage of the coverage interval. This is typically 95%. Since people get tired of saying the same thing over and over again, they often omit the "with 95% confidence" part of the report. This can be dangerous, since sometimes people use confidence levels other than 95%.

The purpose of multiplying the standard error by two is to make the confidence interval approximate a 95% coverage interval of the resampling distribution. (For more information about this, see section 14.9.1.)

Example 14.2: The GPA Confidence Interval The grade-point average in the above example was 3.60 with a standard error of 0.155. This translates to a confidence interval of 3.60 ± 0.31.

Calculating the confidence interval from the regression report is very simple: you just need to remember to multiply the standard error by 2.

Example 14.3: Wage discrimination in trucking? Section 8.1.2 (page 148) looked at data from a trucking company to see how earnings differ between men

DATA FILE truckingjobs.csv

and women.It's time to revisit that example, using confidence intervals to get an idea of whether the data clearly point to the existence of a wage difference.

The model earnings ~ sex ascribed all differences between men and women to their sex itself. Here is the regression report:

	Estimate	Std. Error	t value	Pr(>|t|)
Intercept	35501	2163	16.41	0.0000
sexM	4735	2605	1.82	0.0714

The estimated difference in earnings between men and women is \$4700 ± 5200 — not at all precise.

Another model can be used to take into account the worker's experience, using age as a proxy:

	Estimate	Std. Error	t value	Pr(>|t|)
Intercept	14970	3912	3.83	0.0002
sexM	2354	2338	1.01	0.3200
age	594	99	6.02	0.0000

Earnings go up by \$600 ± 200 for each additional year of age. The model suggests that part of the difference between the earnings of men and women at this trucking company is due to their age: women tend to be younger than men. The confidence interval on the earnings difference is very broad — \$2350 ± 5700 — so broad that the sample doesn't provide much evidence for any difference at all.

One issue is whether the age dependence of earnings is just a mask for discrimination. To check out this possibility, fit another model that looks at the age dependence separately for men and women:

	Estimate	Std. Error	t value	Pr(>|t|)
Intercept	17178	8026	2.14	0.0343
sexM	-443	9174	-0.05	0.9615
age	530	225	2.35	0.0203
sexM:age	79	251	0.32	0.7530

The coefficient on the interaction term between age and sex is 79 ± 502 — no reason at all to think that it's different from zero. So, evidently, both women and men show the same increase in earnings with age.

Notice that in this last model the coefficient on sex itself has reversed sign from the previous models: Simpson's paradox. Of course the confidence interval is so broad — $-$\$443 ± 18348 — that the sex coefficient is not distinguishable from zero. This huge inflation in the width of the confidence interval is the result of the collinearity between age and sex and their interaction term. As discussed in Section 14.7, sometimes it is necessary to leave out model terms in order to get reliable results.

Example 14.4: SAT Scores and Spending, revisited Example 8.1.3 used data from 50 US states to try to see how teacher salaries and student-teacher ratios are related to test scores.The model used was sat ~ salary+ratio + frac, where frac is a covariate — what fraction of students in each state took the SAT. To interpret the results properly, it's important to know the confidence interval of the coefficients:

	Estimate	Std. Error	t value	Pr(>|t|)
Intercept	1057.9	44.3	23.86	0.0000
salary	2.5	1.0	2.54	0.0145
ratio	-4.6	2.1	-2.19	0.0339
frac	-2.9	0.3	-12.76	0.0000

The confidence interval on salary is 2.5 ± 2.0, leaving little doubt that the data support the idea that higher salaries are associated with higher test scores. Higher student-faculty ratios seem to be associated with lower test scores. You can see this because the confidence interval on ratio, -4.6 ± 4.2, doesn't cover zero. The important role of the covariate frac is also confirmed by its tight confidence interval away from zero.

Collecting more data would allow more precise estimates to be made. For instance, collecting data over 4 years would reduce the standard errors in half, although what the estimates would be with the new data cannot be known until the analysis is done.

Aside. 14.1 Confidence intervals for very small samples.

When you have data with a very small n, say $n < 20$, a multiplier of 2 can be misleading. The reason has to do with how well a small sample can be assumed to represent the population. The correct multiplier depends on the difference between the sample size n and the number of model coefficients m:

$n - m$	1	2	3	5	10	15
Multiplier for 95% confidence level	12.7	4.3	3.2	2.6	2.2	2.1

For example, if you fit the running time versus age and sex model to 4 cases, the appropriate multiplier should be 12.7, not even close to 2.

14.4 Interpreting the Confidence Interval

Take a typical confidence interval, perhaps something like "17 ± 6 with 95% confidence." Calculating the confidence interval is easy. Interpreting it is hard.

A 95% confidence interval is intended to reflect a 95% coverage interval of the sampling distribution, as approximated by the resampling distribution. So, it's tempting to say something like this, "The true coefficient will be in the range 11 to 23 with 95 percent probability." One question this raises what "true" means. Statisticians are more comfortable talking about **population parameters** — the value of the model if it could be fit to the entire population — than "truth." Those who interpret probabilities according to relative frequencies can point out that unlike the coefficients from random samples, the population parameter is not actually random, so you shouldn't talk about the probability of it being this or that.

Another tempting statement is, "If I repeated our study with a different random sample, the new result would be within ± 6 of the original result." But that statement isn't correct mathematically, unless your point estimate happens to align perfectly with the population parameter — and there's no reason to think this is the case.

Treat the confidence interval just as an indication of the precision of the measurement. If you do a study that gets a coefficient of 17 ± 6 and someone else does a study that gives 23 ± 5, then there is little reason to think that the two studies are inconsistent. On the other hand, if your study gives 17 ± 2 and the other study is 23 ± 1, then something seems to be going on.

Now return back to the first interpretation offered of the interval 17 ± 6 with 95% confidence: "The true coefficient will be in the range 11 to 23 with 95 percent probability." Taking "true" to mean "population parameter," you can get around the frequentist's objection if you consider probability from a different angle. It's not the population parameter that's random; it is your study that is a random sample from all the possible studies that could have been done. From this perspective, restate the interpretation like this:

> Of all the studies that have computed 95% confidence intervals properly, 95% of them will have captured the population parameter relevant to their study within their confidence interval.

Why use a 95% confidence level? Why not 100%? Because a 100% confidence interval would be too broad to be useful. In theory, 100% confidence intervals tend to look like $-\infty$ to ∞ — that doesn't give any information. Certainty comes at the cost of ignorance.

The 95% confidence level is standard in contemporary science; it's a convention. For that reason alone, it is a good idea to use 95% so that the people reading your work will tend to interpret things right.

14.5 Confidence in Predictions

When a model is used for making a prediction, the coefficients themselves aren't of direct interest; it's the prediction that counts. The logic of confidence intervals can be extended to prediction. The idea is to take the precision of the coefficients and propagate that through the model formula.

It's important to distinquish two kinds of prediction confidence intervals. One kind is the interval on the model value itself; this reflects the uncertainty in the coefficients as reflected by the standard errors. The second kind — the prediction interval — is the interval of the likely outcome according to the model. This outcome interval incorporates the uncertainty due to the precision of the coefficients and also due to the residuals of actual values around the model value.

DATA FILE
galton.csv

To illustrate, consider making a prediction of a child's adult height when you know the heights of the mother and father and the child's sex. Using Galton's data from the 19th century, a simple and appropriate model is height ~ sex + mother + father.

	Estimate	Std. Error	t value	Pr(>|t|)
Intercept	15.3448	2.7470	5.59	0.0000
sexM	5.2260	0.1440	36.29	0.0000
mother	0.3215	0.0313	10.28	0.0000
father	0.4060	0.0292	13.90	0.0000

Now consider a hypothetical man — call him Bill — whose mother is 67 inches tall and whose father is 69 inches tall. According to the model formula, Bill's predicted height is $15.3448 + 5.226 + 0.3215 \times 67 + 0.406 \times 69 = 70.13$ inches.

Calculating a confidence interval on the model value is more involved and usually done using software. For Bill, the 95% confidence interval on this model value is 70.13 ± 0.27: precise to about a quarter of an inch. This high precision reflects the small standard errors of the coefficients which arise in turn from the large amount of data used to fit the model.

It is wrong to interpret this interval as saying something about the actual range of heights of men like Bill, that is, men whose mother is 67 inches and father 69 inches. The model-value confidence interval should not be interpreted as saying that 95 percent of such men will have heights in the interval 70.13 ± 0.27. Instead, this interval means that if you were to fit a model based on the entire population — not just the 898 cases in Galton's data — the model you would fit would likely produce a model value for Bill close to 70.13. In other words, the model-value confidence interval is not so much about the uncertainty in Bill's height as in what the model has to say about the average for men like Bill.

If you are interested in what the model has to say about the uncertainty in Bill's height, you need to ask a different question and compute a different confidence interval. The prediction confidence interval takes into account the spread of the cases around their model values: the residuals. For the model given above, the standard deviation of the residuals is 2.15 inches — a typical person varies from the model value by that amount. This suggests that

The prediction confidence interval takes into account this case-by-case residual to give an indication of the range of heights into which the actual value is likely to fall. For men like Bill, the 95% prediction interval is 70.13 ± 4.24 inches. This is much larger than the interval on the model values, and reflects mainly the size of the residual of actual cases around their model values.

Example 14.5: Catastrophe in Grand Forks In April 1997, there was massive flooding on the Red River in Minnesota and North Dakota, states in the northern US, due to record setting winter snows. The towns of Grand Forks and East Grand Forks were endangered and the story was in the news. I remember hearing a news report saying that the dikes in Grand Forks could protect against a flood level of 50 feet and that the National Weather Service predictions were for the river to reach a maximum of 47.5 to 49 feet. To the reporter and the city planners, this was good news. The city had never been better prepared and the preparations were paying off. To me, evening knowing nothing about the area, the report was a sign of trouble. What kind of confidence interval was this 47.5 to 49? No confidence level was reported. Was it at a 50% level, was it at a 95% level? Was it a confidence interval on the model values alone or did it include the residuals from the model values? Did it take into account the extrapolation involved in handling record-setting conditions? Nothing in the news stories gave any insight into who precise previous predictions had been.

In the event, the floods reached 54.11 feet in Grand Forks, overtopping the dikes and innundating both towns. Damage was estimated to be more than $1 billion, a huge amount given the small population of the area. In the aftermath of the flood, the mayor of East Grand Forks said, understandably, "They [the National Weather Service] missed it, and they not only missed it, they blew it big." The Grand Forks city engineer lamented, "with proper advance notice we could have protected the city to almost any elevation . . . if we had known, I'm sure that we could have protected a majority of the city."

But all the necessary information was available at the time. The forecast would have been accurate if a proper prediction confidence interval had been given. It turns out that the quoted 47.5 to 49 foot interval was not a confidence interval at all — it was the range of predictions from a model under two different scenarios, with and without ongoing precipitation.

Looking back on the history of predictions from the National Weather Service, the typical residual was about 11% of the prediction. Thus, a reasonable 95% confidence interval might have been $\pm 22\%$, or, translated to feet, 48 ± 10 feet. Had this interval been presented, the towns might have been better prepared for the actual level of 54.11 feet, well within the confidence band.

Whether a 95% confidence level is appropriate for disaster planning is an open question and reflects the balance between the costs of preparation and the potential damage. If you plan using a 95% level, the upper boundary of the interval will be exceeded something like 2.5% of the time. This might be acceptable, or it might not be.

What shouldn't be controversial is that confidence intervals need to come with a clear statement of what they mean. For disaster planning, a model-value confidence interval is not so useful — it's about the quality of the model rather than the uncertainty in the actual outcome.

[Much of this example is drawn from the account by Roger Pielke [24].]

14.6 Finding the Resampling Distribution

When you fit a model to a dataset consisting of random samples from a population, the resulting coefficients will be somewhat random. The sampling distribution describes this variation.

One can in principle find the sampling distribution by doing the work of actually drawing many random samples from the population. In practice, there is little need to do this. That's fortunate, since samples are typically collected with great difficulty and expense and there is often no real possibility of collecting much more just to find the sampling distribution.

Instead, the same sample to which the model was fitted is used to construct a different but closely related distribution, called the **resampling distribution**, which approximates the important properties of the sampling distribution, particularly the standard error or the 95% coverage interval.

This section shows two methods to construct resampling distributions: resampling and simulation. These methods are rooted in simple notions of randomness. Both of the methods can and are used in practice. For linear models, it's possible to build a simple theory of standard errors and to give formulas for them. Such formulas are used by statistical software in constructing the regression report.

14.6.1 Resampling

Here's a simple idea: use the sample itself to stand for the population. The sample is already in hand, in the form of a data frame, so it's easy to draw cases of it. Such new samples, taken from your original sample, not from the population, are called **resamples**. Sampling from the sample.

Figure 14.3 illustrates how resampling works. There is just one sample drawn (with the concordant expense) from the real population. Thereafter, the sample itself is used as a stand-in for the population and new samples are drawn from the sample.

Will such resamples capture the sampling variation that expected if you were genuinely drawing new samples from the population? An objection might come to mind: If you draw n cases out of a sample consisting of n cases, the resample will look exactly like the sample itself. No variability. As you'll see, this problem is easily overcome by **sampling with replacement**.

To illustrate, consider a very small sample of size $n = 5$, although in practice one uses resampling only for samples that are much bigger, say $n > 20$.

state	net	age	sex
MD	5701	53	M
VA	5610	28	F
DC	5621	24	F
MA	5541	45	F
DC	4726	34	M

Taking this sample as exactly representative of the population, imagine that the population looks like this:

state	net	age	sex
MD	5701	53	M
MD	5701	53	M
MD	5701	53	M
and so on ...			
VA	5610	28	F
VA	5610	28	F
VA	5610	28	F
and so on ...			
DC	5621	24	F
DC	5621	24	F
DC	5621	24	F
and so on ...			
MA	5541	45	F
MA	5541	45	F
MA	5541	45	F
and so on ...			
DC	4726	34	M
DC	4726	34	M
DC	4726	34	M
and so on ...			

The simulated population looks just like the sample, but every case in the sample is repeated over and over again. Thus, the population is very large, but the cases in it look just like the sample.

The resampling process is arranged to make it seem that the sample itself is infinite in size. This is accomplished by sampling with replacement: whenever a case is drawn from the sample to put in a resample, the case is put back so that it is available to be used again. This is not something you would do when collecting the original sample; in sampling (as opposed to resampling) you don't use a case more than once.

The resamples in Figure 14.3 may seem a bit odd. They often repeat cases and omit cases. And, of course, any case in the population that was not included in the sample cannot be included in any of the resamples. Even so, the resamples do they job; the simulate the variation in model coefficients introduced by the process of random sampling.

It's important to emphasize what the resamples do not and cannot do: they don't construct the sampling distribution. The resamples merely show what the sampling distribution would look like *if the population looked like your sample.* The center of the resampling distribution from any given sample is generally not aligned exactly with the center of the sampling distribution. However, in practice, the width of the resampling distribution is a good match to the width

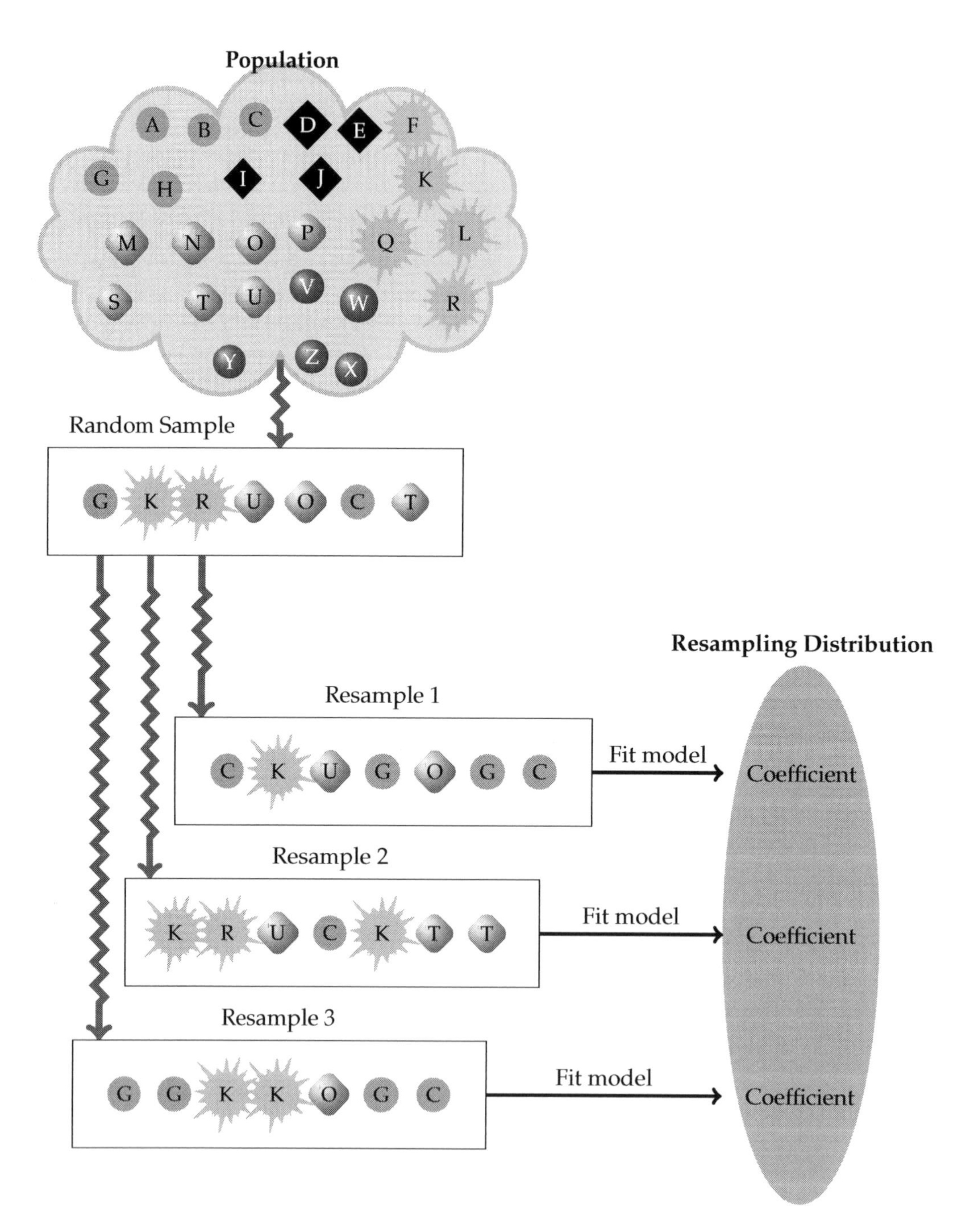

Figure 14.3: Resampling draws randomly from the sample to create new samples.

of the sampling distribution. The resampling distribution is adequate for the purpose of finding standard errors and margins of error.

Figure 14.4 shows an experiment to demonstrate this. At the top of the figure is the sampling distribution on the age coefficient in the running model. This was found by drawing 1000 genuine repeated samples of size $n = 100$ from the population. The remaining panels show resampling distributions, each panel based on a single genuine sample of size $n = 100$ from the population and then drawing 1000 resamples from that sample. The resampling distribution varies from sample to sample. Sometimes it's to the left of the sampling distribution, sometimes to the right, occasionally well aligned. Consistently, however, the resampling distributions have a standard error that is a close match to the standard error of the sampling distribution.

The process of finding confidence intervals directly via resampling is called **bootstrapping**.

14.6.2 Randomizing the Residuals

Models partition variation in the response variable into two parts. In the past, these parts have been called the "explained" and "unexplained," or the "model values" and the "residuals," or the "modeled" or "unmodeled" parts. For this discussion, consider them the *deterministic* and the *random* parts.

The notation used has focussed on the deterministic part of the model. So, a model formula has been written like

$$\underset{\text{Values}}{\text{Model}}\ \text{net} = 5540 + 16.9\ \text{age} - 727\ \text{sexM}.$$

The above formula doesn't include the residuals. Of course, the residuals are defined as the difference between the model values and the actual values of the response variable. So the response values can be written

$$\text{net} = 5540 + 16.9\ \text{age} - 727\ \text{sexM} + \mathcal{E}$$

where $\mathcal{E}$ is the vector of random numbers, one random number for each case.

This process is illustrated in Figure 14.5, which depicts the situation when the response variable A is deterministically set by another variable B, a population coefficient which can be writen c_B, and a random component $\mathcal{E}$ as

$$A = Bc_B + \mathcal{E}.$$

Conventionally, the random component is called the **error**. The word comes from the Latin *errare*, meaning "to wander" or "to stray." After following the determinist path to Bc_B, the system strays off the path to A.

Each case comes with its own random component, so one way to think about random sampling is it is in effect picking a random set of numbers to add to the cases in the sample. If a different sample had been picked, the random numbers would have been different.

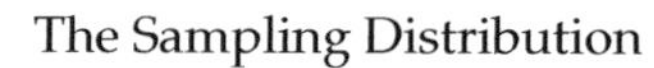

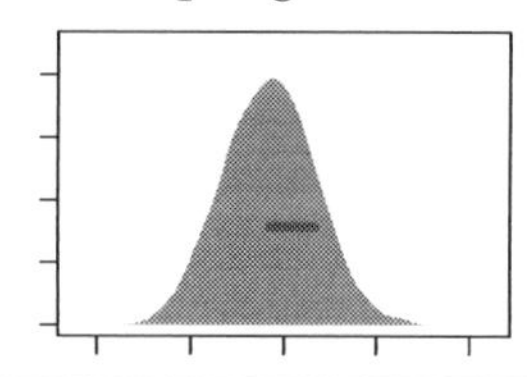

Figure 14.4: Top: Sampling distribution for the coefficient on age for samples of size $n = 100$. The standard error is shown as bar. Below: Several resampling distributions, each from one sample of size $n = 100$. The position of the resampling distributions doesn't always match the sampling distribution, but the standard error is often a good match.

Fitting a model to a sample tries to infer the value of the coefficient from measurements of A and B by projecting A onto B. If the random component $\mathcal{E}$ had been perpendicular to B, this projection would give exactly the right answer. But $\mathcal{E}$ is random, so it won't necessarily be perpendicular to B. (However, if the sample size n is large, $\mathcal{E}$ will almost certainly be clase to perpendicular to B.) Because of the randomness, the projection $A_{\parallel B}$ won't be a perfect match to the deterministic mechanism. That's why the sample doesn't generally give the same coefficients that would be found from the entire population.

To create a resampling distribution take the fitted model values $A_{\parallel B}$ as the deterministic part of the mechanism and add in random values.

But how big should those random values be? A clue is given by the size of the residuals estimated when the model was fit to the data: make the simulated $\mathcal{E}$ just like the residuals.

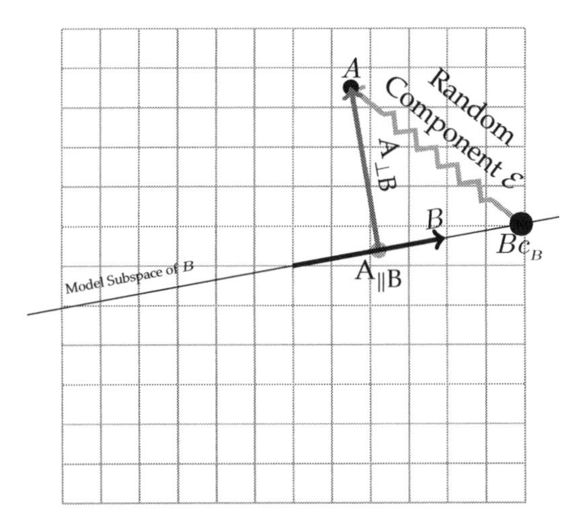

Figure 14.5: The response variable A can be thought of as generated by a deterministic component Bc_B and a random component $\mathcal{E}$, shown as a squiggly line. You can't observe the random component directly. Instead, you find $A_{\perp B}$ and imagine that this approximates $\mathcal{E}$.

Aside. 14.2 Estimating the length of $\mathcal{E}$.

It might seem natural to take the length of $\mathcal{E}$ as the length of the residual vector. But this will systematically understate $\mathcal{E}$. The reason is that the residual vector isn't entirely random: it has no component at all in the directions of the model subspace. How could it? The residual is, by definition, perpendicular to the model subspace. Thus the length of the residual vector is entirely carred in the part of variable space that is perpendicular to the model subspace.
In order to produce a correct length for $\mathcal{E}$, it suffices to multiply the square length of the residual vector by a factor $n/(n-m)$, where where n is the sample size and m is the number of parameters. This makes up for the fact that the residual vector lives in a subspace of dimension $n-m$.

Figure 14.6 shows the process of generating the simulated data, which is denoted A_{sim}. Fitting the model to each of these simulated data sets amounts to projecting A_{sim} onto the model vector B and finding the corresponding coefficient c_{sim}, the multiplier on B to reach the projected point. The collection of these c_{sim} is the resampling distribution.

There is something about Figure 14.6 that is misleading. In that figure, the simulated random components $\mathcal{E}$ point in all different directions. They are just as likely to be aligned with B as to be perpendicular to it. Recall from Chapter 13 that this is wrong. For samples of size $n > 2$, the $\mathcal{E}$ are much more likely to be perpendicular to B than aligned with it.

The probability that random vectors will be almost perpendicular is bigger when the dimension of the space n is bigger. Figure 14.7 attempts to depict this. Since the drawing is restricted to two-dimensional paper, the random vectors look like they are artificially restricted to a fan-shaped area. If you could look at

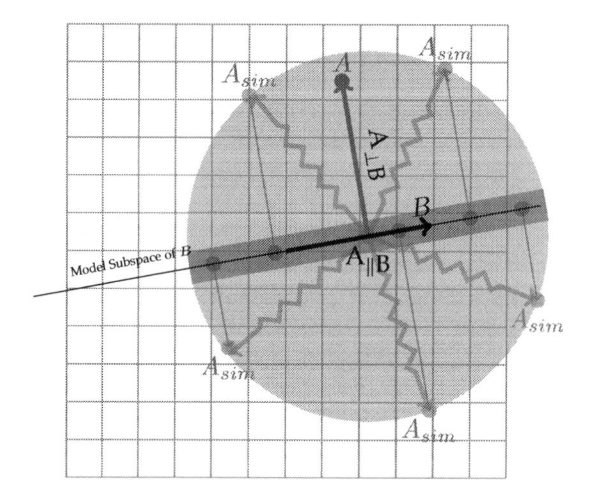

Figure 14.6: Using the residual $A_{\perp B}$ as representing the unknown $\mathcal{E}$, and the fitted model values $A_{||B}$ as representing the true mechanism Bc_B, you can simulate new random draws A_{sim} from the population. The distribution of such draws sets the confidence interval on the estimate of c_B, indicated with the bar running along B which covers a range of coefficients from about -0.6 to 2.0.

things in the full-dimensional space, you wouldn't see any artificial restriction, just that most of the vectors are close to being perpendicular to B.

Projecting the fans in Figure 14.7 onto the model vector B gives the resampling distribution of the model coefficients. The trigonometry of projection is such that the width of the resampling distribution is proportional both to the length of the residual vector and to the spread in the cosine of random angles. As discussed in Chapter 13, for random vectors in n-dimensional space, the mean value of $\cos(\alpha)$ is zero and the standard deviation is $1/\sqrt{n}$.

This approach of resampling by randomizing the residuals applies in the same way to models with more than one explanatory model vector. The projected spread is still proportional both to the length of the residual vector and to the spread in the cosines of the random angles. What's different is the way the coefficients match up with the model vectors.

Figure 14.8 shows how the position of A_{sim} is translated into a coefficient on an explanatory vector B. Each of the contours in the figure shows the set of positions that will project down onto the same place on the subspace due to B. The spacing between the contours indicates how uncertainty in A_{sim} corresponds to uncertainty in the model coefficient; the wider the spacing, the less the uncertainty.

When there is just one explanatory vector, the spacing between the projection contours is set by the length of that vector. But when there are two or more explanatory vectors, the spacing between contours gets smaller if there is any collinearity. This is seen in the right side of Figure 14.8 where the spacing between the projection contours is the distance from the head of B to the subspace

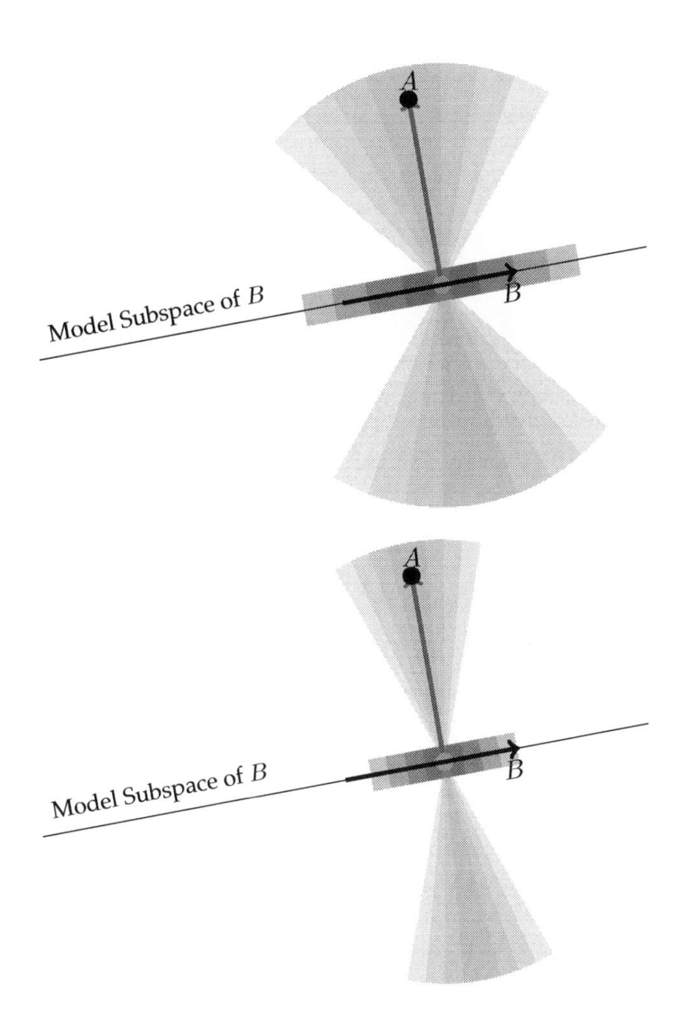

Figure 14.7: For $n > 2$, the distribution of random angles is not uniform but is centered on $90°$. The larger is the number of cases n, the closer the angles tend to be to $90°$. The bottom diagram shows how the interval is narrowed for 4 times as much data as in the top figure.

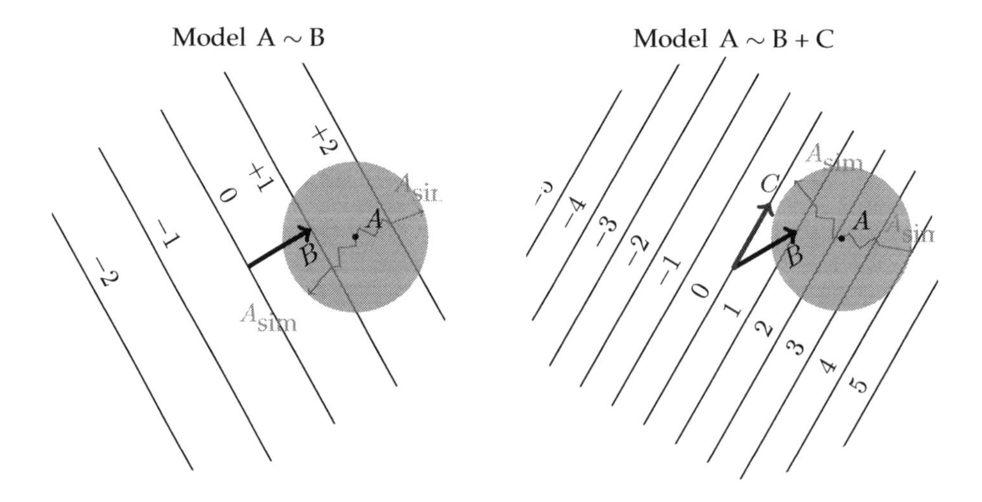

Figure 14.8: Confidence intervals on the B coefficient for two models. The shaded circle shows the range covered by simulated values of the residuals A_{sim}. The contour lines show the values of the coefficient c_B for any possible value of A_{sim}. The confidence interval can be read off as the range of contours covered by the shaded circle. For the model A ∼ B+C that spacing depends on the angle between B and C.

due to C. When the collinearity is very strong, the spacing is very tight and therefore uncertainty in A_{sim} translates into very large uncertainty in the coefficient on B. (Denoting as θ the angle between B and the subspace spanned by the other explanatory vectors, the spacing between contours is proportional to $1/\sin(\theta)$.)

14.6.3 A Formula for the Standard Error

When you do lots of simulations, you start to see patterns. Sometimes these patterns are strong enough that you can anticipate the result of a simulation without doing it. For example, suppose you generate two random vectors A and B and calculate their correlation coefficient: the cosine of the angle between them. You will see some consistent patterns. The distribution of correlation coefficients will have zero mean and it will have standard deviation of $1/\sqrt{n}$.

Traditionally, statisticians have looked for such patterns in order save computational work. Why do a simulation, repeating something over and over again, if you can anticipate the result with a formula?

Such formulas exist for the standard error of a model coefficient. They are implemented in statistical software and used to generate the regression report. Here is a formula for computing the standard error of the coefficient on a model vector B in a model with more than one explanatory vector, e.g., A ∼ B + C + D. You don't need to use this formula to calculate standard errors; the software will do that. The point of giving the formula is to show how the standard error depends on features of the model, so that you can strategize about how to design your models to give suitable standard errors.

$$\text{standard error of B coef.} = \| \text{ residuals } \| \frac{1}{\| \text{ B } \|} \frac{1}{\sin(\theta)} \frac{1}{\sqrt{n}} \sqrt{\frac{n}{n-m}}.$$

There are five components to this formula, each of which says something about the standard error of a coefficient:

$\|$ **residuals** $\|$ The standard error is directly proportional to the size of the residuals.

$1/\|$ **B** $\|$ The length of the model vector reflects the units of that variable or, for the indicator vector for a categorical variable, how many cases are at the corresponding level.

$1/\sin(\theta)$ This is the magnifying effect of collinearity. θ is the angle between vector B and all the other explanatory vectors. More precisely, it is the angle between B and the vector that would be found by fitting B to all the other explanatory vectors. When θ is 90°, then $\sin(\theta) = 1$ and there is no collinearity and no magnification. When $\theta = 45^\circ$, the magnification is only 1.4. But for strong collinearity, the magnification can be very large: when $\theta = 10^\circ$, the magnification is 5.8. When $\theta = 1^\circ$, magnification is 57. For $\theta = 0$, the alignment becomes a matter of redundancy, since B can be exactly modeled by the other explanatory variables. Magnification would be infinite, but statistical software will identify the redundancy and eliminate it.

$1/\sqrt{n}$ This reflects the amount of data. Larger samples give more precise estimates of coefficients, that is, a smaller standard error. Because of the square-root relationship, quadrupling the size of the sample will half the size of the standard error. To improve the standard error by ten-fold requires a 100-fold increase in the size of the sample.

$\sqrt{n/(n-m)}$ This is a bias correction factor. (See Aside 14.2.) It reflects how the length of the residual vector is biased to underestimate the length of the random sampling component $\mathcal{E}$. n is, as always, the number of cases in the sample and m is the number of model vectors. So long as m is much less than n, which is typical of models, the bias correction factor is close to 1.

The formula suggests some ways to build models that will have coefficients with small standard errors.

- Use lots of data: large n.
- Include model terms that will account for lots of variation in the response and produce small residuals.
- But avoid collinearity.

The next section emphasizes the problems that can be introduced by collinearity.

14.7 Confidence and Collinearity

One of the key decisions that a modeler makes is whether to include an explanatory term; it might be a main term or it might be an interaction term or a transformation term.

The starting point for most people new to statistics might be summarized as follows:

Methodologically Ignorant Approach. Include just one explanatory variable.

Most people who work with data can understand tabulations of group means (in modeling language, A ~ 1 + G, where G is a categorical variable) or simple straight-line models (A ~ 1 + B, where B is a quantitative variable.)

At this point, the reader of this book should understand how to include in a model multiple explanatory variables and terms such as interactions. When people first learn this, they adopt a different approach, one that might be called the

Greedy Approach. Include everything, just in case.

The idea here is that you don't want to miss any detail; the fitting technology will untangle everything.

Perhaps the word "greedy" is already signalling the problem in a metaphorical way. The poet Horace (65BC-8BC) wrote, "He who is greedy is always in want." Or Seneca (1st century AD), "To greed, all nature is insufficient."

The problem with the greedy approach to statistical modeling isn't moral. If adding a term improves your model, go for it! The problem is that, sometimes, adding a term can hurt the model by dramatically inflating standard errors.

To illustrate, consider again the example from Chapter 11: modeling wages using the Current Population Survey dataset.

A relatively simple model might take just the worker's education into account: wage ~ 1 + educ:

	Estimate	Std. Error	t value	Pr(>|t|)
Intercept	-0.746	1.045	-0.71	0.4758
educ	0.750	0.079	9.53	0.0000

The coefficient on educ says that a one-year increase in education is associated with a wage that is higher by 75 cents/hour. (Remember, these data are from the mid-1980s.) The standard error is 8 cents/hour per year of education, so the confidence interval is 75 ± 16 cents/hour per year of education.

The coefficient from this model might be misleading because the availability of education has expanded over the decades. Due to this, younger workers typically have higher levels of education than older workers. A simple adjustment will take that into account: add age as an explanatory variable. wage ~ 1 + educ + age:

	Estimate	Std. Error	t value	Pr(>|t|)
Intercept	-5.534	1.279	-4.33	0.0000
educ	0.821	0.077	10.66	0.0000
age	0.105	0.017	6.11	0.0000

This has increased the coefficient on educ slightly to 82 cents/hour per year of education while leaving the standard error more or less the same.

Of course, wages also depend on experience. Every year of additional education tends to cut out a year of experience. Here is the regression resport from the model wage $\sim$ 1 + educ + age + exper:

	Estimate	Std. Error	t value	Pr(>|t|)
Intercept	-4.770	7.042	-0.68	0.4985
educ	0.948	1.155	0.82	0.4121
age	-0.022	1.155	-0.02	0.9845
exper	0.128	1.156	0.11	0.9122

The standard errors have exploded! The confidence interval on educ is now 95 ± 231 cents/hour per year of education — a very imprecise estimate. Indeed, the confidence interval now includes zero, so the model is consistent with the skeptic's claim that education has no effect on wages, once you hold constant the level of experience and age of the worker.

A better interpretation is that greed has gotten the better of the modeler. The large standard errors in this model reflect a deficiency in the data: the multi-collinearity of exper with age and educ as described in Example 6.3.

A simple statement of the situation is this:

> Collinearity and multi-collinearity cause model coefficients to become less precise — they increase standard errors.

This doesn't mean that you should never include model terms that are collinear or multi-collinear. If the precision of your estimates is good enough, even with the collinear terms, then there is no problem. You, the modeler, can judge if the precision is adequate for your purposes. For instance, the age-adjusted coefficient on educ is 82 ± 15 cents/hour per year of education. Suppose that your purpose was to compare the effect of education in these workers with a group of workers who went through a different sort of educational system. If that different group had an educ coefficient of 69 ± 18, you would be hard pressed to say you had found a difference; the precision of your estimates would be inadequate. But if your purpose were to show that education is indeed correlated with wage, even when holding age constant, then 82 ± 15 cents/hour per year of education is a perfectly adequate level of precision.

The estimate 95 ± 231 cents/hour is probably not precise enough for any reasonable purpose. Including the exper term in the model has made the entire model useless, not just the exper term, but the other terms as well! That's what happens when you are greedy.

Aside. 14.3 Redundancy and Multi-collinearity

Redundancy can be seen as an extreme case of multi-collinearity. So extreme that it renders the coefficients completely meaningless. That's why statistical software is written to delete redundant vectors. With multi-collinearity, things are more of a judgment call. The judgment needs to reside in you, the modeler.

The good news is that you don't need to be afraid of trying a model term. Use the standard errors to judge when your model is acceptable and when you need to think about dropping terms due to collinearity.

14.8 Confidence and Bias

The margin of error tells you about the **precision** of a coefficient: how much variation is to be anticipated due to the random nature of sampling. It does not, unfortunately, tell you about the **accuracy** of the coefficient; how close it is to the "true" or "population" value.

A coefficient makes sense only within the context of a model. Suppose you fit a model $A \sim B + C$. Then you think better of things and look at the model $A \sim B + C + D$. Unless variable D happens to be orthogonal to B and C, the coefficients on B and C are going to change from the first model to the second.

If you take the point of view that the second model is "correct," then the coefficients you get from the first model are wrong. That is, the omission of D in the first model has **biased** the estimates of B and C.

On the other hand, if you think that the model $A \sim B + C$ reflects the "truth," then the coefficients on B and C from $A \sim B + C + D$ are biased.

The margin of error from any model tells you nothing about the potential bias. How could it? In order to measure the bias you would have to know the "correct" or "true" model.

Avoiding bias is a difficult matter and it's hard to know when you have succeeded. Care in sampling is important. Adopting formal random sampling techniques can help — "formal" meaning "really random" rather than "just giving it a try." When the sample is collected in a genuinely random way, you will make unmeasured covariates approximately orthogonal to the variables you can measure.

If your sampling procedure is not demonstrably random, you should suspect that your coefficients are biased even though you may not know how. But even if the sample is random, you still have to decide which of the measured variables to include in a model. Ultimately the issue of bias comes down to which is the correct set of model terms to include. This is not a statistical issue — it depends on how well your model corresponds to reality. Since people disagree about reality, perhaps it would be best to think about bias with respect to a given theory of how the world works. A model that produces unbiased coefficients according to your theory of reality might give biased coefficients with respect to someone else's theory.

14.9 Computational Technique

Regression reports are generated using software you have already encountered: `lm` to fit a model and `summary` to construct the report from the fitted model. To illustrate:

```
> swim = ISMdata("swim100m.csv")
> mod = lm( time ~ year + sex, data=swim)
> summary(mod)
Coefficients:
            Estimate Std. Error t value Pr(>|t|)
(Intercept) 555.7168    33.7999   16.44  < 2e-16
year          -0.2515     0.0173  -14.52  < 2e-16
sexM          -9.7980     1.0129   -9.67  8.8e-14

Residual standard error:
  3.98 on 59 degrees of freedom
Multiple R-Squared: 0.844,
  Adjusted R-squared: 0.839
F-statistic:  160 on 2 and 59 DF,
  p-value: <2e-16
```

14.9.1 Confidence Intervals from Standard Errors

Given the coefficient estimate and the standard error from the regression report, the confidence interval is easily generated. For a 95% confidence interval, you just multiply the standard error by 2 to get the margin of error. For example, in the above, the margin of error on **sexM** is $2 \times 1.013 = 2.03$, or, in computer notation:

```
> 2 * 1.0129
[1] 2.026
```

If you want the two endpoints of the confidence interval, rather than just the margin of error, do these simple calculations: (1) subtract the margin of error from the estimate; (2) add the margin of error to the estimate. So,

```
> -9.798 - 2*1.0129
[1] -11.82
> -9.798 + 2*1.0129
[1] -7.772
```

The key thing is to remember the multiplier that is applied to the standard error. A multiplier of 2 is for a 95% confidence level.

Technical Note: The $t^\star$ multiplier

It's very standard to use a confidence level of 95% in constructing a confidence interval. You may never need to use anything else. Still, here is how to find the multiplier sanctioned by conventional statistical practice, often called $t^{\star}$.

1. Suppose your desired confidence level is called α, say $\alpha = 0.90$. Then compute the "tail probability" $(1-\alpha)/2$. For $\alpha = 0.90$, this is $0.1/2 = 0.05$.

2. Look at the regression report to find the "degrees of freedom" of the residual. In the above model of swim times, this is reported as 59. You can, if you want, calculate it yourself by subtracting the number of coefficients in the report from the number of cases. For the swim data, there are $n = 62$ cases and $m = 3$ coefficients in the model, so the calculation is $62 - 3 = 59$.

3. The appropriate multipler will be the quantile from the t-distribution with the number of degrees of freedom from (2) and the tail probability from (1). For this example:

```
> qt( 0.05, df=59)
[1] -1.671
```

Following this algorithm, the 90-percent confidence interval on the `sexM` coefficient will be

```
> -9.7980 - 1.671*1.0129
[1] -11.49
> -9.7980 + 1.671*1.0129
[1] -8.105
```

This algorithm is consistent with the multiplier 2 for a 95% confidence interval when the degrees of freedom is larger than, say, 50:

```
> qt( 0.025, df=50)
[1] -2.009
```

There is a convenient operator, `confint`, that does the tedious part of the calculation for you, calculating the correct multiplier and applying it to the standard error to produce the confidence intervals.

```
> mod = lm( time ~ year + sex, data=swim)
> confint(mod)
                2.5 %   97.5 %
(Intercept) 488.0833 623.3503
year         -0.2861  -0.2168
sexM        -11.8247  -7.7712
```

Traditionally, introductory statistics courses have spent a lot of time and effort on the t-distribution and, perhaps for this reason, expected students to do the calculation by hand, generally using a printed table instead of software such as the `pt` operator.

14.9.2 Repeating Trials

You have already seen the `resample` operator which, when combined with `lm` and other software, makes it easy to see how coefficients vary due to sampling variability. For example, here are models fit to the `swim` data and to resampled versions of it.

```
> lm( time~year+sex, data=swim)
(Intercept)        year        sexM
   555.7168     -0.2515     -9.7980

> lm( time~year+sex, data=resample(swim))
(Intercept)        year        sexM
   661.2486     -0.3048    -12.2551

> lm( time~year+sex, data=resample(swim))
(Intercept)        year        sexM
   597.0033     -0.2727     -9.7788

> lm( time~year+sex, data=resample(swim))
(Intercept)        year        sexM
    542.774      -0.245      -9.757
```

Even from this small set of trials, you can get a rough idea of the amount of sampling variability in the coefficients.

It's helpful to be able to automate this process of generating trials. The `do` operator does this. To show how it works, this section gives some simple examples that don't use model fitting. The next section shows how to use repeated resampling to generate the resampling distribution of model coefficients.

As an example, consider playing a board game in which you roll two dice and add up the results to determine your next move. The `resample` operator can be used to generate a single move — in statistical language a single **trial** in which two dice are rolled and the results summed up.

```
> die = c(1,2,3,4,5,6)
> sum( resample( die, 2 ) )
[1] 5
```

In order to perform more trials, you could give the same command over and over:

```
> sum(resample(die,2))
[1] 9
> sum(resample(die,2))
[1] 7
> sum(resample(die,2))
[1] 6
> sum(resample(die,2))
[1] 8
```

The do operator automates this process. It takes a single arguments: the number of times to repeat a statement. Just put it before the statement to be repeated with a multiplication sign.

```
> do(25) * sum(resample(die,2))
 [1] 12  6 12 11  9 10  8  7  3 12  3  4 11 11  6
[16]  6  11 3  7  8  5  9  6  2  5
```

You can read it like this "do 25 times ..."

The do operator will repeat the statement over and over again, n times. It collects the results into a vector. It is just as if you typed in the statement yourself, over and over, and then collected the results yourself.

The do part of the command must always come first, before the statement to be repeated.

Usually, you will want to do some further calculation on the results, so it is worthwhile to save them into an object. Here I'll do 1000 trials, each of which involves summing the score from two dice. Then I'll count up how often each outcome occurs:

```
> samps = do(1000) * sum(resample(die,2))
> table(samps)
samps
  2   3   4   5   6   7   8   9  10  11  12
 26  50  84 112 133 168 134 115  81  69  28
```

Evidently 7 is the most likely total score from rolling two dice.

In constructing a command involving do, you may find it worthwhile first to construct and test the statement that's to be repeated. When you have this working, then use the command editor to preceed the statement with do(25)*. Of course, you might want to use some other value for the number of trials than $n = 25$.

Occasionally, you might need to include more than one statement to be repeated by do. One way to do this is to group the statements into a function. For example:

```
f = function() {
  s = resample( die, 2 )
  sum(s)
}
```

Then hand the name of the function to do as the first argument:

```
> do(10) * f()
 [1]  8 12  7  8  8  3  7  8  6  9
```

For convenience, a number of related operators have been defined in terms of do: five, ten, dozen, hundred, thousand. For instance

```
> five*f()
[1] 4 9 7 7 9
> dozen*f()
 [1]  5  7  3  6  3  9  8  9  9  6  7 11
```

14.9.3 Bootstrapping Confidence Intervals

The `do` operator can be used along with `resample` to explore the sampling variability in model coefficients (or any statistic computed from data). The generation of the resampling distribution is called **bootstrapping**.

From a programming point of view, the only difficulty is making sure to make clear to the computer that you want to look at the coefficients and not at other descriptions of the model such as the R^2 or sum of squares of the residuals or whatever.

To look at the model coefficients, just use the `coef` operator applied to the output of `lm`. To illustrate, I'll show an evolution of statements leading up to the overall bootstrapping calculation.

- First, a statement to fit the model and extract the coefficients:

```
> coef( lm( time~year+sex, data=swim))
(Intercept)          year          sexM
   555.7168       -0.2515       -9.7980
```

- Next, resampling the data frame to give a single trial that reflects the variability introduced by random sampling:

```
> coef( lm( time~year+sex, data=resample(swim)))
(Intercept)          year          sexM
   520.4456       -0.2336       -9.3447
```

- Constructing many trials and collecting the results

```
> samps = five*
  coef( lm( time~year+sex, data=resample(swim)))

> samps
  (Intercept)     year   sexM
1       562.8 -0.2550 -10.35
2       612.2 -0.2802 -10.10
3       487.6 -0.2172  -8.81
4       635.5 -0.2916 -11.61
5       612.2 -0.2791 -12.05
```

  Here I generated just five trials so that it's easy to look at the results and confirm that the output is as expected. Each row is one trial and it's clear that the coefficients are varying a bit from trial to trial.

- Conduct many trials in order to find the standard deviation of the resampling distribution, in other words, the "standard error." For the purposes of finding a standard error, 500 trials is enough to give reliable results:

```
> samps = do(500)*
    coef( lm( time~year+sex, data=resample(swim)))
> sd(samps)
(Intercept)        year        sexM
   45.59395     0.02303     0.95315
```

- From the standard errors, you can find a confidence interval in the usual way: multiplying by 2 for a 95% confidence interval. For instance, the 95% confidence interval on **sexM** is $-9.798 \pm 2 \times 0.95315$ or, -11.7 to 7.9.

14.9.4 Prediction Confidence Intervals

When a model is used to make a prediction, it's helpful to be able to describe how precise the prediction is. To illustrate, return to an example used in Section 8.4.2 where model values were calculated based on new inputs. In that example, a model of foot width was constructed and the model value constructed for new inputs stored in a file called `nancy.csv`.

The sequence of commands was this:

```
# Construct the model
feet = ISMdata("kidsfeet.csv")
mod = lm( width ~ length + sex, data=feet)
# Apply the model to new inputs
nancy = read.csv("nancy.csv")
predict(mod, newdata=nancy)
[1] 9.1553
```

In order to generate a confidence interval, the `predict` operator needs to be told what type of interval is wanted. There are two types of prediction confidence intervals:

Interval on the model value which reflects the sampling distributions of the coefficients themselves. To calculate this, use the `interval="confidence"` named argument:

```
> predict( mod, newdata=nancy, interval="confidence")
        fit   lwr   upr
[1,] 9.155 8.909 9.402
```

The components names `lwr` and `upr` are the lower and upper limits of the confidence interval, respectively.

Interval on the prediction which includes the variation due to the uncertainty in the coefficients as well as the size of a typical residual. To find this interval, use the `interval="prediction"` named argument:

```
> predict( mod, newdata=nancy,
    interval="prediction")
```

```
        fit   lwr   upr
[1,] 9.155 8.337 9.974
```

The prediction interval is larger than the model-value confidence interval because the residual always gives additional uncertainty around the model value.

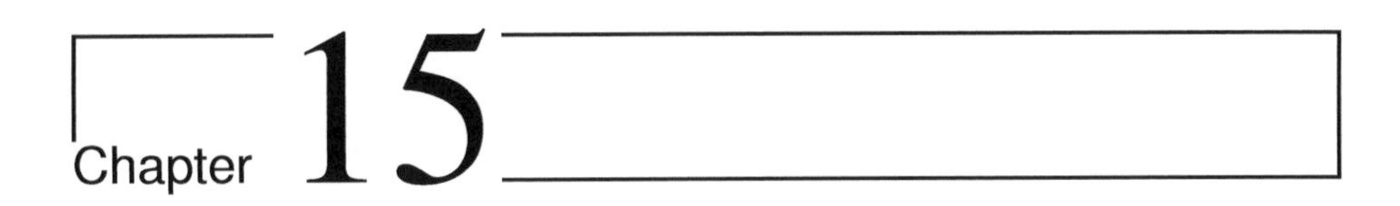

The Logic of Hypothesis Testing

Extraordinary claims demand extraordinary evidence. — Carl Sagan

A **hypothesis test** is a standard format for assessing statistical evidence. It is ubiquitous in scientific literature, most often appearing in the form of statements of **statistical significance** and notations like "$p < 0.01$" that pepper scientific journals.

Hypothesis testing involves a substantial technical vocabulary: null hypotheses, alternative hypotheses, test statistics, significance, power, p-values, and so on. The last section of this chapter lists the terms and gives definitions.

The technical aspects of hypothesis testing arise because it is a highly formal and quite artificial way of reasoning. This isn't a criticism. Hypothesis testing is this way because the "natural" forms of reasoning are inappropriate. To illustrate why, consider an example.

The stock market's ups and downs are reported each working day. Some people make money by investing in the market, some people lose. Is there reason to believe that there is a trend in the market that goes beyond the random-seeming daily ups and downs?

Figure 15.1 shows the closing price of the Dow Jones Industrial Average stock index for a period of about 10 years. It's evident that the price is going up and down in an irregular way, like a random walk. But it's also true that the price at the end of the period is much higher than the price at the start of the period.

Is there a trend or is this just a random walk? It's undeniable that there are fluctuations that look something like a random-walk, but is there a trend buried under the fluctuations?

As phrased, the question contrasts two different possible hypotheses. The first is that the market is a pure random walk. The second is that the market has a systematic trend in addition to the random walk.

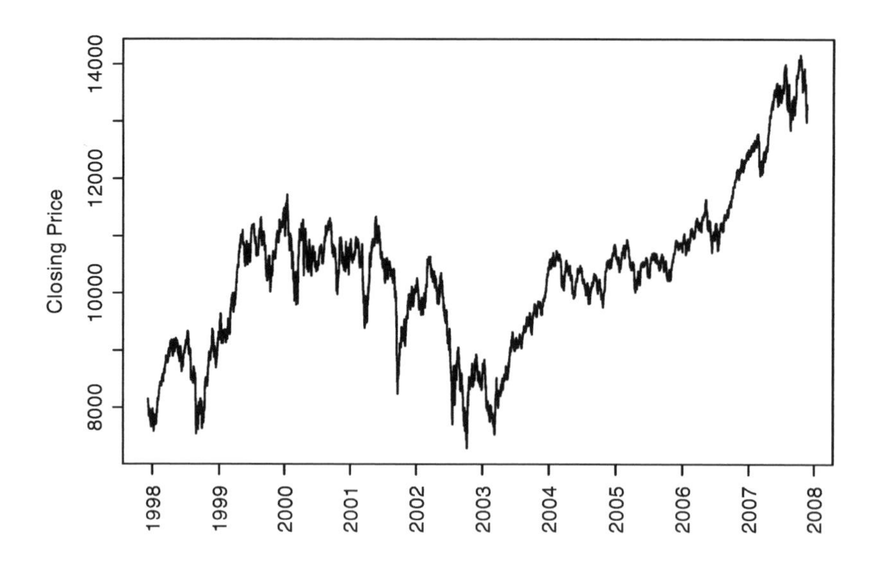

Figure 15.1: The closing price of the DJIA each day over 2500 trading days — a roughly 10 year period from the close on Dec. 5, 1997 to the close on Nov. 14, 2007.

The natural question to ask is this: Which hypothesis is right?

Each of the hypotheses is actually a model: a representation of the world for a particular purpose. But each of the models is an incomplete representation of the world, so each is wrong.

It's tempting to rephrase the question slightly to avoid the simplistic idea of right versus wrong models: Which hypothesis is a better approximation to the real world? That's a nice question, but how to answer it in practice? To say how each hypothesis differs from the real world, you need to know already what the real world is like: Is there a trend in stock prices or not? That approach won't take you anywhere.

Another idea: Which hypothesis gives a better match to the data? This seems a simple matter: fit each of the models to the data and see which one gives the better fit. But recall that even junk model terms can lead to smaller residuals. In the case of the stock market data, it happens that the model that includes a trend will almost always give smaller residuals than the pure random walk model, even if the data really do come from a pure random walk.

The logic of hypothesis testing avoids these problems. The basic idea is to avoid having to reason about the real world by setting up a hypothetical world that is completely understood. The observed patterns of the data are then compared to what would be generated in the hypothetical world. If they don't match, then there is reason to doubt that the data support the hypothesis.

15.1 An Example of a Hypothesis Test

To illustrate the basic structure of a hypothesis test, here is one using the stock-market data.

The **test statistic** is a number that is calculated from the data and summarizes the observed patterns of the data. A test statistic might be a model coefficient or an R^2 value or something else. For the stock market data, it's sensible to use as the test statistic the start-to-end dollar difference[1] in prices over the 2500-day period. The observed value of this test statistic is \$5074.80 — the DJIA stocks went up by this amount over the 10-year period.

This test statistic can be used to test the hypothesis that the stock market is a random walk. (The reasons to choose the random walk hypothesis instead of the trend hypothesis will be discussed later.)

In a random-walk world the start-to-end price difference would be random. As described in Section 13.3 the price difference is a random variable with a mean of 0 and standard deviation of $s\sqrt{n}$, where s is the typical daily fluctuation. Since the data cover 2500 days, it's safe to set $n = 2500$. But what should the parameter s be? It was not specified by the hypothesis. Such an unknown parameter is called a **nuisance parameter**. You need to know it in order to say what would happen in the hypothetical world, but the hypothesis doesn't state it explicitly.

If the hypothesis were true, you would have a way to find s: measure it from your data. This can be done by taking the standard deviation of day-to-day price changes. For the stock market data, it's \$106.70. That is, the DJIA typically went up or down by about 100 dollars per day for the 10-year period covered by the data.

This gives a complete description of what the test statistic should look like in the hypothesized world: the start-to-end price difference in dollars will be a normal distribution with mean 0 and standard deviation of $106.70 \times \sqrt{2500} = 5335$ dollars. This distribution is drawn in Figure 15.2. Also shown is a tick mark at the observed value of the test statistic, \$5074.80. It seems obvious from the figure that the observed value is quite plausible as an outcome in the world of the hypothesis. In other words, the hypothesis is consistent with the data.

It's tempting to use this result to say, perhaps, "the observations support the hypothesis." In actuality, however, the permitted conclusion is stiff and unnatural:

> *We fail to reject the hypothesis.*

15.2 Inductive and Deductive Reasoning

Hypothesis testing involves a combination of two different styles of reasoning: deduction and induction. In the deductive part, the hypothesis tester makes an

[1] Another way to describe the change in stock prices is by the *proportional increase*, which is 62.3% for the DJIA over the period in Figure 15.1: a rate of 5% per year when compounded. Economists usually prefer to study the proportional change rather than the dollar change.

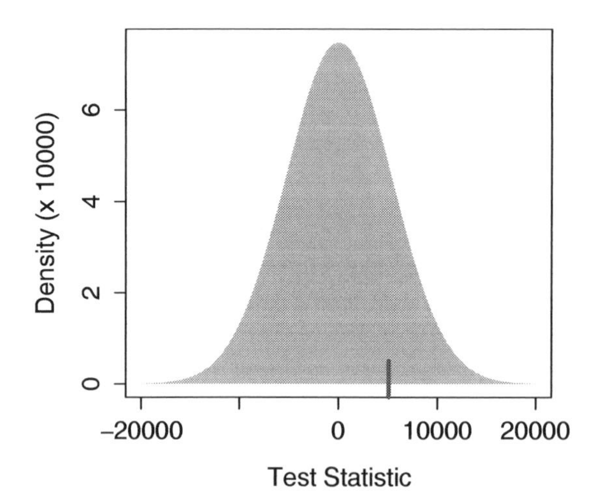

Figure 15.2: The distribution of start-to-end differences in stock price under the hypothesis that day-to-day changes in price are random with no trend. The value observed in the data, $5074.80, is marked with a tick.

assumption about how the world works and draws out, deductively, the consequences of this assumption: what the observed value of the test statistic should be if the hypothesis is true. For instance, the hypothesis that stock prices are a random walk was translated into a statement of the probability distribution of the start-to-end price difference.

In the inductive part of a hypothesis test, the tester compares the actual observations to the deduced consequences of the assumptions and decides whether the observations are consistent with them.

15.2.1 Deductive Reasoning

Deductive reasoning involves a series of rules that bring you from given assumptions to the consequences of those assumptions. For example, here is a form of deductive reasoning called a **syllogism**:

Assumption 1 No healthy food is fattening.

Assumption 2 All cakes are fattening.

Conclusion No cakes are healthy.

The actual assumptions involved here are questionable, but the pattern of logic is correct. If the assumptions were right, the conclusion would be right also.

Deductive reasoning is the dominant form in mathematics. It is at the core of mathematical proofs and lies behind the sorts of manipulations used in algebra. For example, the equation $3x + 2 = 8$ is a kind of assumption. Another assumption, known to be true for numbers, is that subtracting the same amount from both sides of an equation preserves the equality. So you can subtract 2 from both sides to get $3x = 6$. The deductive process continues — divide both sides by 3 — to get a new statement, $x = 2$, that is a logical consequence of the initial assumption. Of course, if the assumption $3x + 2 = 8$ was wrong, then the conclusion $x = 2$ would be wrong too.

The **contrapositive** is a way of recasting an assumption in a new form that will be true so long as the original assumption is true. For example, suppose the original assumption is, "My car is red." Another way to state this assumption is as an statement of implication, an if-then statement:

Assumption If it is my car, then it is red.

To form the contrapositive, you re-arrange the assumption to produce another statement:

Contrapositive If it is **not** red, then it is **not** my car.

Any assumption of the form "if [statement 1] then [statement 2]" has a contrapositive. In the example, statement 1 is "it is my car." Statement 2 is "it is red." The contrapositive looks like this:

Contrapositive If [negate statement 2] then [negate statement 1]

The contrapositive is, like algebraic manipulation, a re-rearrangement: reverse and negate. Reversing means switching the order of the two statements in the if-then structure. Negating a statement means saying the opposite. The negation of "it is red" is "it is not red." The negation of "it is my car" is "it is not my car." (It would be wrong to say that the negation of "it is my car" is "it is your car." Clearly it's true that if it is your car, then it is not my car. But there are many ways that the car can be not mine and yet not be yours. There are, after all, many other people in the world than you and me!)

Contrapositives often make intuitive sense to people. That is, people can see that a contrapostive statement is correct even if they don't know the name of the logical re-arrangement. For instance, here is a variety of ways of re-arranging the two clauses in the assumption, "If that is my car, then it is red." Some of the arrangements are logically correct, and some aren't.

Original Assumption: *If it is my car, then it is red.*

— Negate first statement: *If it is not my car, then it is red.*

Wrong. Other people can have cars that are not red.

— Negate only second statement: *If it is my car, then it is not red.*

Wrong. The statement contradicts the original assumption that my car is red.

— Negate both statements: *If it is not my car, then it is not red.*

Wrong. Other people can have red cars.

— Reverse statements: *If it is red, then it is my car.*

Wrong. Apples are red and they are not my car. Even if "it" is a car, not every red car is mine.

— Reverse and negate first: *If it is red, then it is not my car.*

Wrong. My car is red.

— Reverse and negate second: *If it is not red, then it is my car.*

Wrong. Oranges are not red, and they are not my car.

— Reverse and negate both — the contrapositive: *If it is not red, then it is not my car.*

Correct.

15.2.2 Inductive Reasoning

In contrast to deductive reasoning, **inductive reasoning** involves generalizing or extrapolating from a set of observations to conclusions. An observation is not an assumption: it is something we see or otherwise perceive. For instance, you can go to Australia and see that kangaroos hop on two legs. Every kangaroo you see is hopping on two legs. You conclude, inductively, that all kangaroos hop on two legs.

Inductive conclusions are not necessarily correct. There might be one-legged kangaroos. That you haven't seen them doesn't mean they can't exist. Indeed, Europeans believed that all swans are white until explorers discovered that there are black swans in Australia.

Suppose you conduct an experiment involving 100 people with fever. You give each of them aspirin and observe that in all 100 the fever is reduced. Are you entitled to conclude that giving aspirin to a person with fever will reduce the fever? Not really. How do you know that there are no people who do not respond to aspirin and who just happened not be be included in your study group?

Perhaps you're tempted to hedge by weakening your conclusion: "Giving aspirin to a person with fever will reduce the fever most of the time." This seems reasonable, but it is still not necessarily true. Perhaps the people in you study had a special form of fever-producing illness and that most people with fever have a different form.

By the standards of deductive reasoning, inductive reasoning does not work. No reasonable person can argue about the deductive, contrapositive reasoning concerning the red car. But reasonable people can very well find fault with the conclusions drawn from the study of aspirin.

Here's the difficulty. If you stick to valid deductive reasoning, you will draw conclusions that are correct given that your assumptions are correct. But how can you know if your assumptions are correct? How can you make sure that your assumptions adequately reflect the real world? At a practical level, most knowledge of the world comes from observations and induction.

The philosopher David Hume noted the everyday inductive "fact" that food nourishes us, a conclusion drawn from everyday observations that people who eat are nourished and people who do not eat waste away. Being inductive, the conclusion is suspect. Still, it would be a foolish person who refuses to eat for want of a deductive proof of the benefits of food.

Inductive reasoning may not provide a proof, but it is nevertheless useful.

15.3 The Null Hypothesis

A key aspect of hypothesis testing is the choice of the hypothesis to test. The stock market example involved testing the random-walk hypothesis rather than the trend hypothesis. Why? After all, the hypothesis of a trend is more interesting than the random-walk hypothesis; it's more likely to be useful if true.

It might seem obvious that the hypothesis you should test is the hypothesis that you are most interested in. But this is wrong.

In a hypothesis test one *assumes* that the hypothesis to be tested is true and draws out the consequences of that assumption in a deductive process. This can be written as an if-then statement:

> If hypothesis H is true, then the test statistic S will be drawn from a probability distribution P.

For example, in the stock market test, the assumption that the prices are a random walk led to the conclusion that the test statistic — the start-to-end price difference — would be a draw from a normal distribution with mean 0 and standard deviation 5335 dollars.

The inductive part of the test involves comparing the observed value of the test statistic S to the distribution P. There are two possible outcomes of this comparison:

Agreement S is a plausible outcome from P.

Disagreement S is not a plausible outcome from P.

Suppose the outcome is agreement between S and P. What can be concluded? Not much. Recall the statement "If it is my car, then it is red." An observation of a red car does not legitimately lead to the conclusion that the car is mine. For an if-then statement to be applicable to observations, one needs to observe the if-part of the statement, not the then-part.

A outcome of disagreement gives a more interesting result, because the contrapositive gives logical traction to the observation. "If it is not red, then it is not my car." Seeing "not red" implies "not my car." Similarly, seeing that S is not a plausible outcome from P, tells you that H is not a plausible possibility. In such a situation, you can legitimately say, "I reject the hypothesis."

Ironically, in the case of observing agreement between S and P, the only permissible statement is, "I fail to reject the hypothesis." You certainly aren't entitled to say that the evidence causes you to accept the hypothesis.

This is an emotionally unsatisfying situation. If your observations are consistent with your hypothesis, you certainly want to accept the hypothesis. But that is not an acceptable conclusion when performing a formal hypothesis test. There are only two permissible conclusions from a formal hypothesis test:

- I reject the hypothesis.
- I fail to reject the hypothesis.

In choosing a hypothesis to test, you need to keep in mind two criteria.

Criterion 1 The only possible interesting outcome of a hypothesis test is "I reject the hypothesis." So make sure to pick a hypothesis that it will be interesting to reject.

The role of the hypothesis is to be refuted or nullified, so it is called the **null hypothesis**.

What sorts of statements are interesting to reject? Often these take the form of the **conventional wisdom** or of **no effect**.

For example, in comparing two fever-reducing drugs, an appropriate null hypothesis is that the two drugs have the same effect. If you reject the null, you can say that they don't have the same effect. But if you fail to reject the null, you're in much the same position as before you started the study.

Failing to reject the null may mean that the null is true, but it equally well may mean only that your work was not adequate: not enough data, not a clever enough experiment, etc. Rejecting the null can reasonably be taken to indicate that the null hypothesis is false, but failing to reject the null tells you very little.

Criterion 2 To perform the deductive stage of the test, you need to be able to calculate the range of likely outcomes of the test statistic. This means that the hypothesis needs to be specific.

The assumption that stock prices are a random walk has very definite consequences for how big a start-to-end change you can expect to see. On the other hand, the assumption "there is a trend" leaves open the question of how big the trend is. It's not specific enough to be able to figure out the consequences.

15.4 The p-value

One of the consequences of randomness is that there isn't a completely clean way to say whether the observations fail to match the consequences of the null hypothesis. In principle, this is a problem even with simple statements like "the car is red." There is a continuous range of colors and at some point one needs to make a decision about how orange the car can be before it stops being red.

Figure 15.1 shows the probability distribution for the start-to-end stock price change under the null hypothesis that stock prices are a random walk. The observed value of the test statistic, $5074.80, falls under the tall part of the curve — it's a plausible outcome of a random draw from the probability distribution.

The conventional way to measure the plausibility of an outcome is by a **p-value**. The p-value of an observation is always calculated with reference to a probability distribution derived from the null hypothesis.

P-values are closely related to percentiles. The observed value $5074.80 falls at the 83rd percentile of the distribution. This is plausible because it's in the middle of the distribution. An observation that's at or beyond the extremes of the distribution is implausible. This would correspond to either very high percentiles or very low percentiles. Being at the 83rd percentile implies that 17 percent of draws would be even more extreme, falling even further to the right than $5074.80.

The p-value is the fraction of possible draws from the distribution that are as extreme or more extreme than the observed value. If the concern is only with values bigger than $5074.80, then the p-value is 0.17.

This p-value of 0.17 is called a **one-tailed** p-value, since it considers only events that are extreme to one side of the distribution. Of course, an observation might also be extreme in the other way. For the stock prices, this would be a large negative start-to-end changes, corresponding to a downward trend in stock prices. To take such possible draws into account, you double the p-value. Thus, the p-value for the observed start-to-end stock price change (under the null hypothesis of a random walk) is $2 \times 0.17 = 0.34$.

A small p-value indicates that the actual value of the test statistic is quite surprising as an outcome from the null hypothesis. A large p-value means that the test statistic value is run of the mill, not surprising, not enough to satisfy the "if" part of the contrapositive.

The convention in hypothesis testing is to consider the observation as being implausible when the p-value is less than 0.05. In the stock market example, the p-value is larger than 0.05, so the outcome is to fail to reject the null hypothesis that stock prices are a random walk with no trend.

15.5 Rejecting by Mistake

The p-value for the hypothesis test of the possible trend in stock-price was 0.34, not small enough to justify rejecting the null hypothesis that stock prices are a random walk with no trend. A smaller p-value, one less than 0.05 by convention, would have led to rejection of the null. The small p-value would have indicated that the observed value of the test statistic was implausible in a world where the null hypothesis is true.

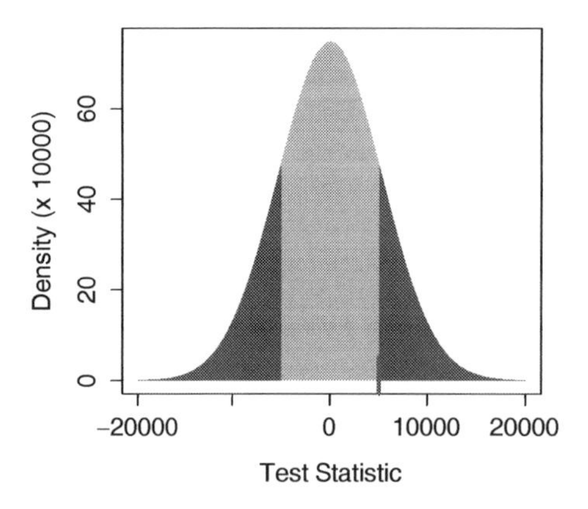

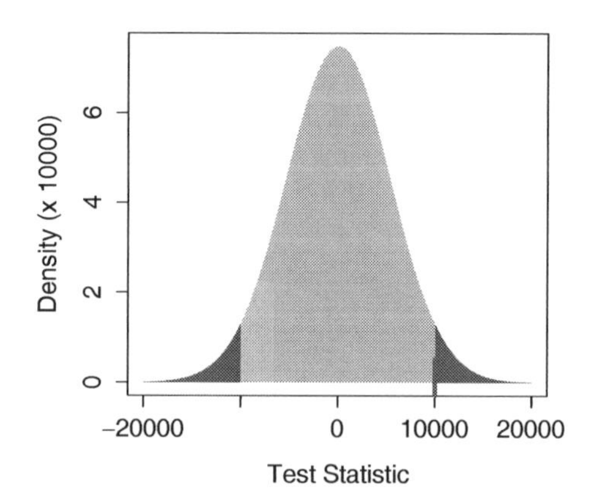

Figure 15.3: The p-value is a probability of observing a test statistic as extreme as the actual value, or more extreme. Like percentiles, this corresponds to the area under the probability distribution for observations assuming that the null hypothesis is true. Left: the shaded area corresponds to the p-value for the actual observation of a start-to-end change of $5074.80. This p-value is 0.34. Right: If the observation had instead been $10,000.00, the p-value would have been smaller: about 0.06.

Now turn this around. Suppose the null hypothesis really were true; suppose stock prices really are a random walk with no trend. In such a world, it's still possible to see an implausible value of the test statistic. But, if the null hypothesis is true, then seeing an implausible value is misleading; rejecting the null is a mistake. This sort of mistake is called a **Type I error**.

Such mistakes are not uncommon. In a world where the null is true — the only sort of world where you can falsely reject the null — they will happen 5% of the time so long as the threshold for rejecting the null is a p-value of 0.05.

The way to avoid such mistakes is to lower the p-value threshold for rejecting the null. Lowering it to, say, 0.01, would make it harder to mistakenly reject the null. On the other hand, it would also make it harder to correctly reject the null in a world where the null ought to be rejected.

The threshold value of the p-value below which the null should be rejected is a probability: the probability of rejecting the null in a world where the null hypothesis is true. This probability is called the **significance level** of the test.

It's important to remember that the significance level is a **conditional probability**. It is the probability of rejecting the null in a world where the null hypothesis is actually true. Of course that's a hypothetical world, not necessarily the real world.

15.6 Failing to Reject

In the stock-price example, the large p-value of 0.34 led to a failure to reject the null hypothesis that stock prices are a random walk. Such a failure doesn't mean that the null hypothesis is true, although it's encouraging news to people who want to believe that the null hypothesis is true.

You never get to "accept the null" because there are reasons why, even if the null were wrong, it might *not* have been rejected:

- You might have been unlucky. The randomness of the sample might have obscured your being able to see the trend in stock prices.

- You might not have had enough data. Perhaps the trend is small and can't easily be seen

A helpful idea in hypothesis testing is the **alternative hypothesis**: the pet idea of what the world is like if the null hypothesis is wrong. The alternative hypothesis plays the role of the thing that you would like to prove. In the hypothesis-testing drama, this is a very small role, since the only possible outcomes of a hypothesis test are (1) reject the null and (2) fail to reject the null. The alternative hypothesis is not directly addressed by the outcome of a hypothesis test.

The role of the alternative hypothesis is to guide you in interpreting the results if you do fail to reject the null. The alternative hypothesis is also helpful in deciding how much data to collect.

To illustrate, suppose that the stock market really does have a trend hidden inside the random day-to-day fluctuations with a standard deviation of \$106.70. Imagine that the trend is \$2 per day: a pet hypothesis. If this were true, the start-to-end change in the price over $N = 2500$ days would be a draw from a normal distribution with mean $\$2 \times 2500 = \$5000$ and standard deviation $\$106.70 \times \sqrt{2500} = \$5335$.

Suppose the world really were like the alternative hypothesis. What is the probability that, in such a world, you would end up rejecting the null hypothesis? (Such a mistake, where you fail to reject the null in a world where the alternative is actually true, is called a **Type II error**.)

The probability of rejecting the null in a world where the alternative is true is called the **power** of the hypothesis test. Of course, if the alternative is true, then it's completely appropriate to reject the null, so a large power is desirable.

A power calculation involves considering both the null and alternative hypotheses. Aside 15.1 shows the logic applied to the stock-market question. It results in a power of 15%.

The power of 15% for the stock market test means that even if the pet theory of the \$2 trend were correct, there is only a 15% chance of rejecting the null. In other words, the study is quite weak.

When the power is small, failure to reject the null can reasonably be interpreted as a failure in the modeler (or in the data collection or in the experiment). The study has given very little information.

Aside. 15.1 Calculating a Power

Here are the steps in calculating the power of the hypothesis test of stock market prices. The null hypothesis is that prices are a pure random walk. The alternative hypothesis is that they have a trend of $2 per day.

1. Go back to the null hypothesis world and find the thresholds for the test statistic that would cause you to reject the null hypothesis. Referring to Figure 15.3, you can see that a test statistic of $10,000.00 would have produced a p-value of 0.061, close to the rejection threshold. So if the test statistic were a little larger, you would have reached the threshold for rejection. The exact threshold value turns out to be $10,456.41, which is the 97.5th percentile for the null hypothesis distribution (the normal distribution with mean zero and standard deviation of $5335). So, you would have rejected the null at a significance level of 0.05 if the test statistic had been bigger than $10,456.41.
2. Now return to the alternative hypothesis world. In this world, what is the probability that the test statistic would have been bigger than $10,456.41? This is a straightforward probability calculation, since the distribution of the test statistic in the alternative hypothesis world is normal with mean $5000 and standard deviation $5335. Doing the calculation gives a probability of 0.15.

Section 17.7 discusses power calculations for models.

Just because the power is small is no reason to doubt the null hypothesis. Instead, you should think about how to conduct a better, more powerful study.

One way a study can be made more powerful is to increase the sample size. Rather than studying the trend over $N = 2500$ days, perhaps it should have been studied over twice or three times as long. To illustrate, consider what would happen in a study with $N = 5000$ days of data.

Null Hypothesis In the world where the null is true, the start-to-end change would be normal with mean $0 and standard deviation $106.70 $\times \sqrt{5000} =$ $7544.83. The threshold for rejection will be a start-to-end change of $14,787.66 or bigger.

Alternative Hypothesis In the alternative hypothesis world, the start-to-end change in price will be normal with mean 2\times 5000 =$$10,000 and standard deviation $106.70 $\times \sqrt{5000}$ =$7544.83. The probability of the start-to-end change being above the threshold for rejecting the null hypothesis is 26%.

So, even a sample size of $n = 5000$ doesn't give a very powerful study: there is little reason to think anyone could reject the null even if there is a trend in stock prices.

The logic of the power calculation can be used to decide how big a study is needed. Repeating the power calculation for different values of n gives the following powers:

n	Power
2500	15%
5000	26%
10000	47%
20000	76%
30000	90%

Reliably detecting a $2 per day trend requires a lot of data. For instance $n = 20000$ is about 80 years of data. This long historical period is probably not relevant to today's investor. Indeed, it's just about all the data that is actually available: the DJIA was started in 1928.

When the power is small for realistic amounts of data, the phenomenon you are seeking to find may be undetectable.

15.7 A Glossary of Hypothesis Testing

Null Hypothesis A statement about the world that you are interested to disprove. The null is almost always something that is clearly relevant and not controversial: that the conventional wisdom is true or that there is no relationship between variables. Examples: "The drug has no influence on blood pressure." "Smaller classes do not improve school performance."

The allowed outcomes of the hypothesis test relate only to the null:

- Reject the null hypothesis.
- Fail to reject the null hypothesis.

Alternative Hypothesis A statement about the world that motivates your study and stands in contrast to the null hypothesis. "The drug will reduce blood pressure by 5 mmHg on average." "Decreasing class size from 30 to 25 will improve test scores by 3%."

The outcome of the hypothesis test is not informative about the alternative. The importance of the alternative is in setting up the study: choosing a relevant test statistic and collecting enough data.

Test Statistic The number that you use to summarize your study. This might be the sample mean, a model coefficient, or some other number. Later chapters will give several examples of test statistics that are particularly appropriate for modeling.

Type I Error A wrong outcome of the hypothesis test of a particular type. Suppose the null hypothesis were really true. If you rejected it, this would be an error: a type I error.

Type II Error A wrong outcome of a different sort. Suppose the alternative hypothesis were really true. In this situation, failing to reject the null would be an error: a type II error.

Significance Level A conditional probability. In the world where the null hypothesis is true, the significance is the probability of making a type I error. Typically, hypothesis tests are set up so that the significance level will be less than 1 in 20, that is, less than 0.05. One of the things that makes hypothesis testing confusing is that you do not know whether the null hypothesis is correct; it is merely assumed to be correct for the purposes of the deductive phase of the test. So you can't say what is the probability of a type I error. Instead, the significance level is the probability of a type I error *assuming* that the null hypothesis is correct.

Ideally, the significance level would be zero. In practice, one accepts the risk of making a type I error in order to reduce the risk of making a type II error.

p-value This is the usual way of presenting the result of the hypothesis test. It is a number that summarizes how atypical the observed value of the test statistic would be in a world where the null hypothesis is true. The convention for rejecting the null hypothesis is $p < 0.05$.

The p-value is closely related to the significance level. It is sometimes called the **achieved significance level**.

Power This is a conditional probability. But unlike the significance, the condition is that the alternative hypothesis is true. The power is the probability that, in the world where the alternative is true, you will reject the null. Ideally, the power should be 100%, so that if the alternative really were true the null hypothesis would certainly be rejected. In practice, the power is less than this and sometimes much less.

In science, there is an accepted threshold for the p-value: 0.05. But, somewhat strangely, there is no standard threshold for the power. When you see a study which failed to reject the null, it is helpful to know what the power of the study was. If the power was small, then failing to reject the null is not informative.

15.8 Computational Technique

The computational techniques described here relate to finding p-values and illustrating how power can be estimated.

15.8.1 Computing p-values

The p-value always involves the position of an observed value of the test statistic within a sampling distribution. Once you have the observed value, finding the p-value involves three steps:

1. Finding the sampling distribution under the null hypothesis.
2. Finding the percentile of the observed value of the test statistic within the sampling distribution.

3. Translating the percentile to a p-value.

Later chapters will deal with ways to find the sampling distribution of model coefficients and other test statistics (such as R^2). Here I'll assume that you already know the form of the sampling distribution either as a probability model or as a sample of values from the distribution.

To illustrate, consider the example of the start-to-end difference in the stock price. Under the null hypothesis that there is no systematic trend in prices, the sampling distribution described in the chapter is that the price difference will have a mean of zero and a standard deviation of 5335 dollars. The observed value of the test statistic was $5074.80.

Since the sampling distribution in this case is a normal distribution, the percentile of the observed value in the sampling distribution can be found using the `pnorm` operator:

```
> pnorm( 5074.80, mean=0, sd=5335)
[1] 0.8293
```

This is not yet a p-value. The value 0.8293 says that the sampling distribution will generate a value that is less than the observed value (5074.80) on 82.9 percent of trials. Thus, there is a $= 100 - 82.9 = 17.1$ percent chance that a randomly generated value would be greater than the observed value. The p-value is the chance that the randomly generated value would be *more extreme* than the observed value. "More extreme" might mean "bigger than" or "less than," depending on the context. In order to judge, it helps to have a picture of the situation. There happens to be one in figure 15.3.

In this case, a value *bigger* than 5074.80 will be more extreme, so the p-value is 17.1 percent. This is a one-tailed p-value, since it considers only one way to be more extreme. The two-tailed p-value includes the possibility that the test statistic might have been extreme but on the other side of the sampling distribution. The two-tailed p-value is, for the nicely symmetric normal distribution, twice the one-tailed value: $2 \times 17.1 = 34.2$ percent, or 0.342.

It can be helpful to sketch your own pictures of sampling distributions in order to judge which tails of the distribution should be included. Here are the commands for making a simple plot to compare the sampling density to the observed value of the test statistic. In the example, the sampling density will be poisson with rate parameter `lambda=100` and the observed value will be 110.

```
samps = rpois( 10000, lambda=100 )
densityplot( samps, panel=hypothesis.test.panel,
  observed=110,
  plot.points=FALSE,lwd=3 )
```

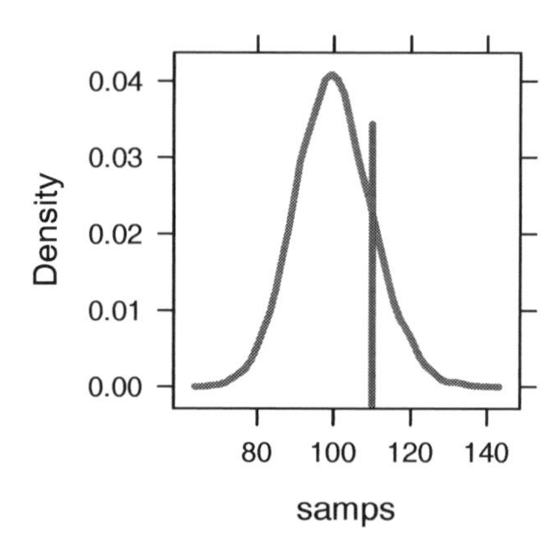

From the graph, you can see that, in this case, "more extreme than" means "greater than. " In making your own plots, you would need to change the first line (`samps=...`) to match your own sampling distribution and the observed value in the `panel.rug` line. The `trellis.focus` and `trellis.unfocus` commands instruct R to add the rug plot to the existing graph on the correct scale.

In later chapters, you'll see situations where you have only a sample from the sampling distribution. Translating this into a percentile can be done with `table` and `prop.table`. To illustrate, I'll use simulation to generate a sample from the stock market price-change distribution.

Recall that the null hypothesis model of the stock market was a simple random walk in prices with no systematic trend. The standard deviation of price changes on each day was estimated from the data to be \$106.70. Over a period of $N = 2500$ days the random walk can be simulated as the sum of 2500 normal random numbers with a mean of zero (no trend) and a standard deviation of \$106.70. Here's one trial:

```
> sum( rnorm( 2500, mean=0, sd=106.70))
[1] -3996
```

The simulated market went down by \$3996 over the 2500 days.

To draw a sizeable sample from this random process, use the `do` operator.

```
> samps = do(500)*sum(rnorm(2500,mean=0,sd=106.70))
```

This command produces a sample of size $n = 500$ from the random process — as if there were 500 different copies of the world in which the null hypothesis was active.

Now to the main point. Given a sample like this from the null hypothesis, the probability of the process generating a value that's smaller than the observed test statistic can be found like this:

```
> pdata( 5074.80, samps)
[1] 0.834
```

This is more or less the same as what was found using the theoretically derived distribution and pnorm.

Power calculations involve finding rejection thresholds. To do this, you need to find the quantile at the appropriate level of the sampling distribution. For a 5% significance level, the appropriate quantile levels for a two-tailed test are 0.025 and 0.975. To find these from the sampling distribution, use the appropiate probability model or the samples:

```
> qnorm( c(.025, .975), mean=0, sd=5335)
[1] -10456  10456
> qdata( c(.025, .975), samps )
      2.5%      97.5%
-10369.931   9794.801
```

As always, the approach based on a sample from the distribution is subject to random fluctuations due to the small size of the sample. These are particularly acute when looking at quantiles near the extremes of the distributions. When the sampling distribution is normally shaped, as it often is, a better estimate can be had by taking the standard deviation, multiplying it out to reach the 95% coverage interval:

```
> sd(samps)*2
[1] 10511
```

Chapter 16

Hypothesis Testing on Whole Models

A wise man ... proportions his belief to the evidence. — David Hume (1711 – 1776)

Fitted models describe patterns in samples. Modelers interpret these patterns as indicating relationships between variables in the population from which the sample was drawn. But there is another possibility. Just as the constellations in the night sky are the product of human imagination applied to the random scattering of stars within range of sight, so the patterns indicated by a model might be the result of accidental alignments in the sample.

Deciding how seriously to take the patterns identified by a model is a problem that involves judgment. Are the patterns consistent with well established understanding of how the system works? Are the patterns corroborated by other sources of data? Are the model results sensitive to trivial changes in the model design?

Before undertaking that judgment, it helps to apply a much simpler standard of evidence. The conventional interpretation of a model such as A ~ B+C+... is that the variables on the right side of the modeler's tilde explain the response variable on the left side. The first question to ask is whether the explanation provided by the model is stronger than the "explanation" that would be arrived at if the variables on the right side were random — explanatory variables in name only without any real connection with the response variable A.

It's important to remember that in a hypothesis test, the null hypothesis relates to the **population**, that in the population the explanatory variables are unlinked to the response variable. Such a hypothesis does not rule out the possibility that, in the sample, the explanatory variables are aligned with the response variable. The hypothesis merely claims that any such alignment is accidental, due to the randomness of the sampling process.

16.1 The Permutation Test

The null hypothesis is that the explanatory variables are unlinked with the response variable. One way to see how big a test statistic will be in a world where the null hypothesis holds true is to randomize the explanatory variables in the sample to destroy any relationship between them and the response variable. To illustrate how this can be done in a way that stays true to the sample, consider a small data set:

A	B	C
3	37.1	M
4	17.4	M
5	26.8	F
7	44.3	F
5	19.7	F

Imagine that the table has been cut into horizontal slips with one case on each slip. The response variable — say, A — has been written to the left of a dotted line. The explanatory variables B and C are on the right of the dotted line, like this:

A=3	B=37.1	C=M
A=4	B=17.4	C=M
A=5	B=26.8	C=F
A=7	B=44.3	C=F
A=5	B=19.7	C=F

To randomize the cases, tear each sheet along the dotted line. Place the right sides — the explanatory variables — on a table in their original order. Then, randomly shuffle the left halves — the response variable — and attach each to a right half.

A=4	B=37.1	C=M
A=5	B=17.4	C=M
A=5	B=26.8	C=F
A=3	B=44.3	C=F
A=7	B=19.7	C=F

None of the cases in the shuffle are genuine cases, except possibly by chance. Yet each of the shuffled explanatory variables is true to it's distribution in the original sample and the relationships among explanatory variables — collinearity and multi-collinearity — are also authentic.

Each possible order for the bottom halves of the cards is called a **permutation**. A hypothesis test conducted in this way is called a **permutation test**.

The logic of a permutation test is straightforward. To set up the test, you need to choose a test statistic that reflects some aspect of the system of interest to you.

Here are the steps involved in permutation test:

Step 1 Calculate the value of the test statistic on the original data.

Step 2 Permute the data and calculate the test statistic again. Repeat this many times, collecting the results. This gives the distribution of the test statistic under the null hypothesis.

Step 3 Read off the p-value as the fraction of the results in (2) that are more extreme than the value in (1).

To illustrate, consider a model of heights from Galton's data and a question Galton didn't consider: Does the number of children in a family help explain the eventual adult height of the children? Perhaps in families with large numbers of children, there is competition over food, so children don't grow so well. Or, perhaps having a large number of children is a sign of economic success, and the children of successful families have more to eat.

The regression report indicates that larger family size is associated with shorter children:

	Estimate	Std. Error	t value	Pr(>|t|)
(Intercept)	67.800	0.296	228.96	0.0000
nkids	**-0.169**	0.044	-3.83	0.0001

For every additional sibling, the family's children are shorter by about 0.17 inches on average. The confidence interval is -0.169 ± 0.088.

Now for the permutation test, using the coefficient on nkids as the test statistic:

Step 1 Calculate the test statistic on the data without any shuffling. As shown above, the coefficient on nkids is -0.169.

Step 2 Permute and re-calculate the test statistic, many times. One set of 10000 trials gave values of -0.040, 0.037, -0.094, -0.069, 0.062, -0.045, and so on. The distribution is shown in Figure 16.1.

Step 3 The p-value is fraction of times that the values from (2) are more extreme than the value of -0.169 from the unshuffled data. As Figure 16.1 shows, few of the permutations produced an nkid coefficient anywhere near -0.169. The p-value is very small, $p < 0.001$.

Conclusion, the number of kids in a family accounts for somewhat more of the children's heights than is likely to occur with a random explanatory variable.

The idea of a permutation test is almost a century old. It was proposed originally by the brilliant statistician Ronald Fisher (1890-1960). Permutation tests

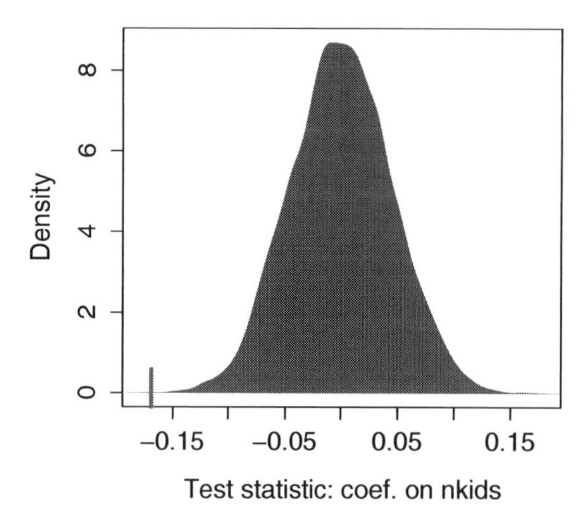

Figure 16.1: Distribution of the coefficient on nkids from the model height $\sim$ 1+nkids from many permutation trials. Tick mark: the coefficient -0.169 from the unshuffled data.

were infeasible for even moderately sized data sets until the 1970s when inexpensive computation became a reality. In Fisher's day, when computing was expensive, permutation tests were treated as a theoretical notion and actual calculations were performed using algebraic formulas. Such formulas could be derived for a narrow range of test statistics such as the sample mean, differences between group means, and the coefficient of determination R^2. Fisher himself derived the sampling distributions of these test statistics.[25]

Ronald Fisher

16.2 R^2 and the F Statistic

The coefficient of determination R^2 measures what fraction of the variance of the response variable is "explained" or "accounted for" or — to put it simply — "modeled" by the explanatory variables. R^2 is a comparison of two quantities: the variance of the fitted model values to the variance of the response variable.

R^2 is a single number that puts the explanation in the context of what remains unexplained. It's a good test statistic for a hypothesis test.

Using R^2 as the test statistic in a permutation test would be simple enough. There are advantages, however, to thinking about things in terms of a closely related statistic invented by Fisher and named in honor of him: the **F statistic**.

Like R^2, the F statistic compares the size of the fitted model values to the size of the residuals. But the notion of "size" is somewhat different. Rather than measuring size directly by the variance or by the sum of squares, the F statistic takes into account the number of model vectors.

To see where F comes from, consider a special sort of random walk: the **random model walk**. In a regular random walk (Chapter 13), each new step is taken in a random direction. In a random model walk, each "step" consists of adding a new random explanatory term to a model. The "position" is measured as the R^2 from the model.

The starting point of the random model walk is the simple model A ~ 1 with just $m = 1$ model vector. This model always produces $R^2 = 0$ because the all-cases-the-same model can't account for any variance. Taking a "step" means adding a random model vector, x_1, giving the model A ~ 1+x_1. Each new step adds a new random vector to the model:

m	Model
1	A ~ 1
2	A ~ 1+x_1
3	A ~ 1+x_1+x_2
4	A ~ 1+x_1+x_2+x_3
⋮	
n	A ~ 1+x_1+x_2+x_3 + ⋯ + x_{n-1}

Figure 16.2 shows R^2 versus m for several random model walks in data with $n = 50$ cases. Each successive step adds in its own individual random explanatory term.

All the random walks start at $R^2 = 0$ for $m = 1$. All of them reach $R^2 = 1$ when $m = n$. Adding any more vectors beyond $m = n$ simply creates redundancy; $R^2 = 1$ is the best that can be done.

Notice that each step increases R^2 — none of the random walks goes down in value as m gets bigger.

The R^2 from a fitted model gives a single point on the model walk that divides the overall walk into two segments, as shown in Figure 16.3. The slope of each of the two segments has a straightforward interpretation. The slope of the segmented labeled "Model" describes the rate at which R^2 is increased by a typical model vector. The slope can be calculated as $R^2/(m-1)$.

The slope of the segment labeled "Residuals" describes how adding a random vector to the model would increase R^2. From the figure, you can see that a typical model vector increases R^2 much faster than a typical random vector. Numerically, the slope is $(1 - R^2)/(m - n)$.

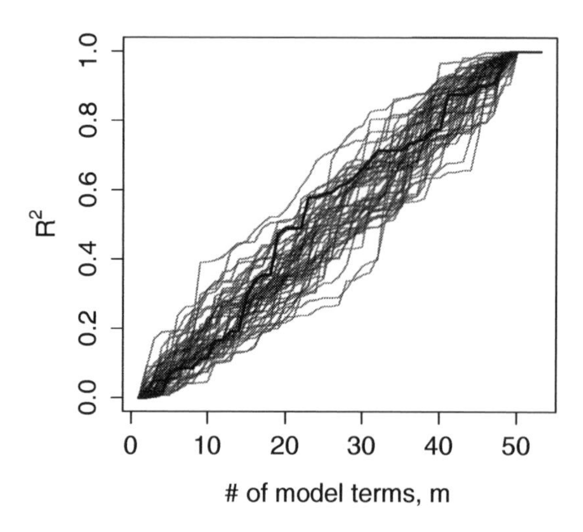

Figure 16.2: Random model walks for $n = 50$ cases. The "position" of the walk is R^2. This is plotted versus number of model vectors m. The heavy line shows one simulation. The light lines show other simulations. All of the simulations reach $R^2 = 1$ when $m = n$.

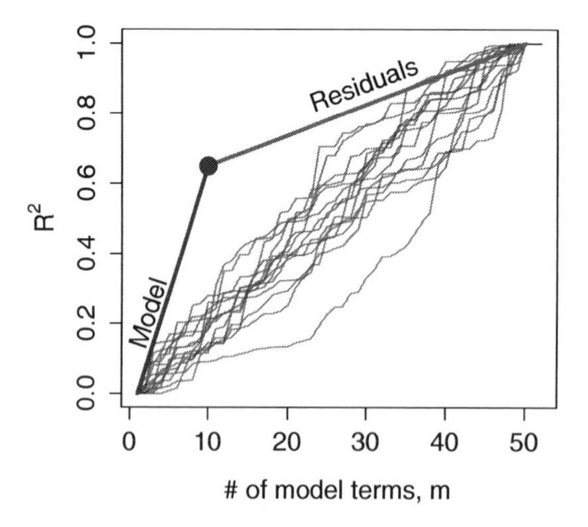

Figure 16.3: The R^2 from a fitted model defines a model walk with two segments. The F statistic is the ratio of the slopes of these segments. Here a model with $m = 10$ has produced $R^2 = 0.65$ when fitted to a data set with $n = 50$ cases.

The F statistic is the ratio of these two slopes:

$$F = \frac{\text{Slope of model segment}}{\text{Slope of residual segment}} = \frac{R^2}{m-1} \Bigg/ \frac{1-R^2}{n-m}. \tag{16.1}$$

In interpreting the F statistic, keep in mind that if the model vectors were no better than random, F should be near 1. When F is much bigger than 1, it indicates that the model terms explain more than would be expected at random. The p-value provides an effective way to identify when F is "much bigger than 1." Calculating the p-value involves knowing the exact shape of the F distribution. In practice, this is done with software.

The number $m-1$ in the numerator of F counts how many model terms there are other than the intercept. In standard statistical nomenclature, this is called the **degrees of freedom of the numerator**. The number $n-m$ in the denominator of F counts how many random terms would need to be added to make a "perfect" fit to the response variable. This number is called the **degrees of freedom of the denominator**.

Example 16.1: Marriage and Astrology

Does your astrological sign predict when you will get married? To test this possibility, consider a small set of data collected from an on-line repository of marriage licenses in Mobile County, Alabama in the US.The licenses contain a variety of information: age at marriage, years of college, date of birth, date of the wedding and so on. Combining the date of the wedding and date of birth gives the age of the person when they got married. Date of birth can also be converted to an astrological sign, resulting in a data set that looks like this:

Sign	Age	College
Aquarius	22.6	4
Sagittarius	25.1	0
Taurus	39.6	1
Cancer	45.8	6
Leo	26.4	1

... and so on for 98 cases altogether.

There are, of course, 12 different astrological signs, so the model Age ~ Sign has $m = 12$. With the intercept term, that leaves 11 vectors to represent Sign. Fitting the model produces $R^2 = 0.070$. To generate a p-value, this can be compared to the distribution of R^2 for random vectors with $n = 98$ and $m = 12$, giving $p = 0.885$ — quite consistent with Sign being unlinked with Age.

The pair $m = 12, R^2 = 0.070$ gives one point on a model walk. This is compared to random model walks in the Figure 16.4 The 11 indicator vectors that arise from the 12 levels of Sign do not contribute more to R^2 than would be expected from 11 random vectors.

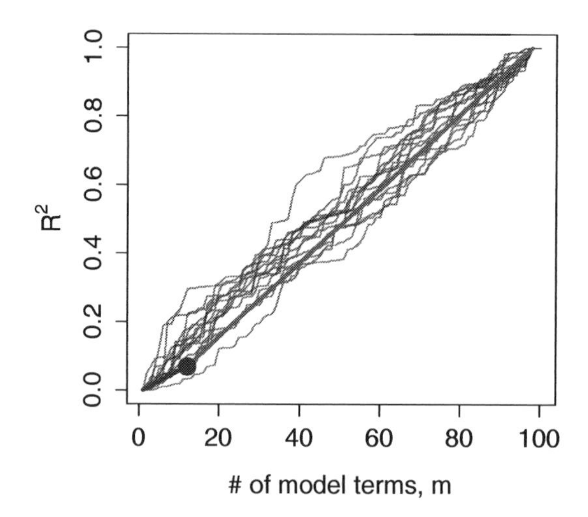

Figure 16.4: Random model walks compared to the model Age ~ Sign. The model does not stand out from the random model works.

Now consider a different model, Age ~ College. This model undertakes to explain age at marriage by whether the person when to college. The R^2 from this model is only 0.065 — the College variable accounts for somewhat less of the variability in Age than astrological Sign. Yet this small R^2 is statistically significant — $p = 0.017$. The reason is that College accomplishes its explanation using only a single model vector, compared to the 11 vectors in Sign. The model walk for the College model starts with a steep increase because the whole gain of 0.065 is accomplished with just one model vector, rather than being achieved using 11 vectors as in Sign.

Aside. 16.1 The shape of F

The shape of the F distribution depends on both n and m. Some of the distributions are shown for various m and n in Figure 16.5. Despite slight differences, the F distributions are all centered on 1. (In contrast, distributions of R^2 change shape substantially with m and n.) This steadiness in the F distribution makes it easy to interpret an F statistic by eye since a value near 1 is a plausible outcome from the null hypothesis. (The meaning of "near" reflects the width of the distribution: When n is much bigger than m, the the F distribution has mean 1 and standard deviation that is roughly $\sqrt{2/m}$.)
Presenting F with m and n as the parameters violates a convention. The F distribution is always specified in terms of the degrees of freedom of the numerator $m - 1$ and the degrees of freedom of the denominator $n - m$.

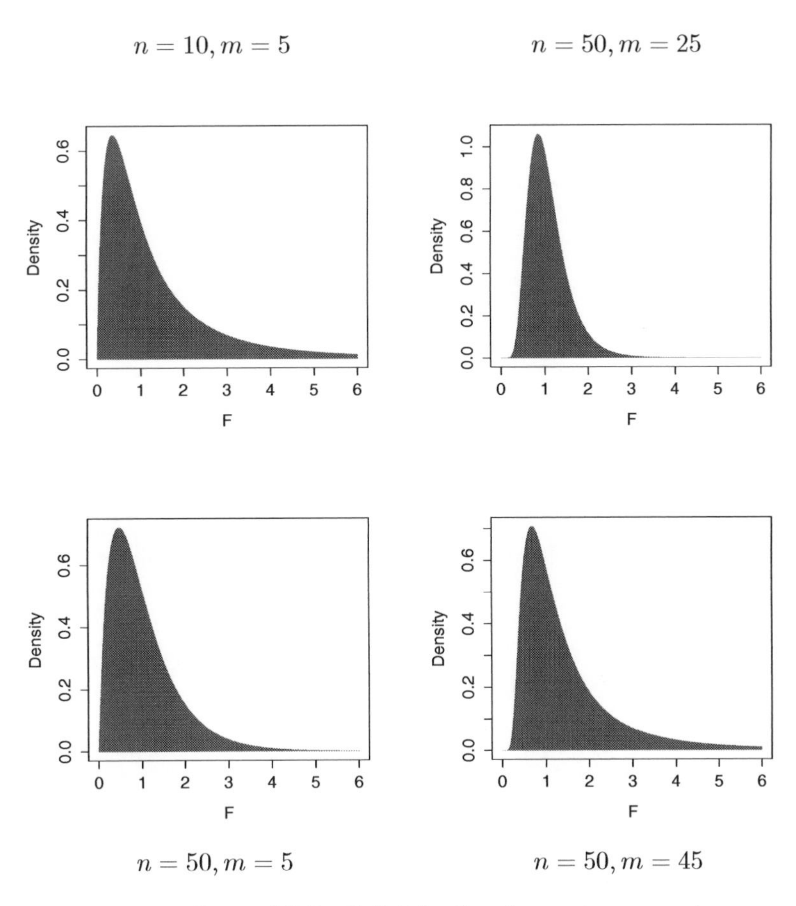

Figure 16.5: F distribution for various n and m.

Example 16.2: F and Astrology Returning to the attempt to model age at marriage by astrological sign ...

The model Age $\sim$ Sign produced $R^2 = 0.070$ with $m = 12$ model vectors and $n = 98$ cases. The F statistic is therefore $\frac{0.070/11}{0.930/86} = 0.585$. Since F is somewhat less than 1, there is no reason to think that the Sign vectors are more effective than random vectors in explaining Age. Finding a p-value involves looking up the value F$= 0.585$ in the F distribution with $m - 1 = 11$ degrees of freedom in the numerator and $n - m = 86$ degrees of freedom in the denominator. Figure 16.6 shows this F distribution.

The p-value is the probability of seeing an F value from this distribution that is larger than the observed value of $F = 0.585$. From the figure, it's easy to see that p is more than one-half since a majority of the distribution falls to the right of 0.585. Calculating it exactly using software gives $p = 0.836$.

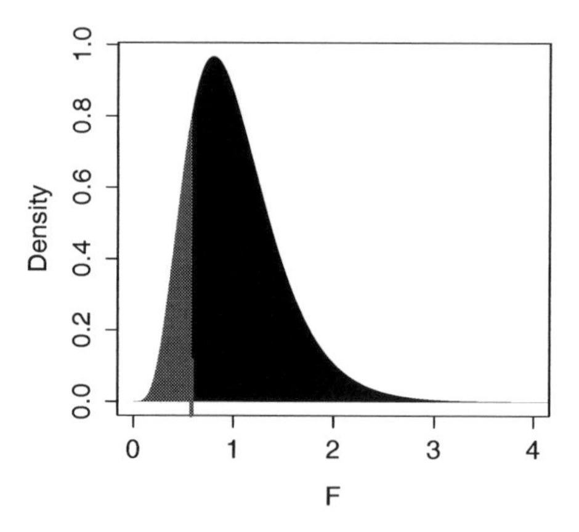

Figure 16.6: Finding a p-value from an F distribution always involves the area to the right of the test statistic, in this case marked by the tick at $F = 0.585$.

16.3 The ANOVA Report

The F statistic compares the variation that is explained by the model to the variation that remains unexplained. Doing this involves taking apart the variation; partitioning it into a part associated with the model and a part associated with the residuals.

Such partitioning of variation is fundamental to statistical methods. When done in the context of linear models, this partitioning is given the name **analysis of variance** (ANOVA, for short). This name stays close to the dictionary definition of "analysis" as "the process of separating something into its constituent elements." [26]

The analysis of variance is often presented in a standard way: the **ANOVA report**. For example, here is the ANOVA report for the marriage-astrology model Age ~ Sign:

	Df	Sum Sq	Mean Sq	F value	Pr(>F)
Sign	11	1402	127	0.59	0.8359
Residuals	86	18724	218		

In this ANOVA report, there are two rows. The first refers to the explanatory variable, the second to the residual.

Since ANOVA is about variance, the mean of the response variable is subtracted out before the analysis is performed, just as it would be when calculating a variance.

The column labeled "Sum Sq" gives the sum of squares of the fitted model values and the residual vector. The column headed "Df" is the degrees of freedom of the term. This is simply the number of model vectors associated with that term or, for the residual, the total number of cases minus the number of explanatory vectors. For a categorical variable, the number of explanatory vectors is the number of levels of that variable minus one for each redundant vector. (For instance, every categorical variable has one degree of redundancy with the intercept). Because there are 12 levels in Sign, the 12 astrological signs, the Sign variable has 11 degrees of freedom.

The degrees of freedom of the residuals is the number of cases n minus the number of model terms m.

Aside. 16.2 F and R^2

The R^2 doesn't appear explicitly in the ANOVA report, but it's easy to calculate since the report does give the square lengths of the two legs of the model triangle. In the above ANOVA report for Age ~ Sign, the square length of the hypotenuse is, through the pythagorean relationship, simply the sum of the two legs: in this report, $1402 + 18724 = 20126$. The model's R^2 is the square length of the fitted-model-value leg divided into the square length of the hypotenuse: $1402/20126 = 0.070$.

The **mean square** in each row is the sum of squares divided by the degrees of freedom. The table's F value is the ratio of the mean square of the fitted model values to the mean square of the residuals: $\frac{127}{218} = 0.585$ in this example. This is the same as given by Equation 16.1, but the calculations are being done in a different order.

Finally, the F value is converted to a p-value by look-up in the appropriate F distribution with the indicated degrees of freedom in the numerator and the denominator.

Example 16.3: Is height genetically determined?

Francis Galton's study of height in the late 1800s was motivated by his desire to quantify genetic inheritance. In 1859, Galton read Charles Darwin's *On the Origin of Species* in which Darwin first put forward the theory of natural selection of heritable traits. (Galton and Darwin were half-cousins, sharing the same grandfather, Erasmus Darwin.)

The publication of *Origin of Species* preceded by a half-century any real understanding of the mechanism of genetic heritability. Today, of course, even elementary-school children hear about DNA and chromosomes, but during Darwin's life these ideas were unknown. Darwin himself thought that traits were transferred from the parents to the child literally through the blood, with individual traits carried by "gemmules." Galton tried to confirm this experimentally by doing blood-mixing experiments on rabbits; these were unsuccessful at transferring traits.

By collecting data on the heights of children and their parents, Galton to sought quantify to what extent height is determined genetically. Galton faced the challenge that the appropriate statistical methods had not yet been developed — he had to start down this path himself. In a very real sense, the development of ANOVA by Ronald Fisher in the early part of the twentieth century

Charles Darwin (1809-1882)

Francis Galton (1822-1911)

was a direct outgrowth of Galton's work.

Had he been able to, here is the ANOVA report that Galton might have generated. The report summarizes the extent to which height is associated with inheritance from the mother and the father and by the genetic trait sex, using the model height ~ 1 + sex + mother + father.

	Df	Sum Sq	Mean Sq	F value	Pr(>F)
Genetic Terms	3	7366	2455	**529**	**0.000**
Residuals	894	4159	5		

The F value is huge and correspondingly the p-value is very small: the genetic variables are clearly explaining much more of height than random vectors.

If Galton had access to modern statistical approaches — even those from as long ago as the middle of the twentieth century — he might have wondered how to go further, for example how to figure out whether the mother and father each contribute to height, or whether it is a trait that comes primarily from one parent. Answering such questions involves partitioning the variance not merely between the model terms and the residual but *among* the individual model terms. This is the subject of Chapter 17.

16.4 Visualizing p-values

To help develop some intuition about p-values, here is a hypothesis test on a ridiculously simple setting, the model A ~ B with these data:

	A	B
case 1	4.5	1
case 2	2.1	1

As a test statistic, use the angle θ between A and B. When θ is small, B is closely aligned with A, the sort of behavior you want for an effective explanatory variable.

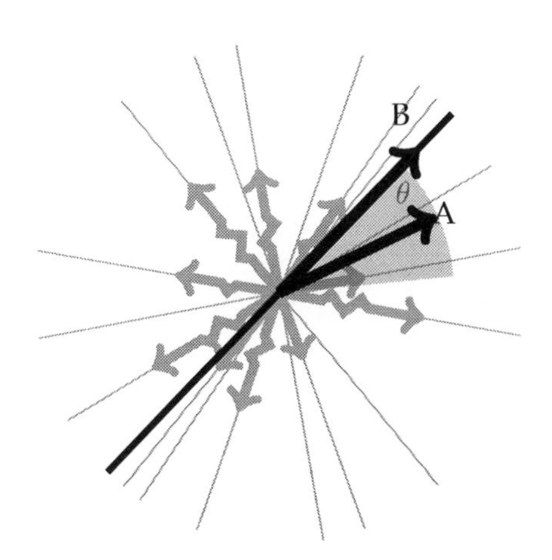

When $n = 2$ the p-value is $p = \theta/90°$.

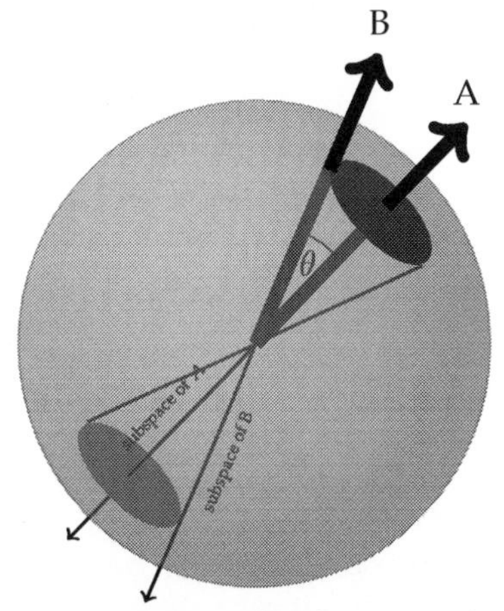

For $n = 3$ the p-value is proportional to the area of the polar cap: $p = 1 - \cos(\theta)$.

Figure 16.7: The p-value is the probability that a random vector will be closer to A than is B.

The null hypothesis is that, in the population, B is unrelated to A. If a different random sample had been selected from the population rather than these two cases, B would have pointed in some other, random direction.

Figure 16.7 (left) shows A and B and some of the random possibilities for different samples of B.

The angle between A and B can be measured with a protractor or calculated from the data values. It is $\theta = 20°$. The p-value is the probability that a random vector would fall closer to A than is B. This would happen if the subspace of the vector falls into the shaded wedges in the figure that indicate an angle of θ on either side of A. The random vectors in the plane are equally likely to point in any direction, so the probability that one of them has a subspace in the wedges is $2\theta/180 = \theta/90 = 0.22$

It's also possible to visualize the p-value for the model A $\sim$ B when there are $n = 3$ cases. (Figure 16.7, right side) In this setting, A and B are both vectors in 3-dimensional space. The angle between them, θ still makes a nice test statistic. (θ is equivalent to using R^2, since $\cos(\theta)^2 = R^2$). Again, the p-value is the probability that the subspace of a randomly oriented vector would be closer to A than B is. This is proportional to the area of the "polar" cap that surrounds A.

Visual intuition fails when $n > 3$, but it's still reasonable to think of the null hypothesis as meaning that the explanatory vectors point in random directions and that the p-value gives the probability that a random vector will align more closely with the response variable A than the actual explanatory vector.

16.5 Tests of Simple Models

Often, only one explanatory variable is of interest. The question is whether that one explanatory variable is related to the response variable. In such situations, the other explanatory variables are not of direct interest, they play the role of covariates. Chapter 17 shows how covariates can be incorporated into ANOVA.

A very simple situation is when there are no covariates; just the one explanatory variable of interest. Often this situation arises because the person analyzing the data doesn't know how to handle covariates or doesn't appreciate how they can influence the interpretation of the variable of interest. If covariates are likely to be important in the system you are studying, use the appropriate methods, those introduced in the next chapter.

Historically, most students of statistics have been taught methods that involve only a single explanatory variable. There's nothing magic about these methods; they are powerless to adjust for covariates. Still, you will encounter them often since they have been so strongly emphasized in statistics education. You need to know what they are, when they are appropriately used (for example, when there are no covariates), and how you can replace them with more appropriate techniques.

16.5.1 Group Means

A simple and commonly asked question is whether a quantity is different for two or more different groups. For example, you might want to know if wages are different for men and women or for people who work in different sectors of the economy. In a clinical experiment, you might want to know if a medication has a different effect than a placebo.

Suppose that A is the response variable and G is a categorical variable that describes which group each case belongs to. The model A ~ 1 + G gives model values that are the groupwise means. The null hypothesis is that, in the population, G has no link to A.

This null hypothesis is often described in another way, that the population means of the different groups are the same. The whole-model hypothesis test on the model amounts to asking whether the sample provides compelling evidence that, over the whole population, the means of the groups are different from one another.

Depending on how many groups there are, people use different names to refer to the hypothesis test on group-wise means:

Two groups The hypothesis test is called a **t-test**.

More than two groups The test is called a **one-way analysis of variance**.

This distinction is hardly worthwhile, but you need to be aware of it so you can work with people who make it; they are in the big majority. Both of the tests are just ANOVA. (There is one kind of t-test, the **unequal variance t-test** that is somewhat different from ANOVA. But there is hardly any advantage to this test, and it incurs a large disadvantage by not allowing the incorporation of covariates. See Section 16.7.3.)

Example 16.4: Heights of the two sexes Do Galton's data give good reason to believe that males and females have different group means in the population? The model is height ~ 1+sex. Here is the ANOVA report:

	Df	Sum Sq	Mean Sq	F value	Pr(>F)
sex	1	5875	5875	**933.0**	0.0000
Residuals	896	5640	6.3		

The sex variable has only one degree of freedom because of the redundancy of its two indicator vectors with the intercept vector. Note that the F value is huge, 933. Correspondingly, the p-value is practically zero. The data provide good evidence that the population means of the two sexes are different.

Example 16.5: Wage and the sector of the economy Does the CPS data provide evidence that wages are systematically different in different sectors of the economy? The model is wage ~ 1+sector

	Df	Sum Sq	Mean Sq	F value	Pr(>F)
sector	7	2572	367	**16.8**	0.0000
Residuals	526	11505	22		

The F value, 16.8, is large. Correspondingly the p-value is very small. This leads to the rejection of the null hypothesis.

16.5.2 Slopes

Exactly the same ideas of the model triangle, square lengths, and R^2 apply to models where the explanatory variable is quantitative. The null hypothesis is still that the explanatory variable is not linked to the response variable in the population, the hypothesis test explores whether the degree of alignment in the sample is such as to call the null into question.

In the model A ~ 1 + B, the coefficient on B is the slope of the straight line model of A versus B. If B is not linked to A, that slope will be zero — A is independent of B. So, the conventional way to describe the null hypothesis is that it says the slope is zero in the population.

Example 16.6: Mother and Child Heights

Is there a relationship between the height of the mother and of her adult child? A simple model to capture this is height ~ mother. Here is the ANOVA report on that model:

	Df	Sum Sq	Mean Sq	F value	Pr(>F)
mother	1	468	468	38	0.0000
Residuals	896	11047	12		

The F value is 38, large enough to produce a tiny p-value of about 10^{-9}. Notice that the ANOVA report doesn't give information about the direction of the effect. For this, you need to look at the regression report.

	Estimate	Std. Error	t value	Pr(>|t|)
(Intercept)	46.6908	3.2587	14.33	0.0000
mother	0.3132	0.0508	6.16	0.0000

It's worth noting that the regression report also gives a p-value on the mother term: it's the same as that from the ANOVA.

For some reason, the term ANOVA has come to be associated by many people only with a test of the difference in group-wise means. So you may encounter people who say that a test of a slope is not ANOVA.

16.5.3 The Grand Mean

Occasionally, you may be interested in knowing whether the sample mean of a variable is consistent with some hypothetical value. For example, the average height of adult males nowadays is about 72 inches. Was it different in Galton's time?

This involves comparing the sample mean height for Galton's data to a hypothetical population mean of 72 inches. The conventional way to perform such a test is called the **one-sample t-test**. Usually, this test is presented in terms of a formula or a software implementation of the formula.

However, the one-sample t-test is really just a form of analysis of variance, where the model is $A \sim 1$. In the spirit of showing hypothesis tests in a unified way, here's how to perform a one-sample t-test using ANOVA.

1. Subtract the null hypothesis value from the variable of interest. Since the hypothesis is that Galton's men were randomly selected from a population with mean height 72 inches, subtract 72 from the actual heights. That is, A = height-72.

2. Using the data from (1), perform ANOVA on the model $A \sim 1$.

Here's the ANOVA report on the model $A \sim 1$.

	Df	Sum Sq	Mean Sq	F value	Pr(>F)
(Intercept)	1	24650.4	24650.4	**1920.2**	0.0000
Residuals	897	11515.1	12.8		

The F value is much bigger than 1, so the null hypothesis can be rejected.

16.6 Interpreting the p-value

The p-value is a useful format for summarizing the strength of evidence that the data provide. But statistical evidence, like any other form of evidence, needs to be interpreted in context.

16.6.1 Multiple Comparisons

Keep in mind that a hypothesis test, particularly when p is near the conventional $p < 0.05$ threshold for rejection of the null, is a very weak standard of evidence. Failure to reject the null may just mean that there isn't enough data to reveal the patterns, or that important explanatory variables have not been included in the model.

But even when the p-value is below 0.05, it needs to be interpreted in the context within which the tested hypothesis was generated. Often, models are used to explore data, looking for relationships among variables. The hypothesis tests that stem from such an exploration can give misleadingly low p-values.

To illustrate, consider an analogous situation in the world of crime. Suppose there were 20 different genetic markers evenly distributed through the population. A crime has been committed and solid evidence found on the scene shows that the perpetrator has a particular genetic marker, M, found in only 5% of the population. The police fan out through the city, testing passersby for the M marker. A man with M is quickly found and arrested.

Should he be convicted based on this evidence? Of course not. That this specific man should match the crime marker is unlikely, a probability of only 5%. But by testing large numbers of people who have no particular connection to the crime, it's a guarantee that someone who matches will be found.

Now suppose that eyewitnesses had seen the crime at a distance. The police arrived at the scene and quickly cordoned off the area. A man, clearly nervous and disturbed, was caught trying to sneak through the cordon. Questioning by the police revealed that he didn't live in the area. His story for why he was there did not hold up. A genetic test shows he has marker M. This provides much stronger, much more credible evidence. The physical datum of a match is just the same as in the previous scenario, but the context in which that datum is set is much more compelling.

The lesson here is that the p-value needs to be interpreted in context. The p-value itself doesn't reveal that context. Instead, the story of the research project come into play. Have many different models been fitted to the data? If so, then it's likely that one or more of them may have $p < 0.05$ even if the explanatory variables are not linked to the response.

One of the gravest misinterpretations of hypothesis testing treats the p-value as the probability that the rejection of the null is wrong. With this misconception, a researcher who has just rejected the null with a p-value of 0.05 treats this as strong evidence: only a one-in-twenty chance that the conclusion is mistaken. This is utterly wrong, but unfortunately the difference from the true interpretation is rather subtle.

To see that it's incorrect to interpret the p-value as the probability that the null is wrong, consider the unfortunate researcher who happens to work in a world where the null hypothesis is always true. In this world, each study that the researcher performs will produce a p-value that is effectively a random number equally likely to be anywhere between 0 and 1. If this researcher performs many studies, it's highly likely that one or more of them will produce a p-value less than 0.05 even though the null is true.

Rejecting the null hypothesis on the basis of that nominally "significant" finding would be wrong. Statistician David Freedman writes, "Given a significant finding ... the chance of the null hypothesis being true is ill-defined — especially when publication is driven by the search for significance." [17]

One approach to dealing with multiple tests is to adjust the threshold for rejection of the null to reflect the multiple possibilities that chance has to produce a small p-value. A simple and conservative method for adjusting for multiple tests is the **Bonferroni correction**. Suppose you perform k different tests. The Bonferroni correction adjusts the threshold for rejection downward by a factor of $1/k$. For example, if you perform 15 hypothesis tests, rather than rejecting the null at a level of 0.05, your threshold for rejection should be $0.05/15 = 0.0033$.

Another strategy is to treat multiple testing situations as "hypothesis generators." Tests that produce low p-values are not to be automatically deemed as significant but as worthwhile candidates for further testing.[27]. Go out and collect a new sample, independent of the original one. Then use this sample to perform exactly one hypothesis test: re-testing the model whose low p-value originally prompted your interest. This common-sense procedure is called an **out-of-sample** test, since the test is performed on data outside the original sample used to form the hypothesis. In contrast, tests conducted on the data used to form the hypothesis are called **in-sample** tests. It's appropriate to be skeptical of in-sample tests.

When reading work from others, it can be hard to know for sure whether a test is in-sample or out-of-sample. For this reason, researchers value **prospective studies**, where the data are collected *after* the hypothesis has been framed. Obviously, data from a prospective study must be out-of-sample. In contrast, in **retrospective studies**, where data that have already been collected are used to test the hypothesis, there is a possibility that the same data used to form the hypothesis are also being used to test it. Retrospective studies are often disparaged for this reason, although really the issue is whether the data are in-sample or out-of-sample. (Retrospective studies also have the disadvantage that the data being analyzed might not have been collected in a way that optimally addresses the hypothesis. For example, important covariates might have been neglected.)

Example 16.7: Multiple Jeopardy

You might be thinking: Who would conduct multiple tests, effectively shopping around for a low p-value? It happens all the time, even in situations where wrong conclusions have important implications.

To illustrate, consider the procedures adopted by the US Government's Office of Federal Contract Compliance Programs when auditing government contractors to see if they discriminate in their hiring practices. The OFCCP requires contractors to submit a report listing the number of applicants and the number of hires broken down by sex, by several racial/ethnic categories, and by job group. In one such report, there were 9 job groups (manager, professional, technical, sales workers, office/clerical, skilled crafts, operatives, laborers, service workers) and 6 discrimination categories (Black, Hispanic, Asian/Pacific Islander, American Indian/American Native, females, and people with disabilities).

According to the OFCCP operating procedures,[28] a separate hypothesis test with a rejection threshold of about 0.05 is to be undertaken for each job group and for each discrimination category, for example, discrimination against Black managers or against female sales workers. This corresponds to 54 different hypothesis tests. Rejection of the null in any one of these tests triggers punitive action by the OFCCP. Audits can occur annually, so there is even more potential for repeated testing.

Such procedures are not legitimately called hypothesis tests; they should be treated as **screening tests** meant to identify potential problem areas which can then be confirmed by collecting more data. A legitimate hypothesis test would require that the threshold be adjusted to reflect the number of tests conducted, or that a pattern identified by screening in one year has to be confirmed in the following year.

16.6.2 Significance vs Substance

It's very common for people to mis-interpret the p-value as a measure of the strength of a relationship, rather than as a measure for the *evidence* that the data provide. Even when a relationship is very slight, as indicated by model coefficients or an R^2, the evidence for it can be very strong so long as there are enough cases.

Suppose, for example, that you are studying how long patients survive after being diagnosed with a particular disease. The typical survival time is 10 years, with a standard deviation of 5 years. In your research, you have identified a genetic trait that explains some of the survival time: modeling survival by the genetics produces an R^2 of 0.01. Taking this genetic trait into account makes the survival time more predictable; it reduces the standard deviation of survival time from 5 years to 4.975 years. Big deal! The fact is, an R^2 of 0.01 does not reduce uncertainty very much; it leaves 99 percent of the variance unaccounted for.

To see how the F statistic stems from both $R^2 = 0.01$ and the sample size n, re-write the formula for F in Equation 16.1:

$$F = \frac{R^2}{m-1} \Big/ \frac{1-R^2}{n-m} \tag{16.2}$$

$$= \left(\frac{n-m}{m-1}\right) \frac{R^2}{1-R^2} \tag{16.3}$$

$$= \text{“Sample size”} \times \text{“Substance”}. \tag{16.4}$$

The ratio $\frac{R^2}{1-R^2}$ is a reasonable way to measure the substance of the result: how forceful the relationship is. The ratio $\frac{n-m}{m-1}$ reflects the amount of data. Any relationship, even one that lacks practical substance, can be made to give a large F value so long as n is big enough. For example, taking $m = 2$ in your genetic trait model, a small study with $n = 100$ gives an F value of approximately 1: not significant. But if $n = 1000$ cases were collected, the F value would become a hugely significant $F = 10$.

When used in a statistical sense to describe a relationship, the word "significant" does not mean important or substantial. It merely means that there is evidence that the relationship did not arise purely by accident from random variation in the sample.

Example 16.8: The Significance of Finger Lengths

Example 7.3 (page 134) commented on a study that found $R^2 = 0.044$ for the relationship between finger-length ratios and aggressiveness. The relationship was hyped in the news media as the "key to aggression." (BBC News, 2005/03/04) The researchers who published the study[14] didn't characterize their results in this dramatic way. They reported merely that they found a statistically signficant result, that is, p-value of 0.028. The small p-value stems not from a forceful correlation but from number of cases; their study, at $n = 134$, was large enough to produce a value of F bigger than 4 even though the "substance", $R^2/(1 - R^2)$, is only 0.046.

The researchers were careful in presenting their results and gave a detailed presentation of their study. They described their use of four different ways of measuring of aggression: physical, hostility, verbal, and anger. These four scales were each used to model finger-length ratio for men and women separately. Here are the p-values they report:

Sex	Scale	R^2	p-value
Males	Physical	0.044	0.028
	Hostility	0.016	0.198
	Verbal	0.008	0.347
	Anger	0.001	0.721
Females	Physical	0.010	0.308
	Hostility	0.001	0.670
	Verbal	0.001	0.778
	Anger	0.000	0.887

Notice that only one of the eight p-values is below the 0.05 threshold. The others seem randomly scattered on the interval 0 to 1, as would be expected if there were no relationship between the finger-length ratio and the aggressiveness scales. A Bonferroni correction to account for the eight tests gives a rejection threshold of $0.05/8 = 0.00625$. None of the tests satisfies this threshold.

16.7 Computational Technique

The chapter presents two ways of doing hypothesis tests on whole models: (1) permutation tests where the connection is severed between the explanatory and response variables, (2) tests such as ANOVA where the sampling distribution is calculated from first principles. In practice, first-principle tests are used most of the time. Still, the permutation test is useful for developing intuition about hypothesis testing — our main purpose here — and for those occasions where the assumptions behind the first-principle tests are dubious.

16.7.1 The Permutation Test

The idea of a permutation test is to enforce the null hypothesis that there is no connection between the response variables and the explanatory variables. An effective way to do this is to randomize the response variable in a way that is consistent with sampling variability. The resampling operation lets you do this. But there is a new twist.

In constructing confidence intervals, resampling was applied to the entire data frame. This means that the cases were a random selection of cases from the frame but that each case was itself authentic. Now, you'll use resampling in a different way, to assign a random value of the response variable to each case.

I'll illustrate with a model that explores whether sex and mother's height are related to the height of the child:

```
> galton = ISMdata("galton.csv")
> mod = lm( height ~ sex + mother, data=galton)
> coefficients(mod)
(Intercept)        sexM      mother
    41.4495      5.1767      0.3531
```

The coefficients indicate that typical males are taller than typical females by about 5 inches and that for each inch taller the mother is, a child will typically be taller by 0.35 inches. The test statistic that I'll use to summarize the whole model is R^2:

```
> Rsquared(mod)
[1] 0.5618
```

By applying resampling to the entire data frame, we get an idea of the sampling variability in these coefficients. To highlight typographically where the resampling occurs, it's printed in bold type:

```
> lm(height~sex+mother,data= resample(galton) )
(Intercept)          sexM       mother
    40.0057        5.2317       0.3745
> lm(height~sex+mother,data= resample(galton) )
(Intercept)          sexM       mother
    44.1076        5.3368       0.3114
> lm(height~sex+mother,data= resample(galton) )
(Intercept)          sexM       mother
    40.1458        5.0296       0.3752
```

The sexM coefficients are tightly grouped near 5 inches, the mother coefficients are around 0.3 to 0.4.

In order to a permutation test, do not randomize the whole data frame. Instead, shuffle just the response variable:

```
> lm( shuffle(height) ~ sex+mother, data=galton)
(Intercept)          sexM       mother
   63.69634      -0.33714      0.05174
> lm( shuffle(height) ~ sex+mother, data=galton)
(Intercept)          sexM       mother
   69.66573      -0.00719     -0.04485
> lm( shuffle(height) ~ sex+mother, data=galton)
(Intercept)          sexM       mother
    64.9935        0.1456       0.0264
```

Now the sexM and mother coefficients are close to zero, as would be expected when there is no relationship between the response variable and the explanatory variables.

To construct a sample from the sampling distribution of the test statistic under the null hypothesis, use do to do the simulation over and over again:

```
> samps = do(500)*
    Rsquared(lm( shuffle(height) ~ sex + mother,
                 data=galton))
```

Notice that the Rsquared operator has been used to calculate the test statistic R^2 from the model.

```
> samps
  [1] 4.78e-04 8.43e-05 1.08e-03 9.24e-05 1.59e-03
  [6] 1.95e-03 7.33e-04 5.02e-04 2.64e-03 2.75e-03
      ... and so on ...
[496] 3.47e-04 1.25e-03 4.81e-03 3.69e-03 3.05e-03
```

Naturally, all of the R^2 values are close to zero since there is no relation between the response variable (after randomization with shuffle and the explanatory variables.

The p-value can be calculated directly from the sample. The observed value of the test statistic R^2 was 0.5618 and, as it happens, none of the simulations reached that level:

```
pdata( 0.5618, samps )
[1] 1
```

Since the observed test statistic $R^2 = 0.5618$ is less than in all 500 of the simulations of the null hypothesis, it's reasonable to say that the p-value is $p < 1/500$.

16.7.2 First-Principle Tests

On modern computers, the permutation test is entirely practical. But a few decades ago, it was not. Great creativity was applied to finding test statistics where the sampling distribution could be estimated without extensive calculation. One of these is the F statistic. This is still very useful today. It is presented in the regression report.

Here is the regression report from the height ~ sex+mother:

```
> mod = lm( height ~ sex + mother, data=galton)
> summary(mod)

 Coefficients:
             Estimate Std. Error t value Pr(>|t|)
 (Intercept) 41.4495      2.2095    18.8   <2e-16
 sexM         5.1767      0.1587    32.6   <2e-16
 mother       0.3531      0.0344    10.3   <2e-16

 Residual standard error: 2.37 on 895 degrees of freedom
 Multiple R-Squared: 0.562, Adjusted R-squared: 0.561
 F-statistic: 574 on 2 and 895 DF,  p-value: <2e-16
```

The report shows an F statistic of 574 based on an R^2 of 0.562 and translates this to a p-value that is practically zero: <2e-16.

By way of showing that the regression report is rooted in the same approach shown in the chapter, you can confirm the calculations. There are $m = 3$ coefficients and $n = 898$ cases, producing $n - m = 895$ degrees of freedom in the denominator and $m - 1 = 2$ degrees of freedom in the numerator. The calculation of the F statistic from R^2 and the degrees of freedom follows the formula given in the chapter.

$$F = \frac{R^2}{m-1} \bigg/ \frac{1-R^2}{n-m}$$

Plugging the values into the formula

```
> (0.562/2) / ((1-.562)/895)
[1] 574.2
```

F is the test statistic. To convert it to a p-value, you need to calculate how extreme the value of F= 574.2 is with reference to the F distribution with 895 and 2 degrees of freedom.

```
> 1 - pf( 574.2, 2, 895)
[1] 0
```

The calculation of p-values from F always follows this form. In the context of the F distribution, "extreme" always means to "bigger than." So, calculate the area under the F distribution to the right of the observed value.

16.7.3 Standard Tests

Applications of the F statistic to very simple models have been widely taught to statistics students. For the reader of this book, those tests will rightly appear to be simple special cases of a general approach. Traditionally, the methods have been given different names. The names — one- and two-sample t-tests, analysis of variance (ANOVA) — belie the underlying unity of the approach.

It's very likely that you will need to communicate with people who use the traditional names. So, it is important that you know how to relate them to the general approach of the F test. Since most people have been taught that the methods are different from one another, don't be surprised to hear something like this: "But I want a t-test, not an F-test." The examples below give, where appropriate, the commands needed to demonstrate that they are the same thing.

Differences in Group-wise Means: One-Way ANOVA

A common question is whether a trait is different between groups. This arises, for example, in an experiment where some people are given a drug and others serve as controls. Perhaps there are three treatment groups: those given the drug, those given a metabolically inert placebo, and those given nothing at all. The treatment is the explanatory variable and the measured trait — call it blood pressure (BP) — is the response variable. Statistically, the issue of group differences often is distilled down to whether the mean response is different among the groups. In the language of models, this would be BP ~ treatment.

An F-test of the sort done when asking whether treatment is associated with BP is traditionally called **one-way analysis of variance**. In conventional language, the statement of the null hypothesis of one-way ANOVA is that all the groups have the same mean BP in the population and that any observed differences between the sample means are due merely to sampling fluctuations.

The calculation follows exactly the pattern seen in this chapter. To illustrate with familiar data, consider the wages from the current population survey . Let the response variable be wage and the explanatory grouping variable be the sector of the economy. In everyday language, the question of the test is whether the data provide significant evidence that wages differ in different sectors of the economy.

To perform the test, first fit the model to the data and then produce the regression report:

```
> mod2 = lm( wage ~ sector, data=cps)
> summary(mod2)
Coefficients:
              Estimate Std. Error t value Pr(>|t|)
(Intercept)      7.423      0.475   15.63  < 2e-16
sectorconst      2.079      1.149    1.81    0.071
sectormanag      5.281      0.789    6.69  5.7e-11
sectormanuf      0.613      0.740    0.83    0.407
sectorother      1.078      0.740    1.46    0.146
sectorprof       4.525      0.659    6.87  1.8e-11
sectorsales      0.170      0.895    0.19    0.849
sectorservice   -0.885      0.699   -1.27    0.206

Residual standard error: 4.68 on 526 degrees of freedom
Multiple R-Squared: 0.183, Adjusted R-squared: 0.172
F-statistic: 16.8 on 7 and 526 DF,  p-value: <2e-16
```

In this whole-model test, the interest is in the F statistic derived from R^2 and the degrees of freedom.

There is a more streamlined way of extracting this information from the test. The format, called an **analysis of variance table** (or **ANOVA table**, for short) will be described in more detail in the next chapter. It's shown here just so that you can see that it is another way of packaging up the information already contained in the regression report:

```
> anova(mod2)
Analysis of Variance Table

Response: wage
             Df Sum Sq Mean Sq F value Pr(>F)
(Intercept)   1  43486   43486  1988.1 <2e-16
sector        7   2572     367    16.8 <2e-16
Residuals   526  11505      22
```

The p-value calculated for this model is practically zero, justifying rejection of the null hypothesis that all the groups have the same mean in the population.

Differences Between Means of Two Groups: the Two-Sample t-test

The one-way ANOVA approach works when comparing the means among groups. It can be applied to any number of groups: 2 groups, 3 groups, 4 groups, and so on.

You might be surprised to hear that there is a special test, called the **two-sample t-test** for testing whether the means are different between two groups. I'll illustrate it using the question of whether the current population survey data give significant evidence that wages are different between men and women — two groups.

Using the modeling approach, the calculations are:

```
> mod3 = lm( wage ~ sex, data=cps)
> summary(mod3)
Coefficients:
            Estimate Std. Error t value Pr(>|t|)
(Intercept)    7.879      0.322   24.50  < 2e-16
sexM           2.116      0.437    4.84  1.7e-06

Residual standard error: 5.03 on 532 degrees of freedom
Multiple R-Squared: 0.0422, Adjusted R-squared: 0.0404
F-statistic: 23.4 on 1 and 532 DF,  p-value: 1.70e-06
```

The p-value is very small, justifying rejection of the null hypothesis. (This conclusion is subject to some legitimate criticism, since no attempt has been made to adjust for covariates such as `age` or the level of education. With a modeling approach, you can add these covariates to the model. But the t-test and one-way ANOVA approaches don't allow this to be done. That's sufficient reason not to use them. For the purpose here, to demonstrate how to perform a t-test, I'll continue on without claiming that the use of the method is justified in anything but a formal sense.)

The streamlined form of the calculation produces the same F and p-value:

```
> anova(mod3)

Analysis of Variance Table

Response: wage
             Df Sum Sq Mean Sq F value  Pr(>F)
(Intercept)   1  43486   43486  1715.8 < 2e-16
sex           1    594     594    23.4 1.7e-06
Residuals   532  13483      25
```

Now to do this in the form of a t-test. The command uses the familiar modeling language to identify the response variable (`wage`) and the explanatory grouping variable (`sex`).

```
> t.test( wage ~ sex, data=cps, var.equal=TRUE)

        Two Sample t-test

data:  wage by sex
t = -4.84, df = 532, p-value = 1.7e-06
alternative hypothesis:
  true difference in means is not equal to 0
95 percent confidence interval:
 -2.975 -1.257
```

```
sample estimates:
mean in group F mean in group M
          7.879           9.995
```

Rather than computing F, the t-test computes a statistic called the t value. The relationship between the two is simple: $t^2 = F$. So, the t of -4.84 corresponds to F $= 23.4$. Similarly, in the t-test there is a quantity called the "degrees of freedom." This corresponds to the degrees of freedom in the denominator in the F test. The p-values from the t-test and the F test are the same!

Technical Note: The Unequal Variance t-test

Actually, the above t-test is not the one that many people would perform in this situation. The full name of the above test is the **equal-variance t-test**. I include this section mainly for the reader who has been told about the **unequal-variance t-test**. The issue of whether to use an equal-variance test or an unequal variance test is one of the places where the approach taken in this book rubs against conventional pedagogical practice.

The term **equal variance** refers to an extension of the null hypothesis: rather than the null hypothesis being that the population means of the response variable are the same in the two groups, the null is expanded to say also that the population variances are equal. The named argument `var.equal=TRUE` is what specifies this additional assumption to the software.

If you suspend the assumption that the group variances are equal, the appropriate t-test test takes on another form, the **unequal variance t-test**. Here it is:

```
> t.test( wage ~ sex, data=cps)


         Welch Two Sample t-test

data:  wage by sex
t = -4.885, df = 530.5, p-value = 1.369e-06
alternative hypothesis:
  true difference in means is not equal to 0
95 percent confidence interval:
 -2.967 -1.265
sample estimates:
mean in group F mean in group M
          7.879           9.995
```

The results of the unequal variance t-test are somewhat different from those of ANOVA or the equal variance t-test. To summarize these differences:

	F-test	equal var. t-test	unequal var. t-test
D.F.	532	532	530.5
t		-4.84	-4.885
t^2 or F	23.4	23.4	23.86
p-value	1.7×10^{-6}	1.7×10^{-6}	1.369×10^{-6}

The values provided by the equal and unequal variance t-tests are different but very close.

Which test to use? Many people will argue that the unequal variance t-test is the appropriate test to use when you don't have a good reason to assume that the variances of the two groups are different. This is reasonable enough if one is trying to choose between the equal-variance and the unequal-variance t-test. But there is an "elephant in the room," an aspect of the situation that dominates things but that, for some reason, people don't talk about. This elephant is the covariates — the other explanatory variables that you should be adjusting for. The t-test approach doesn't provide any capability for such adjustment. Indeed, if you try to add another explanatory variable to the t-test command, you get an error message:

```
> t.test( wage ~ sex + educ, data=cps)
Error in t.test.formula(wage ~ sex + educ, data = cps) :
  'formula' missing or incorrect
```

The same applies to trying to do a t-test with a quantitative explanatory variable:

```
> t.test( wage ~ educ, data=cps)
Error in t.test.formula(wage ~ educ, data = cps) :
  grouping factor must have exactly 2 levels
```

The ANOVA approach easily allows the addition of covariates to the model.

If you find yourself in a situation where there are no covariates and you are comparing two groups with potentially unequal variances, it makes sense to use the unequal variance t-test. I regard it as a special-purpose method that should be used only in such special situations. As it happens, the results you get with the equal-variance t-test and the unequal-variance t-test will be very close ... unless the variances of the two groups are hugely different. And, if the variances are indeed so different, you should be asking yourself deeper questions, such as "Why am I interested in the group mean in the first place? Isn't the difference in variance telling me something, too?"

The Grand Mean: the One-Sample t-test

Occasionally, the question to be asked of a variable is very simple: Is there evidence that the overall mean is different from some specified value. For example, suppose the government had published figures saying that the mean wage was $9.25 per hour at the time the current population survey data were collected. You're interested to know whether the CPS data are consistent with this claim.

The one-sample t-test is designed to deal with this situation.

```
> t.test( cps$wage, mu=9.25)

        One Sample t-test

 t = -1.016, df = 533, p-value = 0.3101
 alternative hypothesis:
   true mean is not equal to 9.25
 95 percent confidence interval:
  8.587 9.461
 sample estimates:
 mean of x
     9.024
```

The named argument `mu=9.25` specifies the population value of the mean under the null hypothesis. In this case, the p-value is large, 0.3101, so there is no reason to reject the null hypothesis.

The one-sample t-test is equivalent to the F test applied to a model where the only explanatory model term is the intercept. As always, $t^2 = F$, so the effective F value from the t-test is $-1.016^2 = 1.032$

```
> mod4 = lm( wage - 9.25 ~ 1, data=cps)
> anova(mod4)

 Analysis of Variance Table

 Response: wage - 9.25
              Df Sum Sq Mean Sq F value Pr(>F)
 (Intercept)   1     27      27    1.03   0.31
 Residuals   533  14077      26
```

The results are identical to the one-sample t-test. (There is no unequal-variance, one-sample t-test for the simple reason that in the one-sample t-test there are no groups to have difference variances!)

Notice that in forming the model, the null hypothesis parameter, 9.25 has been subtracted from the response variable in the modeling statement. This is essential. Otherwise, the test will be whether the sample mean is consistent with a population value of zero. Formally, that's a possible null hypothesis, but it makes no sense in this situation.

One important use for the one-sample t-test occurs in situations where measurements have been made in pairs, for example, before-and-after measurements. When used this way, the one-sample t-test is called a **paired t-test**. The idea is to test the difference within each pair to see if it is non-zero. As you'll see in the next chapter, ANOVA can also be configured to perform such paired tests and, in fact, is able to generalize them to multiple measurements and to adjust for covariates.

Chapter 17

Hypothesis Testing on Parts of Models

Everything should be made as simple as possible, but not simpler. — Albert Einstein

It often happens in studying a system that a single explanatory variable is of direct interest to the modeler. Other explanatory variables may be important in the way the system works, but they are not of primary interest to the modeler. As in previous chapters, call the explanatory variable of interest simply the "explanatory variable." The other explanatory variables, the ones that aren't of direct interest, are the **covariates**. The covariates are ordinary variables. Their designation as covariates reflects the interests of the modeler rather than any intrinsic property of the variables themselves.

The source of the interest in the explanatory variable might be to discover whether it is relevant to a model of a response variable. For example, in studying people's heights, one expects that several covariates, for instance the person's sex or the height of their father or mother, play a role. It would be interesting to find out whether the number of siblings a person has, nkids in Galton's dataset, is linked to the height. Studying nkids in isolation might be misleading since the other variables are reasonably regarded as influencing height. But doing a whole-model hypothesis test on a model that includes both nkids and the covariates would be uninformative: Of course one can account for a significant amount of the variation in height using sex and mother and father. (See Example 16.3 on page 297.) The question about nkids is whether it contributes something to the explanation that goes beyond the covariates. To answer this question, one needs a way of assessing the contribution of nkids on its own, but in the context set by the other variables.

Sometimes the focus on a single explanatory variable comes about because a decision may hang on the role of that variable but that role might depend on the context set by covariates. Those covariates might enhance, mask, or reverse the role of the variable of interest. For instance, it's legitimate that wages might vary according to the type of job and the worker's level of experience and education. It's also possible that wages might vary according to sex or race insofar as those variables happen to be correlated with the covariates, that is, correlated with the type of job or level of education and experience. But, if wages depend on the worker's sex or race *even taking into account* the legitimate factors, that's a different story and suggests a more sinister mechanism at work.

This chapter is about conducting hypothesis tests on a single explanatory variable in the context set by covariates. In talking about general principles, the text will refer to models of the form A ~ B+C or A ~ B+C+D where A is the response variable, B is the variable of interest, and C and D are the variables that you aren't directly interested in: the covariates.

Recall the definition of statistics offered in Chapter 1:

> *Statistics is the explanation of variation in the context of what remains unexplained.*

It's important to include covariates in a hypothesis test because they influence both aspects of this definition:

Explanation of variation When the covariates are correlated with the explanatory variable, as they often are, including the covariates in a model will change the coefficients on the explanatory variable.

What remains unexplained Including covariates raises the R^2 of a model. Or, to put it another way, including covariates reduces the size of the residuals. This can make the explanatory variable look better: smaller residuals generally mean smaller standard errors and higher F statistics. This is not an accounting trick, it genuinely reflects how much of the response variable remains unexplained.

17.1 The Term-by-Term ANOVA Table

The ANOVA table introduced in the Chapter 16 provides a framework for conducting hypothesis tests that include covariates. The whole-model ANOVA table lets you compare the "size" of the explained part of the model with the "size" of the residuals. Here is a whole-model report for the model height ~ nkids + sex + father + mother fit to Galton's data:

	Df	Sum Sq	Mean Sq	F value	Pr(>F)
Model terms	4	7377.9	1845.0	401	0.0000
Residuals	893	4137.1	4.6		

The square length of the residual vector is 4137.1, the square length of the fitted model vector is 7377.9. The F test takes into account the number of vectors involved in each — four explanatory vectors (neglecting the intercept term), leaving 893 residual degrees of freedom. The F value, 401, is the ratio of the mean square of the model terms to the residuals. Since 401 is much, much larger than 1 (which is the expected value when the model terms are just random vectors), it's appropriate to reject the null hypothesis; the p-value is effectively zero since F is so huge.

A term-by-term ANOVA report is much the same, but the report doesn't just partition variation into "modeled" and "residual," it goes further by partitioning the modeled variation among the different explanatory terms. There is one row in the report for each individual model term.

	Df	Sum Sq	Mean Sq	F value	Pr(>F)
nkids	1	185.5	185.5	40.0	0.0000
sex	1	5766.3	5766.3	1244.7	0.0000
father	1	937.6	937.6	202.4	0.0000
mother	1	488.5	488.5	105.5	0.0000
Residuals	893	4137.1	4.6		

Notice that the "Residuals" row is exactly the same in the two ANOVA reports. This is because that row just describes the square length of the residuals — it's the same model with the same residuals being analyzed in the two ANOVA reports. Less obvious, perhaps, is that the sum of squares of the individual model terms adds up across all the terms to give exactly the sum of squares from the whole-model ANOVA report: $185.5 + 5766.3 + 937.6 + 488.5 = 7377.9$. The same is true for the degrees of freedom of the model terms: $1+1+1+1 = 4$.

What's new in the term-by-term ANOVA report is that there is a separate mean square, F value, and p-value for each model term. These are calculated in the familiar way: the mean square for any term is just the sum of squares for that term divided by the degrees of freedom for that term. Similarly, the F value for each term is the mean square for that term divided by the mean square of the residuals. For instance, for nkids, the F value is $185.5/4.6 = 40$. This is translated to a p-value in exactly the same way as in Chapter 16 (using an F distribution with 1 degree of freedom in the numerator and 893 in the denominator — values that come from the corresponding rows of the ANOVA report).

17.2 Covariates Soak Up Variance

An important role of covariates is to account for variance in the response. This reduces the size of residuals. Effectively, the covariates reduce the amount of randomness attributed to the system and thereby make it easier to see patterns in the data. To illustrate, consider a simple question relating to marriage: Do men tend to be older than women when they get married?

The marriage-license dataset contains information on the ages of brides and grooms at the time of their marriage as well as other data on education, race, number of previous marriages, and so on. These data were collected from the on-line marriage license records made available by Mobile County, Alabama in the US. Here's an excerpt:

BookpageID	Person	Age
B230p539	Bride	28.7
B230p539	Groom	32.6
B230p677	Bride	52.6
B230p677	Groom	32.3
B230p766	Bride	26.7
B230p766	Groom	34.8
B230p892	Bride	39.6
B230p892	Groom	40.6
... and so on for 49 couples altogether		

The Person and Age variables are obvious enough. The BookpageID variable records the location of the marriage license in the Mobile County records as a book number and page number. There is only one license on each page, so this serves as a unique identifier of the couple.

It seems simple enough to test whether men tend to be older than women when they get married, just fit the model Age $\sim$ Person. Here it is:

	Estimate	Std. Error	t value	Pr(>|t|)
Intercept	33.239	2.060	16.13	0.0000
PersonGroom	2.546	2.914	0.87	0.3845

The coefficient on person**Groom** indicates that men are typically 2.5 years older than women when they get married. That's roughly what one would expect from experience — husbands and wives tend to be roughly the same age, with the husband typically a bit older. The confidence interval suggests a different interpretation, however. It's 2.5 ± 5.8, giving reason to believe that a different sample of licenses could give a coefficient of zero, indicating no age difference between men and women. ANOVA tells much the same story

	Df	Sum Sq	Mean Sq	F value	Pr(>F)
Person	1	159	159	0.76	0.38
Residuals	96	19967	208.00		

The F value, 0.76, is smaller than one. Correspondingly, the p-value is big: 0.38. Based on these data, it's not appropriate to reject the null hypothesis that men and women are the same age when they get married.

No big deal. Perhaps there just isn't enough data. With some effort, more data could be collected and this might bring the p-value down to the point where the null could be rejected.

Actually, there's much more information in this sample than the hypothesis test suggests. The model Age $\sim$ Person ignores a very important structure in the data: Who is married to whom. A person's age at marriage is very strongly related to the age of their spouse. That is, if you know the age of the wife, you have a lot of information about the age of the husband, and vice versa. This relationship between the husband and wife is carried by the variable BookpageID. Since the model Age ~ Person doesn't include BookpageID as a covariate, the model tends to overstate the unexplained part of age.

To fix things, add in BookpageID as a covariate and examine the model Age ~ Person + BookpageID. In this model the coefficient on Person**Groom** refers to the difference in age between males and females, *holding constant* the family. That is, this model looks at the difference between the husband and wife *within the family*. Here's the term-by-term ANOVA report:

	Df	Sum Sq	Mean Sq	F value	Pr(>F)
Person	1	159.0	159.0	9.07	0.0041
BookpageID	48	19127.0	398.5	22.77	0.0000
Residuals	48	840.0	17.5		

Notice that here the F value of Person is much larger than before, 9.07 versus 0.76 before. Correspondingly the p-value is much smaller: 0.004 versus 0.384. But the sum of squares of Person is exactly the same as it was for the previous model. What's happened is that the inclusion of BookpageID in the model has dramatically reduced the size of the residuals. The residual mean square is now 17.5 compared to 208 in the model that omitted BookpageID as a covariate.

It's not that BookpageID is a variable that's directly of interest. But that doesn't mean that you shouldn't include BookpageID in a model. It accounts for some of the variation in Age and in so doing makes it easier for the role of other variables to be seen above the random variation due to sampling.

The particular hypothesis test conducted here has a specific name, the **paired t-test**. The word "paired" reflects the natural pairing in the data: each family has a wife and a husband. Another context in which pairing arises is before-and-after studies, for example measuring a subject's blood pressure both before and after a drug is given.

A more general term is **repeated measures**. In a paired situation there are two measurements on each subject, but there can also be situations where there are more than two.

A still more general term is **analysis of covariance**, which refers simply to a situation where covariates are used to soak up the residuals.

Example 17.1: Wages and Race The Current Population Survey data (from the mid-1980s) can be analyzed to look for signs that there is discrimination based on race in the wages people earn. That dataset has a variable, race, that encodes the race of each wage earner as either Black or White. The obvious, simple model is wage ~ race. Here's the coefficient report:

	Estimate	Std. Error	t value	Pr(>|t|)
Intercept	8.06	0.63	12.86	0.0000
raceW	1.10	0.67	1.65	0.1001

According to this report, Whites earned an average of \$1.10 per hour more than Blacks. But the standard error is large: the confidence interval on this estimate is \$1.10 $\pm$ 1.34. The ANOVA report on the model suggests that these data don't give enough evidence to reject the null hypothesis that race is a factor in wages:

	Df	Sum Sq	Mean Sq	F value	Pr(>F)
race	1	71.5	71.45	2.71	0.1001
Residuals	532	14005.3	26.33		

The F value is 2.7, bigger than the expected value of one, but not so big as to be compelling evidence. The p-value is 0.10, too large to reject the null.

Remember that the p-value compares the explanatory power of race to a typical random term in order to account for sampling variation. In this model, all of the variation in wage is considered random other than that attributed to race. This is not reasonable. There are other sources of variation in wage such as the education of the worker and the type of job. Consider now a model with educ and sector of the economy as covariates: wage ~ race+educ + sector:

	Df	Sum Sq	Mean Sq	F value	Pr(>F)
race	1	71.45	71.45	3.44	0.0641
educ	1	2017.67	2017.67	97.22	0.0000
sector	7	1112.92	158.99	7.66	0.0000
Residuals	524	10874.66	20.75		

Now the F value on race is increased, simply because the size of the residuals has been reduced by inclusion of the covariates.

The lesson here is, again, that including covariates can make the patterns of the explanatory variable of interest stand out more clearly. Admittedly, the change in F here is not huge, but it has come at a very small price: fitting a new model. Collecting data can be expensive. The modeler seeks to extract the most information from that model.

This example will be continued later in this chapter.

17.3 Measuring the Sum of Squares

At a superficial level, interpretation of the term-by-term ANOVA report is straightforward. Use the p-value on each row to decide whether to reject the null hypothesis for that row. For instance, return to the height-model example from before: height ~ nkids+sex+father+mother

	Df	Sum Sq	Mean Sq	F value	Pr(>F)
nkids	1	185.46	185.46	40.03	**0.0000**
sex	1	5766.33	5766.33	1244.67	0.0000
father	1	937.63	937.63	202.39	0.0000
mother	1	488.52	488.52	105.45	0.0000
Residuals	893	4137.12	4.63		

The very small p-value on nkids (it's actually 0.0000000004) prompts a rejection of the null hypothesis on the nkids term. But there are other ANOVA reports on exactly the same model fitted to exactly the same data. Here's one:

	Df	Sum Sq	Mean Sq	F value	Pr(>F)
sex	1	5874.57	5874.57	1268.03	0.0000
father	1	1001.11	1001.11	216.09	0.0000
mother	1	490.22	490.22	105.81	0.0000
nkids	1	12.04	12.04	2.60	**0.1073**
Residuals	893	4137.12	4.63		

According to this report, nkids is not making a significant contribution to the model.

Why do the two reports give such different results for nkids? Which is right, the second report's p-value on nkids of 0.107 or the first report's p-value of 0.0000000004? Obviously, they lead to different conclusions. But they are *both* right. It's just that there are two different null hypotheses involved. In the first report, the null is that nkids on its own doesn't contribute beyond what's typical for a random vector. In the second report, the null is that nkids doesn't contribute *beyond* the contribution already made by mother, father, and sex. The two reports give us different perspectives on nkids.

In understanding term-by-term ANOVA, it's helpful to start with an analogy. Suppose you have a sports team consisting of players B, C, D, and so on. As a whole, the team scores 30 points in a game. B scored 3 of these points, C 10 of them, D no points, and so on.

The score reflects the team as a whole but you want to be able to identify particularly strong or weak players as individuals. The obvious thing to do is give each player credit for the points that he or she actually scored: B gets 3 points of credit, C 10 points of credit, D no credit, and so on.

This system, though simple, has its flaws. In particular, the system doesn't acknowledge the role of teamwork. Perhaps D's role is primarily defensive; she contributes to the team even though she doesn't actually do the scoring. C scores a lot, but maybe this is because B does a great job of passing to her. In evaluating an individual player's contribution, it's important to take into account the context established by the other players.

Now consider two different ways of assigning credit to individual players. Each involves replaying the game with some changes; this is impractical in real-world game playing but quite feasible when it comes to fitting models.

Take out one player Play the game and record the score. Now take out one player and re-play the game. The player who was taken out gets credit for the difference in scores between the two games.

What's nice about this method is that it gives the player credit for how he or she they contributed to the score **beyond** what the other players could have accomplished on their own.

Take out multiple players Play the game with a subset of the players. Now replay the game, adding in one new player. The new player gets credit for the difference in scores between the two games.

As you can imagine, the result of this method will depend on which subset of players was in the game. In one extreme limit, the new player would be the **only** player: a one person team.

Now to translate this analogy back to ANOVA. First, the game is played with only one player: a single model term, as in the model height ~ 1. The game's score is recorded as the sum of squares of the fitted model values. Next, another player is put in and the game is replayed: as in height ~ 1+nkids. More and more games are played, each successive game adding in one more player: height ~ 1+nkids+sex, then height ~ 1+nkids+sex+father and so on . In each game, the score is measured as the sum of squares.

To illustrate the process, reconstruct term-by-term ANOVA on the model height ~ 1+nkids + sex + father + mother. Think of this as a set of nested models, as shown in the table. For each of the nested models a "score" — that is, a sum of squares — is found.

Model	Sum of Squares	Term added	Additional Sum of Sq.
height ~ 1	4002376.8	1	
height ~ 1+nkids	4002562.3	nkids	185.5
height ~ 1+nkids+sex	4008328.6	sex	5766.3
height ~ 1+nkids+sex+father	4009266.2	father	937.6
height ~ 1+nkids+sex+father+mother	4009754.7	mother	488.5

The "Additional Sum of Squares" assigned to each term reflects how much that term increased the sum of squares compared to the previous model. For instance, adding the nkids term in the second line of the table increased the sum of squares from 4002376.8 to 4002562.3, an increase of 185.5. It's the increase that's attributed to nkids.

Perhaps you're wondering, why did nkids go first? ANOVA can take the terms in any order you want. For example:

Model	Sum of Squares	Term added	Additional Sum of Sq.
height ~ 1	4002376.8	1	
height ~ 1+mother	4002845.1	mother	468.3
height ~ 1+mother+father	4003630.8	sex	785.7
height ~ 1+mother+father+sex	4009742.7	sex	6112.0
height ~ 1+mother+father+sex+nkids	4009754.7	nkids	12.0

Notice that with the new arrangement of terms, the sum of squares assigned to each term is different. In particular, the first report gave nkids a fairly large sum of squares: 185.5. The second report gave nkids only 12.0. This change plays out in the p-value on nkids given in the ANOVA report: it's 0.107 now compared to 0.0000000004.

17.4 ANOVA, Collinearity, and Multi-Collinearity

In sports, it's easy to see why the performance of one player on a team depends on the context set by the other players. Players specialize in offense and defense, they support one another by passing and blocking or by drawing away the opponent's defense. A poor defensive player may hurt his own team by depriving it of opportunities to go on the offense. A good team player improves the performance of other players on the team by creating opportunities for them to perform well.

With model terms, the situation is analogous but obviously lacking the aspect of personality and intensionality that shapes team dynamics. Model terms help or hurt their "teammates" through collinearity.

To illustrate, consider a model A ~ B+C where B and C are collinear. The figure below shows the situation.

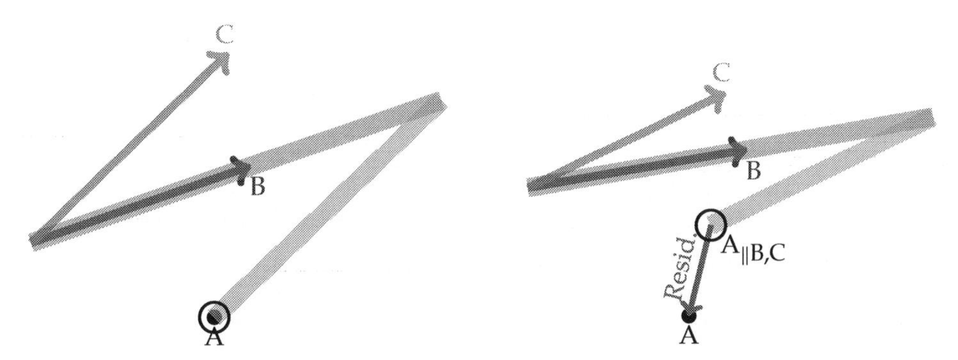

The B,C-plane is on the page. The residual is coming directly out of the page.

Rotated a little about the horizontal axis to show the residual vector.

In the figure, it looks like the linear combination of B and C can reach A exactly. But this is an illusion of perspective. In fact, the linear combination reaches the fitted model values $A_{||B,C}$ drawn as an open circle. The response variable A, drawn as a dot, is really hovering a few cm above the page but because of the viewers perspective it looks like A falls directly on $A_{||B,C}$. To see this, the figure is drawn a second time, rotated it a little so that you can see the residual vector connecting $A_{||B,C}$ to the response variable A.

Because B and C are collinear, the path taken to reach $A_{||B,C}$ is indirect: go 1.86 steps along B and then turn and go -1.15 steps along C. The team's score isn't measured by all this back-and-forth activity, the score is just the sum of squares $\| A_{||B,C} \|^2$.

In whole-model ANOVA, this sum of squares would be compared to the residual sum of squares. In term-by-term ANOVA, the whole-model sum of squares needs to be partitioned among the contributing terms. How to do this?

First, pick an order for the terms. Here, B will come first and then C.

Next, fit a model with just the first of the terms: this is the model $A \sim B$ that is nested within $A \sim B+C$. The nested is analogous to playing the game with only B, leaving C on the bench.

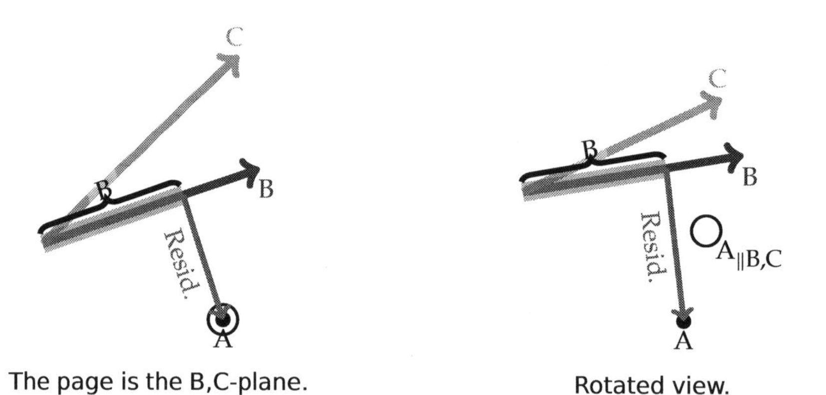

The page is the B,C-plane. Rotated view.

The "credit" that B gets is the sum of squares of the vector $A_{||B}$. This sum of squares is indicated in the figure by drawing a curly brace along the vector. The sum of squares is the square length of the segment marked by the curly brace.

Just so that you can relate the above picture to what is going to happen when C is added to the model, the point $A_{||B,C}$ is drawn even though that doesn't come out of the model $A \sim B$. It looks like $A_{||B,C}$ is the same as A, but this is an illusion of perspective as the rotated view shows.

Now to find the sum of squares to be credited to term C. This is done by adding C to the model, that is, fitting $A \sim B+C$. Of course, this brings the fitted model values to the point $A_{||B,C}$, but neither B nor C get credit for the actual path taken to reach that point. Instead, B gets the credit it got from the nested model $A \sim B$. Since B and C together get credit $\| A_{||B,C} \|^2$, and since B on its own gets credit $\| A_{||B} \|^2$, the amount of credit that can go to C is $\| A_{||B,C} \|^2 - \| A_{||B} \|^2$. Here's the geometry of the situation:

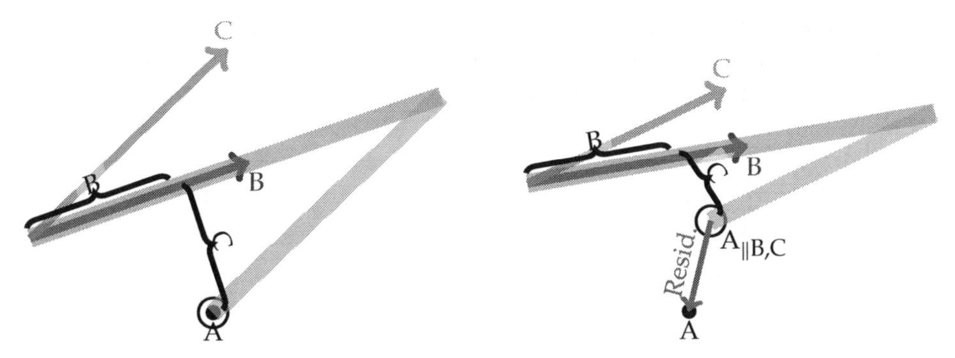

The page is the B,C-plane. Rotated view.

Notice that the credit assigned to each of B and C in the model A ~ B+C has very little to do with the actual path followed by the linear combination of A and B in reaching $A_{||B,C}$.

Instead of placing B first, you could have placed C first. The logic of partitioning the sum of squares is just the same, but the amount of credit assigned to B and C will be different.

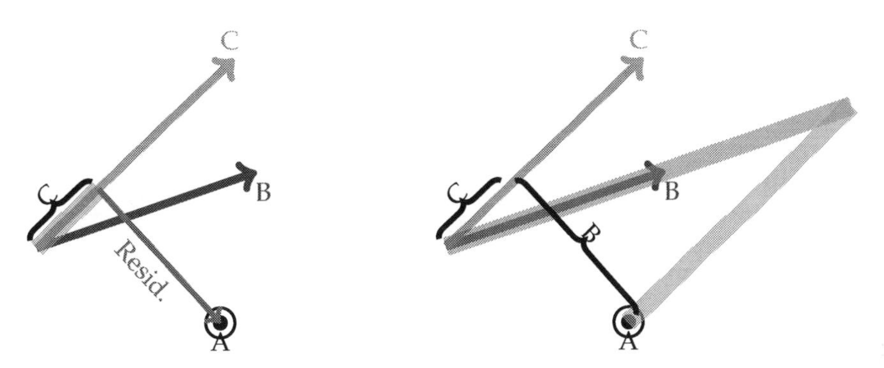

Fitting A ~ C. Then adding the B term.

Example 17.2: Height and Siblings In fitting the model height ~ nkids + sex + father + mother to Galton's data, the p-value on nkids depended substantially on where the term is placed in the ANOVA report. When nkids came first, the p-value was small, leading to rejection of the null hypothesis. But when nkids came last, the p-value was not small enough to justify rejection of the null.

As always, the order dependence of the term-by-term ANOVA results stems from collinearity. You can investigate this directly by modeling nkids as a function of the covariates. Here is the regression report on the model nkids ~ sex + father + mother:

	Estimate	Std. Error	t value	Pr(>|t|)
Intercept	19.2374	3.3772	5.70	0.0000
sexM	-0.3652	0.1770	-2.06	0.0394
father	-0.1751	0.0359	-4.88	0.0000
mother	-0.0123	0.0385	-0.32	0.7488

The model coefficients indicate that boys tend to have fewer siblings than girls. This may sound a bit strange, because obviously within any family every boy has exactly the same number of siblings as every girl. But the model applies *across* families, so this model is saying that families with boys tend to have fewer children than families with girls. This is a well known phenomenon. Depending on your point of view, you might like to interpret in either of two ways. Perhaps it's that parents with boys get worn out and stop having children. An alternative is that, particularly in Galton's time when women were definitely second-class citizens, parents who didn't have boys continued to have more children in the hopes of producing them. Either way, the model indicates that nkids is collinear with sex. It also apears to be related to father, as you will see later in Example 17.4.

A nice way to measure the extent of collinearity is with either the model R^2 or, equivalently but more directly geometrically, with the angle θ between nkids and the fitted model vector. For the model, $R^2 = 0.03$, corresponding to $\theta = 80°$. That's practially $90°$, hardly any alignment at all! How can that trivial amount of collinearity lead to the dramatic order-dependence of the ANOVA result on nkids?

To examine this, imagine splitting nkids into two vectors: a vector that is exactly aligned with the sex, father, mother subspace and the part that is exactly orthogonal to it. These two vectors correspond to the fitted model values and the residuals of the nkids model. Call them aligned and perp and add them to the height model in place of nkids. It's important to remember that neither of these vectors points in exactly the same direction as nkids, but nkids = aligned + perp.

Here's one ANOVA report, where aligned and perp have been placed first:

	Df	Sum Sq	Mean Sq	F value	Pr(>F)
aligned	1	3435.49	3435.49	741.55	0.0000
perp	1	12.04	12.04	2.60	0.1073
sex	1	3529.00	3529.00	761.74	0.0000
father	1	401.41	401.41	86.64	0.0000
Residuals	893	4137.12	4.63		

There are several interesting aspects to this report. First, aligned shows up as very significant, with a huge sum of squares compared to the original nkids — the original was only 185.5. This is because aligned happens to point in a very favorable direction, closely related to the very important variable sex. Second, perp gets a very small sum of squares: only 12. Third, sex is much reduced compared to the original ANOVA. This is because aligned has already grabbed much of the sum of squares originally associated with sex. Finally, the mother term gets absolutely no sum of squares and, indeed, has zero degrees of freedom. This is because there is one degree of redundancy among the set aligned, sex, father, and mother. Since mother came last in the report, it had absolutely nothing to contribute.

Now consider an ANOVA report on the same terms, but in a different order:

	Df	Sum Sq	Mean Sq	F value	Pr(>F)
sex	1	5874.57	5874.57	1268.03	0.0000
father	1	1001.11	1001.11	216.09	0.0000
mother	1	490.22	490.22	105.81	0.0000
aligned	0	0	NA	NA	NA
perp	1	12.04	12.04	2.60	0.1073
Residuals	893	4137.12	4.63		

Coming after father, mother and sex, the aligned vector is the one discarded as being redundant. Indeed, it was constructed to fit exactly into the subspace spanned by father, mother and sex. The term sex has regained is large sum of squares — sex is a strong predictor of height.

Now look at perp. Although it comes at the end of the list of terms, it has exactly the same set of values that it had when it came at the beginning of the list. This is because perp is exactly orthogonal to all the other terms in the report — it was constructed that way as the residual from fitting nkids to the other terms. Whenever a term is orthogonal to the others, the sum of squares for that term will not depend on where the term comes in order with respect to the others.

It's worth pointing out that the sum of squares, F value, and p-value on perp exactly matches that from the original ANOVA report using nkids when nkids came last. This is not a coincidence. It stems from putting father, mother and sex before nkids in the report — those other variables effectively stripped nkids of any component that was aligned with them, leaving only the part of nkids perpendicular to them.

All this may sound like an abstract exercise in geometrical thinking, but it actually carries a strong implication for interpreting the relationship between nkids and height. Earlier, a possible **causal** relationship was suggested; perhaps in families with many kids, competition for food and other resources caused the kids to be stunted in growth. Or it might be that families that had lots of kids back then were healthier or rich in resources that caused the kids to grow well.

The results with aligned and perp suggest something different. If nkids really did **cause** height, then the component of nkids perpendicular to sex and the other covariates ought to have some explanatory power with regard to height. On the other hand, if it's really sex and the other covariates that cause height, and nkids has nothing to do with it, then the perpendicular component of nkids ought to be no better than a random vector. The ANOVA report is consistent with this second scenario. The p-value of 0.107 isn't sufficiently small to justify rejecting the null that perp plays no role.

Even a very small alignment with genuinely important variables — nkids is at an angle of $\theta = 80^\circ$ with respect to the sex, father, mother subspace — can create a sum of squares that's statistically significant. Keep in mind that "significance" in a hypothesis test is defined with respect not to how strong or substantial or authentic the variable's role is, but how it compares in size to a completely random variable. With large n, even a very small alignment can be much larger than expected for a random vector. There are n=898 cases in Galton's data, and in a space of dimension 898 it's very unlikely that a random vector would show even an alignment as far from perpendicular as 80°. (The 95% confidence interval for the angle that a random vector makes with another vector is 86° to 94°.)

17.4.1 Orthogonality Eliminates Order Dependence

If B and C are not collinear, that is, if they are orthogonal to each other, then the order of the terms has no effect. For orthogonal explanatory variables, for instance B and C in the diagram below , the contribution both of each explanatory variable does not depend on the presence of the other in the model.

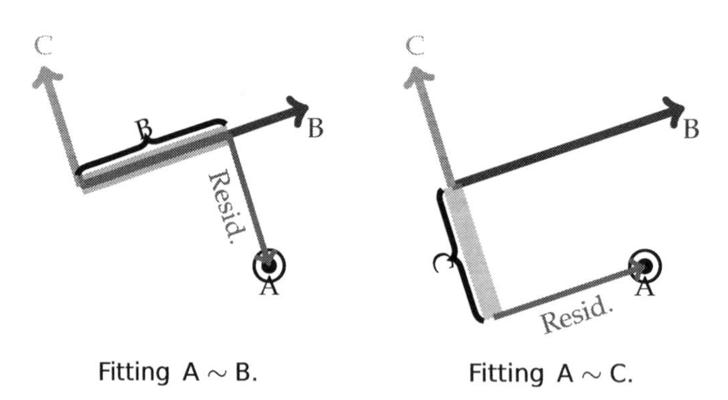

Fitting A $\sim$ B. Fitting A $\sim$ C.

17.4.2 Choosing the Order of Terms in ANOVA

The order-dependence of ANOVA results creates profound problems in interpretation. There is, unfortunately, no generally applicable way to identify a single correct ordering. This is because the order-dependence in ANOVA reflects genuine ambiguities in how to interpret data. Eliminating these ambiguities requires adding new information to the analysis, such as the modeler's assumptions or knowledge about the system under study. An illustration of this is given in Example 17.4, where it was assumed (reasonably enough) that the number of children in a family might be the result of covariates such as the children's sex or the height of the mother and father, but that the number of kids doesn't itself cause those covariates.

The mathematical analysis of the order-dependence of term-by-term ANOVA as it stems from collinearity can illuminate the situation, but doesn't resolve the ambiguity.

Here are some suggestions for dealing with the ambiguity of term-by-term ANOVA. It's important to remember that despite the problems with ANOVA, it offers the important advantage of allowing covariates to soak up unexplained variance. Other methods such as whole-model ANOVA or the regression report (the topic of Section 17.5) may avoid ambiguity but this is only because they don't offer the flexibility of ANOVA. Such methods also impose their own hidden assumptions.

Treat ANOVA as a screening method. In medicine, **screening tests** are ways to get a quick, inexpensive indication of whether a patient has an illness or condition. They tend to have relatively high error rates. Nonetheless, they can be genuinely useful for identifying situations where more invasive or expensive methods can be used to get a more definitive result.

ANOVA provides a quick, powerful, and effective way to assess the strength of evidence for a claim. You can use it to probe whether variables should be included in a model and what covariates might be included as well. The results shouldn't be taken as proof that a hypothesis is correct. Indeed, you should always keep in mind that hypothesis testing never offers a proof of a hypothesis. It can only produce a rejection of the null or a failure to reject the null.

Arrange the covariates so that they are orthogonal to the explanatory variable. Recall that order-dependence is only an issue in the presence of collinearity. If you are collecting data in the context of an experiment, you may well be able to arrange things so that the explanatory variables and covariates are orthogonal to one another. Chapter 20 on **experimental design** covers some of the techniques to do this.

Adopt a skeptical attitude. If you are inclined to be skeptical about a variable, look for the arrangement of terms that produces the smallest sum of squares. If, despite this, the variable still has a significant sum of squares, the ambiguity of order-dependence won't influence your conclusions.

But in interpreting your results, keep in mind that there might be other covariates that weren't included in your models that might reduce your variable to insignificance.

Adopt an enthusiastic attitude of hypothesis formation. Although ANOVA is formally a mechanism for hypothesis testing, it can also be used for hypothesis formation. Explore your data using ANOVA. If you find an arrangement of explanatory variables that produces a significant result, that's a good reason to devote some time and energy to thinking more about that result and how to follow up on it.

But don't be complacent. Once you have formed a hypothesis, you will have to test it skeptically at some point.

Don't forget other ambiguities. The ambiguity of order dependence is readily seen in ANOVA reports. But there are other ambiguities and uncertainties that you won't necessarily be aware of. Why did you choose the particular covariates in your analysis? Why did you choose to measure certain variables and not others? Did you include interaction terms (which often introduce collinearities)? Did you consider transformation terms? For example, Example 17.4, treats the number of kids as a simple linear term. But there is reason to think that it's effect might be nonlinear, that in small families height is associated with having more siblings, but in large families the reverse is true.

Even if there were no ambiguities of order dependence, there would still be ambiguities and uncertainties in your model results.

Think carefully about how your system works, and model the system rather than just the data. The source of order dependence is collinearity among the explanatory variables. Such collinearity is itself evidence that the explanatory variables are related in some way. A model like A ~ B+C+D is about the relationship between A and the explanatory variables; it doesn't describe the relationships among the explanatory variables, even though these may be important to understanding how the system works. If you want to understand the system as a whole, you need to model the system as a whole. This involves a set of models. For example, suppose that in the real system, A is caused by both B and C, and D exerts its influence only because it affects B. For such a system, it's appropriate to consider a set of models: A ~ B+C and B ~ D. Things can quickly get more complicated. For example, suppose C influences D. In that case, the model A ~ B+C would confuse the direct influence of C on A with the indirect influence that C has on A via D and therefore B.

The methods of **structural equation modeling** (SEM) are designed to deal with such complexity by creating structured sets of models. The structure itself is based in the modeler's knowledge or assumptions about how the system works.

Example 17.3: Wages and Race: Part 2

Example 17.2 looked at the Current Population Survey data to see if there is evidence that race is associated with wages. The simple model wage ~ race produced a p-value on race of 0.10, not strong evidence. However, including the covariates educ and sector made the role of race somewhat clearer as shown by the term-by-term ANOVA report:

	Df	Sum Sq	Mean Sq	F value	Pr(>F)
race	1	71.45	71.45	3.44	0.0641
educ	1	2017.67	2017.67	97.22	0.0000
sector	7	1112.92	158.99	7.66	0.0000
Residuals	524	10874.66	20.75		

Since the sum of squares in term-by-term ANOVA can be order dependent, it's worthwhile to look at a diferent ordering of terms. Here's one with race last:

	Df	Sum Sq	Mean Sq	F value	Pr(>F)
educ	1	2053.29	2053.29	98.94	0.0000
sector	7	1136.56	162.37	7.82	0.0000
race	1	12.19	12.19	0.59	0.4439
Residuals	524	10874.66	20.75		

This report suggests that the data provide no evidence that race plays any role at all.

Which conclusion is right? Do the data provide evidence for the importance of race or not? The answer is that it depends on your view of how the system is working.

Suppose, for instance, that race is really a causal influence in determining people's level of education or the sector of the economy that they are permitted to work in. In such a situation, it's appropriate to put race first in order to reflect the claim that race influence educ and sector and therefore race should get credit for any of the way it overlaps with educ and sector in accounting for wage.

Now consider an alternative perspective. It's certainly clear that educ and sector can't possibly be *causing* race. But it is possible that a person's race is correlated with education and sector without having caused them. How? Both education and sector can be influenced by the opportunities and role models that a person had as a child and these in turn are set by the situation in which that person's parents lived, including racial discrimination against the parents. According to this perspective, the person's race doesn't cause their level of education or the sector in which they work. Instead, all these things were causally influenced by other factors. In such a situation, it might be fair to assume in the null hypothesis that educ and sector are primary influences on wage independent of race. To test this null hypothesis, which treats the potential role of race with skepticism, race should come after educ and sector in the ANOVA breakdown.

There's also a third possibility here. Suppose you are looking for evidence that employers discriminate and you suspect that one mechanism for such discrimation is to restrict which sectors people can work in. That's not an unreasonable idea. But it's harder to believe that the employers have influenced the person's level of education. So, put educ *before* race and put sector *after* race:

	Df	Sum Sq	Mean Sq	F value	Pr(>F)
educ	1	2053.29	2053.29	98.94	0.0000
race	1	35.83	35.83	1.73	0.1894
sector	7	1112.92	158.99	7.66	0.0000
Residuals	524	10874.66	20.75		

This set of assumptions leads to failure to reject the null on race: the skeptic's attitude toward race is supported by the analysis.

The core of the matter is whether covariates such as sector or educ are mechanisms through which the effects of race plays out, or whether they provide alternative, non-racial explanations for the variation in wage. Depending on which one of these you believe, you can get very different results from ANOVA.

In this wages and race example, people with different views can each see that the data are consistent with those views. The data provide evidence for the person who thinks that race influences wages through the mechanism of educ and sector. Similarly, for the person who says that educ and sector are the determinants of wage, the data are consistent with the view that race doesn't play a role.

17.5 Hypothesis Tests on Single Coefficients

ANOVA treats a covariate as either previous to or subsequent to the explanatory variable. In contrast, model fitting takes both the explanatory variable and the covariates simultaneously.

The **coefficient t-test** is a hypothesis test on a coefficient in a model. The null hypothesis is that, in the population, that coefficient is zero but all the other coefficients are exactly as given by the model fit. Under this null hypothesis, the sampling distribution of the coefficient of interest is centered on zero. The standard deviation of the sampling distribution is the standard error found in the model regression report.

To illustrate, consider the model height ~ nkids + sex + mother + father fitted to Galton's height data:

	Estimate	Std. Error	t value	Pr(>|t|)
Intercept	16.1877	2.7939	5.79	0.0000
nkids	-0.0438	0.0272	-1.61	0.1073
sexM	5.2099	0.1442	36.12	0.0000
mother	0.3210	0.0313	10.27	0.0000
father	0.3983	0.0296	13.47	0.0000

According to this model, the coefficient on nkids is -0.0438 with a standard error of 0.0272. Under the null hypothesis for nkids, the sampling distribution of the coefficient has a mean of zero and a standard devation of 0.0272. The p-value of the observed coefficient, 0.107, tells how likely it is for that value or a more extreme value to be observed as a random draw from the sampling distribution. In this model, the p-value for nkids is not so small to justify rejecting the null hypothesis.

A nice feature of the regression report is that it is conducting a hypothesis test on each of the coefficients at the same time. This makes it easy to scan for coefficients that are statistically non-zero. On the other hand, this also introduces the need to adjust the rejection threshold to take into account the multiple hypothesis tests that are being conducted simultaneously. For example, in the above report, a Bonferroni adjustment would change the rejection threshold from the conventional 0.05 to 0.0125, since there are four coefficients being tested. (The hypothesis test on the intercept would not be of interest, since it doesn't relate to any of the explanatory variables.)

A not-so-nice feature of the regression report is that collinearity tends to produce high p-values on the coefficients of *all* of the collinear vectors; there's no hint from the report itself in the p-values that any of the collinear vectors might have a low p-value if the others were dropped from the model. To find this out, you need to try various orderings of the terms in the ANOVA report.

As always, the easiest situation for interpretation is when there is no collinearity at all between the explanatory variable and the covariates. In such a case, the regression report p-value is exactly the same as the p-value from ANOVA. However, when there is a correlation between the explanatory variable and the covariates, the regression report suffers from the same sorts of ambiguities as ANOVA, it's just that these are hidden because a simple re-arrangement of terms doesn't change the regression report.

Example 17.4: Wages and Race: Part 3 Earlier in this chapter, the Current Population Survey data was explored using term-by-term ANOVA to investigate the potential role of race. The conclusions reached differed depending on the assumptions made by the modeler as reflected by the order of terms in the ANOVA report.

Here is yet another analysis, using the regression report:

	Estimate	Std. Error	t value	Pr(>|t|)
Intercept	0.2083	1.4298	0.15	0.8842
raceW	0.4614	0.6021	0.77	0.4439
educ	0.5274	0.0965	5.47	0.0000
sectorconst	3.0204	1.1320	2.67	0.0079
sectormanag	4.3935	0.7857	5.59	0.0000
sectormanuf	1.5382	0.7400	2.08	0.0381
sectorother	1.6285	0.7301	2.23	0.0261
sectorprof	3.0602	0.6947	4.41	0.0000
sectorsales	-0.0085	0.8734	-0.01	0.9922
sectorservice	-0.1574	0.6940	-0.23	0.8206

The raceW comes first in this report doesn't signify anything. In a regression report, the values don't depend at all on the order of the terms.

The valuable thing about the regression report is that it gives not just the statistical significance of the term, but a measure of the actual strength of the term. This report indicates that Whites earned \$0.46 $\pm$ 1.20 more than Blacks, after adjusting for the sector of the economy and the educational worker of the worker. The p-value reflects the wide confidence interval: at 0.44 there's no basis for rejecting the null hypothesis that the population coefficient on raceW is zero.

Because the regression report is independent of the order of terms, it's tempting to think that it's somehow more objective than term-by-term ANOVA. That's wrong. The causal model implicit in the regression report has it's own point of view, that each of the terms in the model contributes to the response variable directly, not through the mediation of another variable. Thus, for the person who thinks that `sector` is merely a mechanism for discrimination by `race`, the regression report is not an appropriate form of analysis. Similarly, the regression report isn't appropriate for the person who thinks that it's really `educ` and `sector` that determine wages and that `race` only appears to come into play because it happens to be correlated with those other variables. The regression report is appropriate only for the person for whom causation isn't an issue. This might be because they want to construct a prediction model or because they believe that none of the explanatory variables cause others.

Example 17.5: Testing Universities

University students are used to being given tests and graded as individuals. But who grades the universities to find out how good a job the institutions are doing?

This need not be hard. Most universities require an entrance test — the SAT is widely used in the US for undergraduate admissions, the GRE, LSAT, MCAT, etc. for graduate school. Why not give students the entrance test a second time when they leave and compare the two results? If students at Alpha University show a large increase compared to students at Beta University, then there is evidence that Alpha does a better job than Beta.

There are good and bad reasons why such testing is not widely done. University faculty as a group are notoriously independent and don't like being evaluated. They are suspicious of standardized tests for evaluation of their teaching (although the don't often oppose using them for admissions purposes). They point out that there are many factors outside of their control that influence the results, such as the students' preparedness, motivation, and work ethic. (On the other hand, universities are happy to take credit for the students who succeed).

One legitimate obstacle to meaningful testing of universities is that students are not required to take an exit test and not motivated to perform well on it. A student who is applying to university has to take the entrance exam and has good reason to try to do well on it — favorable admissions decisions and scholarship awards often follow good test results. But the student who is soon to graduate has no such motivation, either to take the exit exam or to do well on it. So what does the difference between the entrance and exit scores show: How much the students learned or that the graduating students have something better to do with their time than take an exit exam?

The Collegiate Learning Assessment (CLA) is a standardized test consisting of several essay questions that probe a student's critical thinking skills.[29] (See `www.cae.org`.) As of 2008, more than 200 universities and colleges have been testing the exam. The CLA is given to one group of first-year students and another group of final-year students. The scores of the two groups are then compared.

Ideally, the same students would be tested in both their first and last years. If this were done, each student could be compared to himself or herself and co-variates such as previous preparation or overall aptitude would be held constant within each entry-to-exit comparison. Unfortunately, such **longitudinal** designs were used at only 15% of the CLA schools. At the remaining schools, for reasons of cost and administrative complexity, a **cross-sectional** design was used, where the two groups consist of different students. When the study is cross-sectional, it can be important to adjust for covariates that may differ from one group to another.

In the case of the CLA, the students' performance on other standardized tests (such as the entrance SAT) is used for the adjustment. As it happens, there is a strong correlation between the SAT score and other explanatory variables. This collinearity tends to increase the standard error of the estimate of school-to-school differences.

Two studies of the CLA adjustment methodology concluded that they so increased standard errors that no clear inference can be made of school-to-school differences.[30] The studies called for longitudinal testing to be done in order to increase the reliability of the results. But, short of requiring students to take both the entrance and exit exams, it's not clear how the longitudinal approach can be implemented.

17.6 Non-Parametric Statistics

It sometimes happens that one or more of your variables have outliers or a highly skew distribution. In this situation, the extreme values of the variable can have an unduly strong influence on the fit of the model. As a result, the p-values calculated from ANOVA can be misleading.

Common sense suggests deleting the outliers or extreme values from the data used to fit the model. Certainly this is an appropriate step if you have good reason to think that the extreme values are bogus.

A more moderate approach is to fit two models: one with and one without the extreme values. If the two models give similar results, then you can reasonably conclude that the extreme values are not having an undue effect.

But if the two models give very different results, you have a quandry. Which model should you use? On the one hand, you don't want some cases to have an "undue" influence. On the other hand, those cases might be telling you something about the system and you need to be able to guard against the criticism that you have altered your data.

There is a simple and effective way to deal with this situation for the purpose of deciding whether the extreme values are dominating p-value calculations. The approach is to transform your data in a way that is genuine to the extreme values by keeping them on the edge of the distribution, but which brings them closer to the mainstream. The easiest such transform is called the **rank transform**. It replaces each value in a variable with the position of the value in the set of values. For example, suppose you have a variable with these values:

$$23.4, 17.3, 26.8, 32.8, 31.3, 34.5, 7352.3.$$

It's obvious that the seventh value is quite different from the other six. Applying the rank transform, this variable becomes

$$2, 1, 3, 5, 4, 6, 7.$$

That is, 23.4 is the 2nd smallest value, 17.3 is the 1st smallest value, 26.8 is the 3rd smallest, and so on. With the rank transform, the extreme value of 7352.3 becomes nothing special: it has rank 7 because it is the seventh smallest value. There are never outliers in a rank transformed variable. Nevertheless, the rank transformed data are honest to the original data in the sense that each value retains its position relative to the other values.

Analysis of rank-transformed data is called **non-parametric statistics**. The coefficients from models of rank-transformed data are hard to interpret, because they describe how the response value changes do to a 1-step increase in rank rather than a one-unit increase the value of the variable. Still, the p-values can be interpreted in a standard way.

When outliers or extreme values are a concern, it's worth comparing the p-values from the original (parametric) model and the non-parametric model fitted to the rank-transformed data. If the p-values lead to the same conclusion, you know that the extreme values are not shaping the results. If the p-values for the two models lead to different conclusions then your work is just beginning and you need to think carefully about what is going on.

17.7 Sample Size and Power

If the p-value on your explanatory variable of interest is small enough, you're entitled to reject the null. If it's not so small, you are left in the ambiguous situation of "failing to reject." This failure might be because the explanatory variable is indeed irrelevant to the response variable; a valuable finding. On the other hand, there can be a much more mundane explanation for the failure: your sample size was not large enough.

In ANOVA, the p-value gets smaller as F gets bigger. In turn, F is a product of both the substance of the effect and the sample size n: larger n produces larger F as described in Equation 16.3 on page 306.

The fundamental issue, however, is not F, but the **power** of your study. In picking a sample size n for your study, your goal should be to produce a study with a large enough power so that, if you do fail to reject the null, you will have good reason to prefer the null hypothesis to the alternative hypothesis as a reasonable representation of your system.

Typically, you start a study with some working hypothesis about how the explanatory variable is related to the response. This working hypothesis is given the name **alternative hypothesis**. (The word "alternative" is simply the statistical vocabulary for distinguishing the working hypothesis from the "null" hypothesis.) The goal of this section is to show how to translate your alternative hypothesis into a decision about the appropriate sample size n to produce good power.

It may seem odd to be thinking about sample size here, at the end of a chapter on analyzing data. After all, you have to have your data already in hand before you can analyze it! But modeling is a process of asking questions. That process often involves going back and collecting more or different data based on the results of the analysis of earlier data.

When deciding on a sample size, think ahead to the analysis you will eventually do on the data you collect. It's likely that your analysis will be analysis of covariance or something like it. So you will end up with an ANOVA table, looking perhaps like this:

Term	Df	Sum Sq	Mean Sq	F value	p value
Intercept	1	500			
Covariates	3	200			
Explanatory Var	1	50	50	2	0.19
Residuals	10	250	25		
TOTAL	15	1000			

From your knowledge of the covariates and the explanatory variable, you should know the degrees of freedom of each model term even before you collect the data. But until you actually have the data, you won't be able to fill in the actual values for the sums of squares. What's been done here is to guess — hopefully an informed guess based on your working hypothesis.

What information might you have to work from? Perhaps you have already done a preliminary study and collected just a bit of data — the $n = 15$ cases here. Then you have the ANOVA report from those data. Or perhaps there is an existing study in the literature from a study of the same explanatory variable or an analogous one in a setting similar to the one you are working on. You'll need to be creative, but realistic.

For the present, assume that the above ANOVA table came from a preliminary study with $n = 15$ cases. The F value of 2 is suggestive, but the p-value is large; you cannot reject the null based on this data set. But perhaps a larger study would be able to reveal what's merely suggested in the preliminary study.

Imagine what the table would look like for larger n. For simplicity here, consider what would happen if you doubled the number of cases, to $n = 30$. (In reality, you might need to consider some much larger than the $n = 30$ used in this example.) Of course, you can't know exactly what the larger n will give until you actually collect the data. But, for planning, it's sensible to assume that your extended study might look something like the preliminary study. If that's so, doubling the number of cases will typically double the total sum of squares. (Why? Because you're adding up twice as many cases. For instance, the sum of squares of $3,\ 4,\ 5$ is 50, the sum of squares of $3,\ 4,\ 5,\ 3,\ 4,\ 5$ is 100.)

Increasing the number of cases doesn't change the degrees of freedom of the explanatory variables. The intercept, the covariates, and the explanatory variable terms still have the same degrees of freedom as before. As a result, for these terms the mean square will double.

In contrast, the degrees of freedom of the residual will increase with the number of cases in the sample. The sum of squares of the residual will also increase, but the net effect is that the mean square of the residual will stay roughly the same.

To illustrate, the following table shows the ANOVA tables side by side for a preliminary study with $n = 15$ and an expanded study with $n = 30$.

	Prelim. Study $n = 15$				New Study $n = 30$			
Term	D.F.	Sum Sq.	Mean Sq.	F	D.F.	Sum Sq.	Mean Sq.	F
Intercept	1	500	500		1	1000	1000	
Covariates	3	200	67		3	400	133	
Explan. Var.	1	50	50	**2**	1	100	100	**5**
Residuals	10	250	25		25	500	20	
TOTAL	15	1000			30	2000		

Overall, by doubling n you can anticipate that the F value will be more than doubled. If n were increased even further, the anticipated F value would similarly be increased further, as in Equation 16.3.

The goal in setting the sample size n is to produce statistical significance — a small p-value. As a rule of thumb for the F distribution, F above 4 or 5 will give a small p-value when the sample size n is around 25 or bigger. (More precisely, when the degrees of freedom of the denominator is around 25 or bigger.)

But targeting such an F value would be a mistake. The reason is that F itself is subject to sampling fluctuations. If your sample size is such to produce a p-value of 0.05 when you hit the target F, then there's a roughly 50% probability that sampling fluctuations will produce an F that's too small and consequently a p-value that's too large.

To avoid suffering from a mishap, it's appropriate to choose the sample size n large enough that sampling fluctuations are unlikely to produce an unsatisfactory p-value. Recall that the **power** is the probability that your study will lead to rejection of the null hypothesis *if the alternative hypothesis were true.* Your goal in choosing n is to produce a large power.

The calculation behind power is somewhat complicated, but the table gives an idea of the power for studies with different sample sizes n and different anticipated F.

	Power			
Anticipated F	$n = 10$	$n = 20$	$n = 100$	$n \to \infty$
3	17%	23%	28%	29%
5	33%	42%	49%	51%
10	64%	76%	83%	85%
20	92%	97%	99%	99%

This table was constructed for an explanatory variable with one degree of freedom. This could be a quantitative variable (for example, family income) or a categorical variable with two levels (for example, sex). As the table indicates the anticipated F must be quite high — 10 or so — to achieve a power of 90% or higher. When the degrees of freedom in the numerator is higher, so long as n is large, high power can be achieved with considerably smaller F. For example, with five degrees of freedom in the denominator, $F = 5$ gives roughly 90% power.

If your study has low power, then failing to reject the null doesn't give much reason to take the null as reasonable: your study is likely to be inconclusive. When you aim for an F of about 5, your power will be only about 50%. If you want a high power, say 90%, your study should be designed to target a high F.

Studies are often expensive and difficult to conduct and so you want to be careful in designing them. The table above gives only a rough idea about power. Since the power calculations are complicated, you may want to consult with a professional to do the calculation that will be appropriate for your specific study design.

17.8 Computational Technique

The basic software for hypothesis testing on parts of models involves the familiar `lm` and `summary` operators for generating the regression report and the `anova` operator for generating an ANOVA report on a model.

17.8.1 ANOVA reports

The `anova` operator takes a model as an argument and produces the term-by term ANOVA report. To illustrate, consider this model of the swimming records data.

```
> swim = ISMdata("swim100m.csv")
> mod1 = lm( time~year+sex+year:sex, data=swim)
> anova(mod1)
```

```
Analysis of Variance Table

Response: time
            Df Sum Sq Mean Sq F value  Pr(>F)
(Intercept)  1 222636  222636 20203.1 < 2e-16
year         1   3579    3579   324.7 < 2e-16
sex          1   1484    1484   134.7 < 2e-16
year:sex     1    297     297    26.9 2.8e-06
Residuals   58    639      11
```

(I've highlighted the p-value on the interaction term because I'll contrast this model with another model later on.)

To change the order of the terms in the report, you can create a new model with the explanatory terms listed in a different order. For example, here I'll put sex first and year second:

```
> mod2 = lm( time ~ sex + year + year:sex, data=swim)
> anova(mod2)
Analysis of Variance Table

Response: time
             Df Sum Sq Mean Sq F value  Pr(>F)
(Intercept)   1 222636  222636 20203.1 < 2e-16
sex           1   1721    1721   156.1 < 2e-16
year          1   3342    3342   303.3 < 2e-16
sex:year      1    297     297    26.9 2.8e-06
Residuals    58    639      11
```

The R package insists on putting interaction terms after the corresponding main terms.

Technical Note

On occasion, you might be interested in answering questions not about individual model terms but about the variables themselves. For instance, you might want to know whether adding year to the model reduces the size of the residuals in a significant way, but you're not particularly interested in just the main term of year but also in how year combines with other variables in interaction terms.

ANOVA can be used to compare complete models to one another. The idea is to see how the second model reduces the residuals compared to what was accomplished in the first model. To illustrate, I'll compare a model with just sex as the explanatory term to a model that includes both sex and year, including the interaction term:

```
> mod3 = lm( time ~ sex, data=swim)
> mod4 = lm( time ~ sex*year, data=swim)
> anova( mod3, mod4)
Analysis of Variance Table

Model 1: time ~ sex
Model 2: time ~ sex * year
  Res.Df  RSS Df Sum of Sq   F Pr(>F)
1     60 4278
2     58  639  2      3639 165 <2e-16
```

Adding the two new terms (year and sex:year) involved two degrees of freedom and reduced the residual sum of squared from 4278 to 639. Thus, these two terms had a mean square of $(4278 - 639)/2 = 1820$. The mean square of the residuals is $639/58 = 11.02$ giving an F value of $1820/11.02 = 165$, as shown in the report.

17.8.2 Non-Parametric Statistics

An effective way to screen for outliers or extreme values in variables is to plot their density. The box-and-whiskers format is handy for marking possible outliers clearly.

```
> bwplot( swim$year)
```

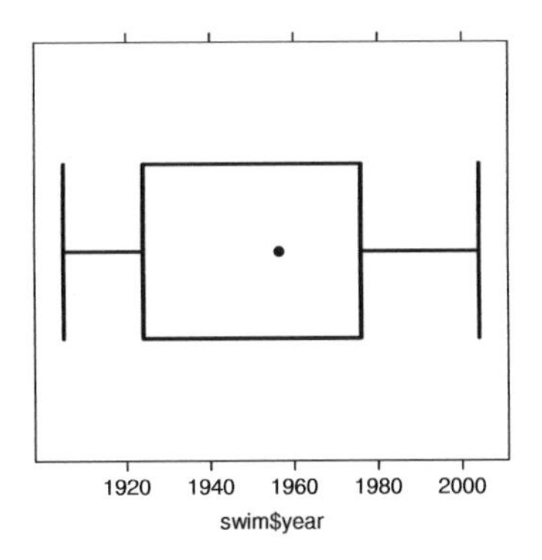

There is no indication of outliers in the year variable.

```
> bwplot( swim$time)
```

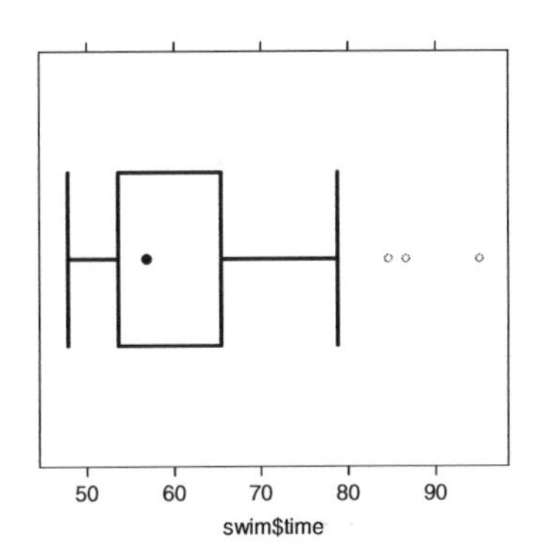

The time variable does have some outlying values. This doesn't mean that the values are bogus — the swimming records are carefully kept and presumably reflect genuine times. Still, those outlying values may have an undue effect on the model.

To perform a hypothesis test non-parametrically, take the rank of any quantitative variables:

```
> mod3 = lm( rank(time) ~ rank(year)*sex, data=swim)
> anova(mod3)
```

```
Analysis of Variance Table

Response: rank(time)
               Df Sum Sq Mean Sq  F value Pr(>F)
(Intercept)     1  61520   61520 16134.45 <2e-16
rank(year)      1  14320   14320  3755.77 <2e-16
sex             1   5313    5313  1393.41 <2e-16
rank(year):sex  1      1       1     0.23   0.63
Residuals      58    221       4
```

Notice the large p-value on the interaction term. In the parametric models, the p-value on the interaction was very small: 2.8×10^{-6}. This discrepancy suggests that the statistical significance of the interaction term may be an effect of the "outliers." These happen to be the very slow world-record times of women from the earliest days of record keeping.

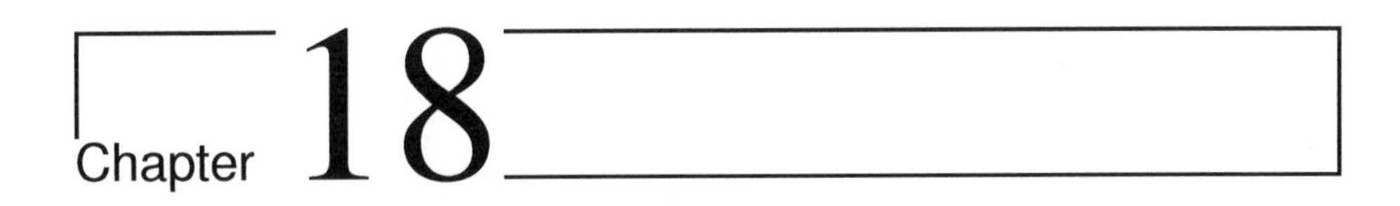

Models of Yes/No Variables

Sometimes the variation that you want to model occurs in a categorical variable. For instance, you might want to know what dietary factors are related to whether a person develops cancer. Or perhaps how income, personality, and education influence what kind of job a person takes.

The techniques studied up until now are intended to model *quantitative* response variables, those where there value is a number, for example height wage, swimming times, foot size. Categorical variables have been used only in the role of explanatory variables.

There are good technical reasons why it's easier to build and interpret models with quantitative response variables. For one, a residual from such a model is simply a number. The size of a typical residual can be described using a variance or a mean square. In contrast, consider a model that tries to predict whether a person will become an businessman or an engineer or a farmer or a lawyer. There are many different ways for the model to be wrong, e.g., it predicts farmer when the outcome is lawyer, or engineer when the outcome is businessman. The "residual" is not a number, so how do you measure how large it is?

This chapter introduces one technique for building and evaluating models where the response variable is categorical. The technique handles a special case, where the categorical variable has only two levels. Generically, these can be called **yes/no** variables, but the two levels could be anything you like: alive/dead, success/failure, engineer/not-engineer, and so on.

18.1 The 0-1 Encoding

The wonderful thing about yes/no variables is that they are always effectively quantitative; they can be naturally encoded as 0 for no and 1 for yes. Given this encoding, you could if you want just use the standard linear modeling approach of quantitative response variables. That's what I'll do at first, showing what's good and what's bad about this approach before moving on to a better method.

To illustrate using the linear modeling approach, consider some data that

relate smoking and mortality. The table below gives a few of the cases from a data frame where women were interviewed about whether they smoked.

Case	Outcome	Smoker	Age
1	Alive	Yes	23
2	Alive	Yes	18
3	Dead	Yes	71
4	Alive	No	67
5	Alive	No	64
6	Alive	Yes	38

... and so on for 1314 cases altogether.

Of course, all the women were alive when interviewed! The outcome variable records whether each woman was still alive 20 years later, during a follow-up study. For instance, case 3 was 71 years old at the time of the interview. Twenty years later, she was no longer alive.

Outcome is the Yes/No variable I'm interested in understanding; it will be the response variable and smoker and age will be the explanatory variables. I'll encode "Alive" as 1 and "Dead" as 0, although it would work just as well do to things the other way around.

The simplest model of the outcome is all-cases-the-same: outcome $\sim$ 1. Here is the regression report from this model:

	Estimate	Std. Error	t value	Pr(>|t|)
Intercept	0.7192	0.0124	57.99	0.0000

The intercept coefficient in the model outcome $\sim$ 1 has a simple interpretation; it's the mean of the response variable. Because of the 0-1 coding, the mean is just the fraction of cases where the person was alive. That is, the coefficient from this model is just the probability that a random case drawn from the sample has an outcome of "alive."

At first that might seem a waste of modeling software, since you could get exactly the same result by counting the number of "alive" cases. But notice that there is a bonus to the modeling approach — there is a standard error given that tells how precise is the estimate: 0.719 ± 0.024 with 95% confidence.

The linear modeling approach also works sensibly with an explanatory variable. Here's the regression report on the model outcome $\sim$ smoker:

	Estimate	Std. Error	t value	Pr(>|t|)
(Intercept)	0.6858	0.0166	41.40	0.0000
smokerYes	0.0754	0.0249	3.03	0.0025

You can interpret the coefficients from this model in a simple way — the group-wise means of outcome for the smokers and the non-smokers. Those means are the proportion of people in each group who were alive at the time of the follow-up study. In other words, it's reasonable to interpret them as the probabilities that the smokers and non-smokers were alive.

Again, you could have found this result in a simpler way: just count the fraction of smokers and of non-smokers who were alive.

The advantage of the modeling approach is that it produces a standard error for each coefficient and a p-value. The intercept says that 0.686 ± 0.033 of non-smokers were alive. The smoker**Yes** coefficient says that an additional 0.075 ± 0.05 of smokers were alive.

This result might be surprising, since most people expect that mortality is higher among smokers than non-smokers. But the confidence interval does not include 0 or any negative number. Correspondingly, the p-value in the report is small: the null hypothesis that smoker is unrelated to outcome can be rejected.

Perhaps it's obvious that a proper model of mortality and smoking should adjust for the age of the person. After all, age is strongly related to mortality and smoking might be related to age. Here's the regression report from the model outcome ~ smoker+age:

	Estimate	Std. Error	t value	Pr(>|t|)
(Intercept)	1.4726	0.0301	48.92	0.0000
smokerYes	0.0105	0.0196	0.54	0.5927
age	**-0.0162**	0.0006	-28.95	0.0000

It seems that the inclusion of age has had the anticipated effect. According to the coefficient -0.016, there is a negative relationship between age and being alive. The effect of smoker has been greatly reduced; the p-value 0.59 is much too large to reject the null hypothesis.

Yet there is a problem. Consider the fitted model value for a 20-year old smoker: $1.47 + 0.010 - 0.016 \times 20 = 1.16$. This value can't be interpreted as a probability that a 20-year old smoker would still be alive at the time of the follow-up study (when she would have been 40). Probabilities must always be between zero and one.

The top panel of Figure 18.1 shows the fitted model values for the linear model along with the actual data. (Since the outcome variable is coded as 0 or 1, it's been jittered slightly up and down so that the density of the individual cases shows up better. Smokers are at the bottom of each band, plotted as small triangles.) The data show clearly that older people are more likely to have died. Less obvious in the figure is that the very old people tended not to smoke.

The problem with the linear model is that it is too rigid, too straight. In order for the model to reflect that the large majority of people in their 40s and 50s were alive, the model overstates survival in the 20-year olds. This isn't actually a problem for the 20-year olds — the model values clearly indicate that they are very likely to be alive. But what isn't known is the extent to which the line has been pulled down to come close to 1 for the 20-year olds, thereby lowering the model values for the middle-aged folks.

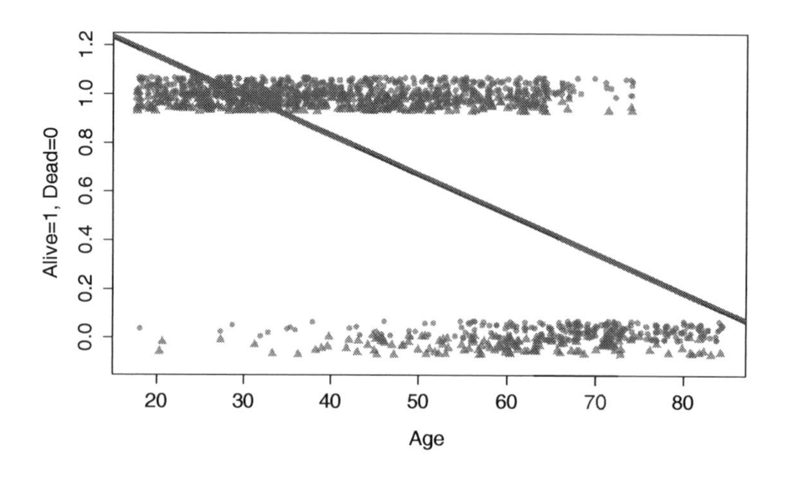

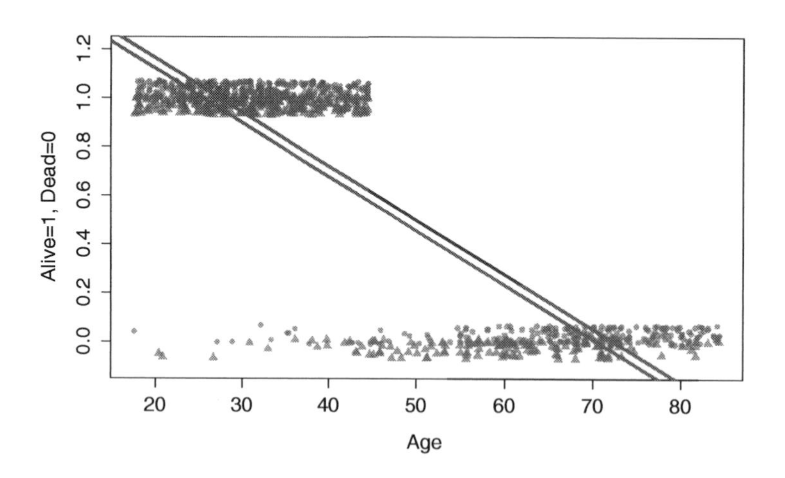

Figure 18.1: Top: The smoking/mortality data along with a linear model. Bottom: A thought experiment which eliminates the "Alive" cases above age 45.

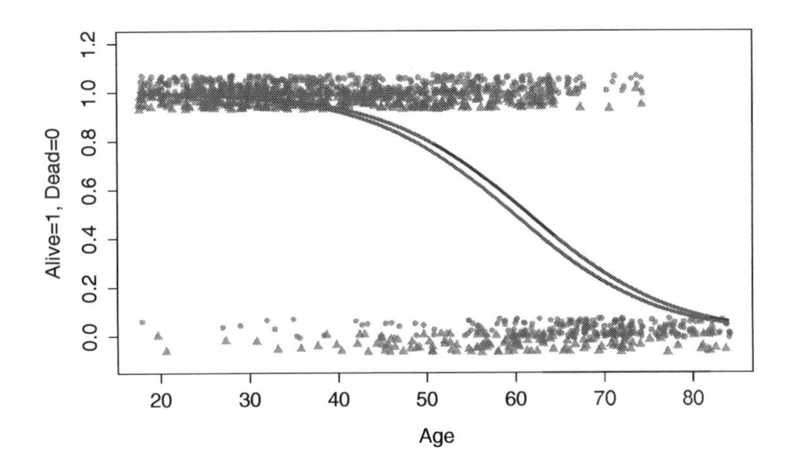

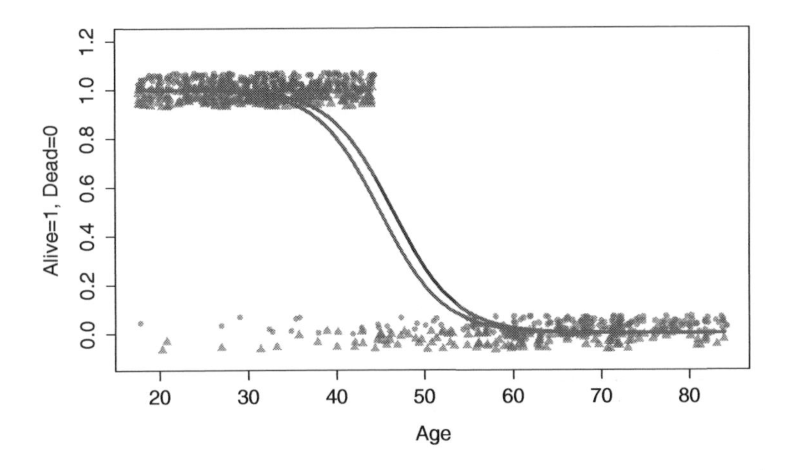

Figure 18.2: Logistic forms for outcome $\sim$ age + smoker. Top: The smoking/mortality data. Bottom: The thought experiment in which the "Alive" cases above age 45 were eliminated.

To illustrate how misleading the straight-line model can be, I'll conduct a small thought experiment and delete all the "alive" cases above the age of 45. The resulting data and the best-fitting linear model are shown in the bottom panel of Figure 18.1.

The thought-experiment model is completely wrong for people in their 50s. Even though there are no cases whatsoever where such people are alive, the model value is around 50%.

What's needed is a more flexible model — something not quite so rigid as a line, something that can bend to stay in the 0-to-1 bounds of legitimate probabilities. There are many ways that this could be accomplished. The one that is most widely used is exceptionally effective, clever, and perhaps unexpected. It consists of a two-stage process:

1. Construct a model value in the standard linear way — multiplication of coefficients times model vectors. The model `outcome` ~ `age+smoker` would involve the familiar terms: an intercept, a vector for `age`, and the indicator vector for `smoker`**`Yes`**. The output of the linear formula — I'll call it Y — is not a probability but rather a **link value**: an intermediary of the fitting process.

2. Rather than taking the link value Y as the model value, transform Y into another quantity P that it is bound by 0 and 1. A number of different transformations could do the job, but the most widely used one is called the **logistic transformation**: $P = \frac{e^Y}{1+e^Y}$.

This type of model is called a **logistic regression** model. The choice of the best coefficients — the model fit — is based not directly on how well the link values Y match the data but on how the probability values P match.

Figure 18.2 illustrates the logistic form on the smoking/mortality data. The logistic transformation lets the model fit the outcomes for the very young and the very old, while maintaining flexibility in the middle to match the data there.

The two-stage approach to logistic regression makes it straightforward to add explanatory terms in a model. Whatever terms there are in a model — main terms, interaction terms, nonlinear terms — the logistic transformation guarantees that the model values P will fall nicely in the 0-to-1 scale.

Figure 18.3 shows how the link value Y is translated into the 0-to-1 scale of P for different "shapes" of models.

18.2 Inference on Logistic Models

The interpretation of logistic models follows the same logic as in linear models. Each case has a model probability value. For example, the model probability value for a 64-year old non-smoker is 0.4215, while for a 64-year old smoker it is 0.3726. These model values are in the form of probabilities; the 64-year old non-smoker had a 42% chance of being alive at the time of the follow-up interview.

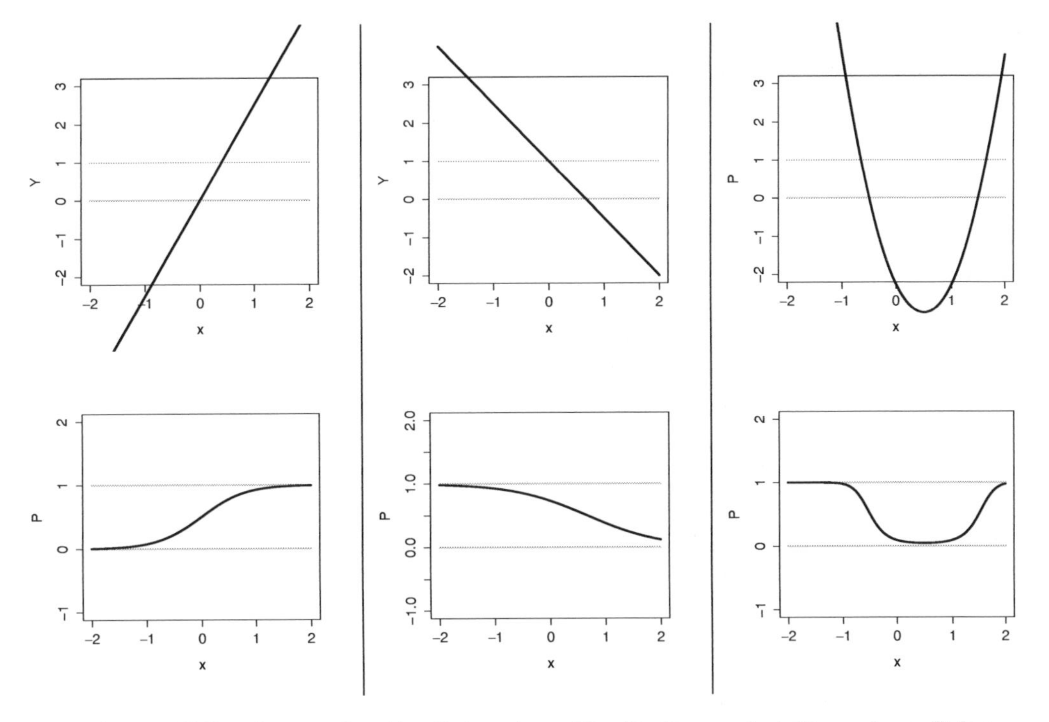

Figure 18.3: Comparing the link values Y with the probability values P for three different models, each with an explanatory variable x. Top: Link values. Bottom: Probability values.

There is a regression report for logistic models. Here it is for outcome ∼ age+smoker:

	Estimate	Std. Error	z value	Pr(>|z|)
Intercept	**7.5992**	0.4412	17.22	0.0000
age	**-0.1237**	0.0072	-17.23	0.0000
smokerYes	-0.2047	0.1684	-1.22	0.2242

The coefficients refer to the link values Y, so for the 64-year old non-smoker the value is $Y = 7.59922 - 0.12368 \times 64 = -0.3163$. This value is not a probability; it's negative, so how could it be? The model value is the output of applying the logistic transform to Y, that is, $\frac{e^{-0.3163}}{1+e^{-0.3163}} = 0.4215$.

The logistic regression report includes standard errors and p-values, just as in the ordinary linear regression report. Even if there were no regression report, you could generate the information by resampling, using bootstrapping to find confidence intervals and doing hypothesis tests by permutation tests or by resampling the explanatory variable.

In the regression report for the model above, you can see that the null hypothesis that age is unrelated to outcome can be rejected. (That's no surprise, given the obvious tie between age and mortality.) On the other hand, these data provide only very weak evidence that there is a difference between smokers and non-smokers; the p-value is 0.22.

An ANOVA-type report can be made for logistic regression. This involves the same sort of reasoning as in linear regression: fitting sequences of nested models and examining how the residuals are reduced as more explanatory terms are added to the model.

It's time to discuss the residuals from logistic regression. Obviously, since there are response values (the 0-1 encoding of the response variable) and there are model values, there must be residuals. The obvious way to define the residuals is as the difference between the response values and the model values, just as is done for linear models. If this were done, the model fit could be chosen as that which minimizes the sum of square residuals.

Although it would be possible to fit a logistic-transform model in this way, it is not the accepted technique. To see the problem with the sum-of-square-residuals approach, think about the process of comparing two candidate models. The table below gives a sketch of two models to be compared during a fitting process. The first model has a one set of model values, and the second model has another set of model values.

	Model Values		
Case	First Model	Second Model	Observed Value
1	0.8	1.0	1
2	0.1	0.0	1
3	0.9	1.0	1
4	0.5	0.0	0
Sum. Sq. Resid.	1.11	1.00	

Which model is better? The sum of square residuals for the first model is $(1-0.8)^2+(1-0.1)^2+(1-0.9)^2+(0-0.5)^2$, which works out to be 1.11. For the second model, the sum of square residuals is $(1-1.0)^2+(1-0.0)^2+(1-1.0)^2+(0-0.0)^2$, or 1.00. Conclusion: the second model is to be preferred.

However, despite the sum of squares, there is a good reason to prefer the first model. For Case 2, the Second Model gives a model value of 0 — this says that it's *impossible* to have an outcome of 1. But, in fact, the observed value was 1; according to the Second Model the impossible has occurred! The First Model has a model value of 0.1 for Case 2, suggesting it is merely *unlikely* that the outcome could be 1.

The actual criterion used to fit logistic models penalizes heavily — infinitely in fact — candidate models that model as impossible events that can happen. The criterion for fitting is called the **likelihood** and is defined to be the probability of the observed values according to the probability model of the candidate. When the observed value is 1, the likelihood of the single observation is just the model probability. When the observed value is 0, the likelihood is 1 minus the model probability. The overall likelihood is the product of the individual likelihoods of each case. So, according to the First Model, the likelihood of the observations $1, 1, 1, 0$ is $0.8 \times 0.1 \times 0.9 \times (1 - 0.5) = 0.036$. According to the Second Model, the likelihood is $1 \times 0 \times 1 \times (1 - 0) = 0$. The First Model wins.

It might seem that the likelihood criterion used in logistic regression is completely different from the sum-of-square-residuals criterion used in linear regression. It turns out that they both fit into a common framework called **maximum likelihood estimation**. A key insight is that one can define a probability model for describing residuals and, from this, compute the likelihood of the observed data given the model.

In the maximum likelihood framework, the equivalent of the sum of squares of the residuals is a quantity called the **deviance**. Just as the linear regression report gives the sum of squares of the residual, the logistic regression report gives the deviance. To illustrate, here is the deviance part of the logistic regression report for the outcome ~ age+smoker model:

```
    Null deviance: 1560.32  on 1313  degrees of freedom
Residual deviance:  945.02  on 1311  degrees of freedom
```

The **null deviance** refers to the simple model outcome ~ 1: it's analogous to the sum of squares of the residuals from that simple model. The reported degrees of freedom, 1313, is the sample size n minus the number of coefficients m in the model. That's $m = 1$ because the model outcome ~ 1 has a single coefficient. For the smoker/mortality data in the example, the sample size is $n = 1314$. The line labelled "Residual deviance" reports the deviance from the full model: outcome ~ age+smoker. The full model has three coefficients altogether: the intercept, age, and smoker**Yes**, leaving $1314 - 3 = 1311$ degrees of freedom in the deviance.

According to the report, the age and smoker**Yes** vectors reduced the deviance from 1560.32 to 945.02. The deviance is constructed in a way so that a random, junky explanatory vector would, on average, consume a proportion of the deviance equal to 1 over the degrees of freedom. Thus, if you constructed the outcome ~ age+smoker + junk, where junk is a random, junky term, you would expect the deviance to be reduced by a fraction $1/1311$.

To perform a hypothesis test, compare the actual amount by which a model term reduces the deviance to the amount expected for random terms. This is analogous to the F test, but involves a different probability distribution called the χ^2 (chi-squared). The end result, as with the F test, is a p-value which can be interpreted in the conventional way, just as you do in linear regression.

18.3 Model Probabilities

The link model values Y in a logistic regression are ordinary numbers that can range from $-\infty$ to ∞. The logistic transform converts Y to numbers P on the scale 0 to 1, which can be interpreted as probabilities. I'll call the values P **model probabilities** to distinguish them from the sorts of model values found in ordinary linear models. The model probabilities describe the chance that the Yes/No response variable takes on the level "Yes."

To be more precise, P is a **conditional probability**. That is, it describes the probability of a "Yes" outcome conditioned on the explanatory variables in the model. In other words, the model probabilities are probabilities *given* the value of the explanatory variables. For example, in the model outcome ~ age + smoker, the link value for a 64-year old non-smoker is $Y = -0.3163$ corresponding to a model probability $P = 0.4215$. According to the model, this is the probability that a person who was 64 and a non-smoker was still alive at the time of the follow-up interview. In other words, the probability of being alive at the follow-up study conditioned on being 64 and a non-smoker at the time of the original interview. Change the values of the explanatory variables — look at a 65-year old smoker, for instance — and the model probability changes.

There is another set of factors that conditions the model probabilities: the situation that applied to the selection of cases for the data frame. For instance, fitting the model outcome ~ 1 to the smoking/mortality data gives a model probability for "alive" of $P = 0.672$. It would not be fair to say that this is the probability that a person will still be alive at the time of the follow-up interview. Instead, it is the probability that a person will still be alive *given* that they were in the sample of data found in the data frame. Only if that sample is representative of a broader population is it fair to treat that probability as applying to that population. If the overall population doesn't match the sample, the probability from the model fitted to the sample won't necessarily match the probability of "alive" in the overall population.

Often, the interest is to apply the results of the model to a broad population that is not similar to the sample. This can sometimes be done, but care must be taken.

To illustrate, consider a study done by researchers in Queensland, Australia on the possible link between cancer of the prostate gland and diet. Pan-fried and grilled meats are a source of carcinogenic compounds such as heterocyclic amines and polycyclic aromatic hydrocarbons. Some studies have found a link between eating meats cooked "well-done" and prostate cancer.[31] The Queensland researchers[32] interviewed more than 300 men with prostate cancer to find out how much meat of various types they eat and how they typically have their meat cooked. They also interviewed about 200 men without prostate cancer to serve as controls. They modeled whether or not each man has prostate cancer (variable pcancer) using both age and intensity of meat consumption as the explanatory variables. Then, to quantify the effect of meat consumption, they compared high-intensity eaters to low-intensity eaters; the model probability at the 10th percentile of intensity compared to the model probability at the 90th percentile. For example, for a 60-year old man, the model probabilities are 69.8% for low-meat intensity eaters, and 82.1% for high-meat intensity eaters.

It would be a mistake to interpret these numbers as the probabilities that a 60-year old man will have prostate cancer. The prevalence of prostate cancer is much lower. (According to one source, the prevalence of clinically evident prostate cancer is less than 10% over a lifetime, so many fewer than 10% of 60-year old men have clinically evident prostate cancer.[33])

The sample was collected in a way that intentionally overstates the prevalence of prostate cancer. More than half of the sample was selected specifically because the person had prostate cancer. Given this, the sample is not representative of the overall population. In order to make a sample that matches the overall population, there would have had to be many more non-cancer controls in the sample.

However, there's reason to believe that the sample reflects in a reasonably accurate way the diet practices of those men with prostate cancer and also of those men without prostate cancer. As such, the sample can give information that might be relevant to the overall population, namely how meat consumption could be linked to prostate cancer. For example, the model indicates that a high-intensity eater has a higher chance of having prostate cancer than a low-intensity eater. Comparing these probabilities — 82.1% and 79.8% — might give some insight.

There is an effective way to compare the probabilities fitted to the sample so that the results can be generalized to apply to the overall population. It will work so long as the different groups in the sample — here, the men with prostate cancer and the non-cancer controls — are each representative of that same group in the overall population.

It's tempting to compare the probabilities directly as a ratio to say how much high-intensity eating increases the risk of prostate cancer compared to low intensity eating. This ratio of model probabilities is 0.821/0.698, about 1.19. The interpretation of this is that high-intensity eating increases the risk of cancer by 19%.

Unfortunately, because the presence of prostate cancer in the sample doesn't match the prevalence in the overall population, the ratio of model probabilities usually will not match that which would be found from a sample that does represent the overall population.

Rather than taking the ratio of model probabilities, it turns out to be better to consider the **odds ratio**. Even though the sample doesn't match the overall population, the odds ratio will (so long as the two groups individually reflect accurate the population in each of those groups).

Odds are just another way of talking about probability P. Rather than being a number between zero and 1, an odds is a number between 0 and ∞. The odds are calculated as $P/(1-P)$. For example, the odds that corresponds to a probability $P = 50\%$ is $0.50/(1-0.50) = 1$. Everyday language works well here: the odds are fifty-fifty.

The odds corresponding to the low-intensity eater's probability of $P = 69.8\%$ is $0.698/0.302 = 2.31$. Similarly, the odds corresponding to high-intensity eater's $P = 82.1\%$ is $0.821/0.179 = 4.60$. The odds ratio compares the two odds: $4.60/2.31 = 1.99$, about two-to-one.

Why use a ratio of odds rather than a simple ratio of probabilities? Because the sample was constructed in a way that doesn't accurately represent the prevalence of cancer in the population. The ratio of probabilities would reflect the prevalence of cancer in the sample: an artifice of the way the sample was collected. The odds ratio compares each group — cancer and non-cancer — to itself and doesn't depend on the prevalence in the sample.

Example 18.1: Log-odds ratios of Prostate Cancer One of the nice features of the logistic transformation is that the link values Y can be used directly to read off the logarithm of odds ratios.

In the prostate-cancer data, the coefficients on the model pcancer $\sim$ age + intensity are these:

	Estimate	Std. Error	z value	Pr(>|z|)
Intercept	4.274	0.821	5.203	0.0000
Age	-0.057	0.012	-4.894	0.0000
Intensity	**0.172**	0.058	2.961	0.0031

These coefficients can be used to calculate the link value Y which corresponds to a log odds. For example, in comparing men at 10th percentile of intensity to those at the 90th percentile, you multiply the intensity coefficient by the difference in intensities. The bottom 10th percentile is 0 intensity and the top 10th percentile is an intensity of 4. So the difference in Y score for the two percentiles is $0.1722 \times (4 - 0) = 0.6888$. This value, 0.6888, is the log odds ratio. To translate this to an odds ratio, you need to undo the log. That is, calculate $e^{0.6888} = 1.99$. So, the odds ratio for the risk of prostate cancer in high-intensity eaters versus low-intensity eaters is approximately 2.

Note that the model coefficient on age is negative. This suggests that the risk of prostate cancer *goes down* as people age. This is wrong biologically as is known from other epidemiological work on prostate cancer, but it does reflect that the sample was constructed rather than randomly collected. Perhaps it's better to say that it reflects a weakness in the way the sample was constructed: the prostate cancer cases tend to be younger than the non-cancer controls. If greater care had been used in selecting the control cases, they would have matched exactly the age distribution in the cancer cases. Including age in the model is an attempt to adjust for the problems in the sample — the idea is that including age allows the model's dependence on intensity to be treated *as if* age were held constant.

18.4 Computational Technique

Fitting logistic models uses many of the same ideas as in linear models.

18.4.1 Fitting Logistic Models

The `glm` operator fits logistic models. (It also fits other kinds of models, but that's another story.) `glm` takes model design and data arguments that are identical to their counterparts in `lm`. Here's an example using the smoking/mortality data:

```
> whickham = ISMdata("whickham.csv")
> mod = glm( outcome ~ age + smoker, data=whickham,
    family="binomial")
```

The last argument, `family="binomial"`, simply specifies to `glm` that the logistic transformation should be used. (`glm` is short for Generalize Linear Modeling, a broad label that covers logistic regression as well as other types of models involving links and transformations.)

The regression report is produced with the `summary` operator, which recognizes that the model was fit logistically and does the right thing:

```
> summary(mod)
Coefficients:
             Estimate Std. Error z value Pr(>|z|)
(Intercept) -7.599221   0.441231 -17.223   <2e-16
age          0.123683   0.007177  17.233   <2e-16
smokerYes    0.204699   0.168422   1.215    0.224

    Null deviance: 1560.32  on 1313  degrees of freedom
Residual deviance:  945.02  on 1311  degrees of freedom
```

Keep in mind that the coefficients refer to the intermediate model values Y. The probability P will be $e^Y/(1+e^Y)$.

In fitting a logistic model, it's crucial that the response variable be categorical, with two levels. It happens that in the `whickham` data, the `outcome` variable fits the bill: the levels are `Alive` and `Dead`.

The `glm` software will automatically recode the response variable as 0/1. The question is, which level gets mapped to 1? In some sense, it makes no difference since there are only two levels. But if you're talking about the probability of dying, it's nice not to mistake that for the probability of staying alive. So make sure that you know which level in the response variable corresponds to 1: it's the second level.

Here is an easy way to make sure which level has been coded as "Yes". First, fit the all-cases-the-same model, outcome ~ 1. The fitted model value P from this model will be the proportion of cases for which the outcome was "Yes."

```
> mod2 = glm( outcome ~ 1, data=whickham, family="binomial")
> summary(mod)
Coefficients:
            Estimate Std. Error z value Pr(>|z|)
(Intercept) -0.94039    0.06139  -15.32   <2e-16
> exp(-.94039)/(1+exp(-.94039))
[1] 0.280822
```

So, 28% of the cases were "Yes." But which of the two levels is "Yes?" Find out just by counting and taking a proportion:

```
> prop.table(table(whickham$outcome))
    Alive      Dead
0.7191781 0.2808219
```

Evidently, by default, "Yes" means Dead.

If you want to dictate which of the two levels is going to be encoded as 1, you can use a comparison operation to do so:

```
> mod3 = glm( outcome=="Alive" ~ 1, data=whickham,
   family="binomial")
```

In this model, "Yes" means Alive.

18.4.2 Fitted Model Values

Logistic regression involves two different kinds of fitted values: the intermediate "link" value Y and the probability P. The fitted operator returns the probabilities:

```
> probs = fitted(mod)
          1           2           3           4
0.010458680 0.005662422 0.800110184 0.665421999
          5           6           7           8
0.578471493 0.063295027 0.138383748 0.858234119
   ... and so on ...
       1309        1310        1311        1312
0.008539232 0.044548261 0.028813443 0.008185469
       1313        1314
0.129003853 0.089194331
```

There is one fitted probability value for each case.

The link values can be gotten via the predict operator

```
> predict(mod, type="link")
          1           2           3           4
-4.54980922 -5.16822509  1.38698315  0.68755138
          5           6           7           8
 0.31650186 -2.69456160 -1.82877938  1.80069995
   ... and so on.
```

Notice that the link values are not necessarily between zero and one.

The predict operator can also be used to calculate the probability values.

```
> predict(mod, type="response")
          1           2           3           4
0.010458680 0.005662422 0.800110184 0.665421999
          5           6           7           8
0.578471493 0.063295027 0.138383748 0.858234119
   ... and so on.
```

This is particularly useful when you want to use predict to find the model values for inputs other than that original data frame used for fitting.

18.4.3 Analysis of Variance

The same basic logic used in analysis of variance applies to logistic regression, although the quantity being broken down into parts is not the sum of squares of the residuals but, rather, the deviance.

The anova software will take apart a logistic model, term by term, using the order specified in the model.

```
> anova(mod, test="Chisq")

Analysis of Deviance Table

        Df Deviance Resid. Df Resid. Dev  P(>|Chi|)
NULL                     1313    1560.32
age      1   613.81      1312     946.51 1.659e-135
smoker   1     1.49      1311     945.02       0.22
```

Notice the second argument, test="Chisq", which instructs anova to calculate a p-value for each term. This involves a slightly different test than the F test used in linear-model ANOVA.

The format of the ANOVA table for logistic regression is somewhat different from that used in linear models, but the concepts of degrees of freedom and partitioning still apply. The basic idea is to ask whether the reduction in deviance accomplished with the model terms is greater than what would be expected if random terms were used instead.

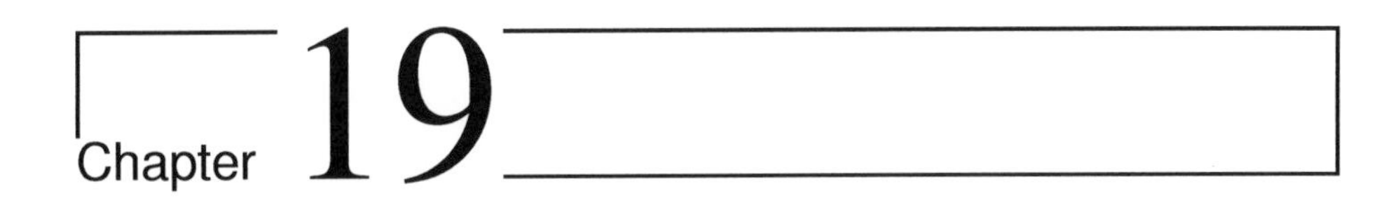

Chapter 19

Causation

If the issues at hand involve responsibilities or decisions or plans, causal reasoning is necessary. — Edward Tufte

Knowing what causes what makes a big difference in how we act. If the rooster's crow causes the sun to rise we could make the night shorter by waking up our rooster earlier and make him crow - say by telling him the latest rooster joke. — Judea Pearl

Starting in the 1860s, Europeans expanded rapidly westward in the United States, settling grasslands in the Great Plains. Early migrants had passed through these semi-arid lands, called the Great American Desert, on the way to territories further west. But unusually heavy rainfall in the 1860s and 1870s supported new migrants who homesteaded on the plains.

The homesteaders were encouraged by a theory that farming itself would increase rainfall. In the phrase of the day, "Rain follows the plow."

The theory that farming leads to rainfall was supported by evidence. As farming spread, measurements of rainfall were increasing. Buffalo grass, a species well adapted to dry conditions, was retreating and grasses more dependent on moisture were advancing. The exact mechanism of the increasing rainfall was uncertain, but many explanations were available. In a phrase-making book published in 1881, Charles Dana Wilber wrote,

> *Suppose now that a new army of frontier farmers ... could, acting in concert, turn over the prairie sod, and after deep plowing and receiving the rain and moisture, present a new surface of green, growing crops instead of the dry, hard-baked earth covered with sparse buffalo grass. No one can question or doubt the inevitable effect of this cool condensing surface upon the moisture in the atmosphere as it moves over by the Western winds. A reduction in temperature must at once occur, accompanied by the usual phenomena of showers. The chief agency in this transformation is agriculture. To be more concise.* Rain follows the plow. [34, p. 68]

Seen in the light of subsequent developments, this theory seems hollow. Many homesteaders were wiped out by drought. The most famous of these, the Dust Bowl of the 1930s, rendered huge areas of US and Canadian prairie useless for agriculture and led to the displacement of hundreds of thousands of families.

Wilber was correct in seeing the association between rainfall and farming, but wrong in his interpretation of the causal connection between them. With a modern perspective, you can see clearly that Wilber got it backwards; there was indeed an association between farming and rainfall, but it was the increased rainfall of the 1870s that lead to the growth of farming. When the rains failed, so did the farms.

The subject of this chapter is the ways in which data and statistical models can and cannot appropriately be used to support claims of causation. When are you entitled to interpret a model as signifying a causal relationship? How can you collect and process data so that such an interpretation is justified? How can you decide which covariates to include or exclude in order to reveal causal links?

The answers to these questions are subtle. Model results can be interpreted only in the context of the researcher's beliefs and prior knowledge about how the system operates. And there are advantages when the researcher becomes a participant, not just collecting observations but actively intervening in the system under study as an experimentalist.

19.1 Interpreting Models Causally

Interpreting statistical models in terms of causation is done for a purpose. It is well to keep that purpose in mind so you can apply appropriate standards of evidence. When causation is an issue, typically you have in mind some intervention that you are considering performing. You want to use your models to estimate what will be the effect of that intervention.

For example, suppose you are a government health official considering the approval of flecainide, a drug intended for the treatment of overly fast heart rhythms such as atrial fibrillation. Your interest is improving patient outcomes, perhaps as measured by survival time. The purpose of your statistical models is to determine whether prescribing flecainide to patients is likely to lead to improved outcomes and how much improvement will typically be achieved.

Or suppose you are the principal of a new school. You have to decide how much to pay teachers but you have to stay within your budget. You have three options: pay relatively high salaries but make classes large, pay standard salaries and make classes the standard size, pay low salaries and make classes small. Your interest is in the effective education of your pupils, perhaps as measured by standardized test scores. The purpose of your statistical models is to determine whether the salary/class-size options will differ in their effects and by how much.

Or suppose you are a judge hearing a case involving a worker's claim of sex discrimination against her employer. You need to decide whether to find for the

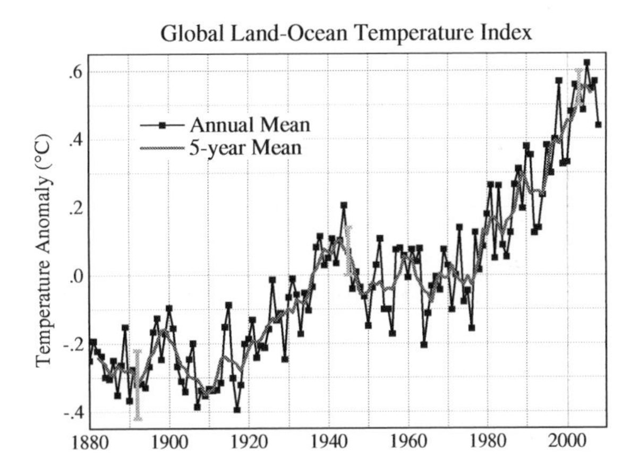

Figure 19.1: Global temperature since 1880 as reported by NASA. http://data.giss.nasa.gov/gistemp/graphs/ accessed on July 7, 2009.

worker or the employer. This situation is somewhat different. In the education or drug examples, you planned to take action to change the variable of interest — reduce class sizes or give patients a drug — in order to produce a better outcome. But you can't change the worker's sex to avoid the discrimination. Even if you could, what's past is past. In this situation, you are dealing with a **counterfactual**: you want to find out how pay or working conditions would have been different if you changed the worker's sex and left everything else the same. This is obviously a hypothetical question. But that doesn't mean it isn't a useful one to answer or that it isn't important to answer the question correctly.

Example 19.1: Greenhouse Gases and Global Warming

Global warming is in the news every day. Warming is not a recent trend, but one that extends over decades, perhaps even a century or more, as shown in Figure 19.1. The cause of the increasing temperatures is thought to be the increased atmospheric concentration of greenhouse gases such as CO_2 and methane. These gases have been emitted at a rapidly growing rate with population growth and industrialization.

In the face of growing consensus about the problem, skeptics note that such temperature data provides support but not proof for claims about greenhouse-gas induced global warming. They point out that climate is not steady; it changes over periods of decades, over centuries, and over much longer periods of time. And it's not just CO_2 that's been increasing over the last century; lots of other variables are correlated with temperature. Why not blame global warming, if it does exist, on them?

Imagine that you were analyzing the data in Figure 19.1 without any idea of a possible mechanism of global temperature change. The data look something like a random walk; a sensible null hypothesis might be exactly that. The typical year-to-year change in temperature is about $0.05°$, so the expected drift from a random walk over the 130 year period depicted in the graph is about $0.6°$, roughly the same as that observed.

The data in the graph do not themselves provide a compelling basis to reject the null hypothesis that global temperatures change in a random way. However, it's important to understand that climatologists have proposed a physical **mechanism** that relates CO_2 and methane concentration in the atmosphere to global climate change. At the core of this mechanism is the increased absorption of infra-red radiation by greenhouse gases. This core is not seriously in doubt: it's solidly established by laboratory measurements. The translation of that absorption mechanism into global climate consequences is somewhat less solid. It's based on computer models of the physics of the atmosphere and the ocean. These models have increased in sophistication over the last couple of decades to incorporate more detail: the formation of clouds, the thermohaline circulation in the oceans, etc. It's the increasing confidence in the models that drives scientific support for the theory that greenhouse gases cause global warming.

It's not data like Figure 19.1 that lead to the conclusion that CO_2 is causing global warming. It's the data insofar as they support models of mechanisms.

19.2 Causation and Correlation

It's often said, "Correlation is not causation." True enough. But it's an odd thing to say, like saying, "A movie is not a train."

In the earliest days of the cinema, a Lumière brothers film showing a train arriving at a station caused viewers to rise to their feet as if the train were real. Reportedly, some panicked from fear of being run over.[35, p. 222]

Modern viewers would not be fooled; we know that a movie train is incapable of causing us harm. The Lumière movie authentically represented a real

Figure 19.2: A frame from the 1895 film, "L'Arrivée d'un train en gare de La Ciotat"

train — the movie is a kind of model, a representation for a purpose — but the representation is not the mechanical reality of the train itself. Similarly, correlation is a representation of the relationship between variables. It captures some aspects of that relationship, but it is not the relationship itself and it doesn't fully reflect the mechanical realities, whatever they may be, of the real relationships that drive the system.

Correlation, along with the closely related idea of model coefficients, is a concept that applies to data and variables. The correlation between two variables depends on the data set, and how the data were collected: what sampling frame was used, whether the sample was randomly taken from the sampling frame, etc.

In contrast, causation refers to the influence that components of a system exert on one another. I write "component" rather than "variable" because the variables that are measured are not necessarily the active components themselves. For example, the score on an IQ test is not intelligence itself, but a reflection of intelligence.

As a metaphor for the differences between correlation and causation, consider a chain hanging from supports. (Figure 19.3.) Each link of the chain is mechanically connected to its two neighbors. The chain as a whole is a collection of such local connections. Its shape is set by these mechanical connections together with outside forces: the supports, the wind, etc. The overall system — both internal connections and outside forces — determines the global shape of

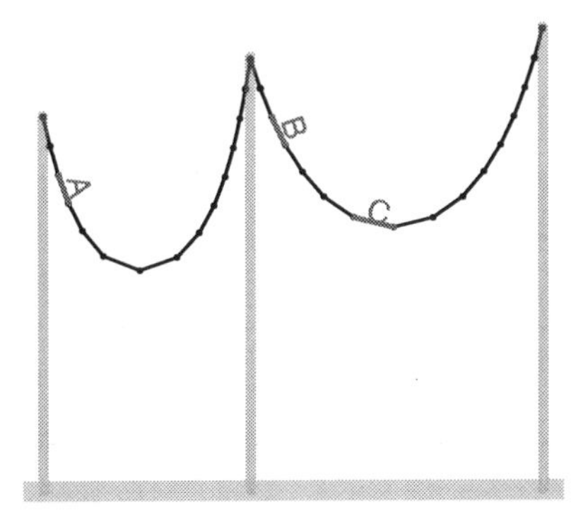

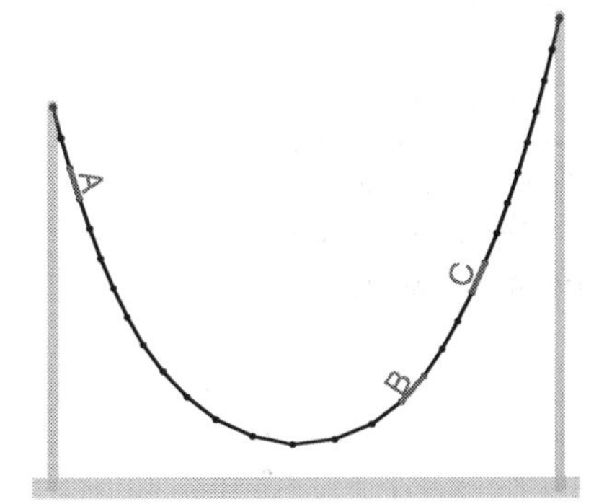

Figure 19.3: A metaphor for causation and correlation: the links of a chain with external supports. It's the same chain in both pictures but supported differently. The relationship between link orientations — links A and B are aligned in the picture on the left, but not in the picture on the right — depends both on the mechanical connections between links and on the external forces at work. So the alignment itself (correlation) is not a good signal for the mechanical connections (causation).

the chain. That overall shape sets the correlation between any two links in the chain, whether they be neighboring or distant.

In this metaphor, the shape of the chain is analogous to correlation; the mechanical connections between neighboring links is causation.

One way to understand the shape of the chain is to study its overall shape: the correlations in it. But be careful; the lessons you learn may not apply in different circumstances. For example, if you change the location of the supports, or add a new support in the middle, the shape of the chain can change completely as can the correlation between components of the chain.

Another way to understand the shape of the chain is to look at the local relationships between components: the causal connections. This does not directly tell you the overall shape, but you can use those local connections to figure out the global shape in whatever circumstances may apply. It's important to know about the mechanism so that you can anticipate the response to actions you take: actions that might change the overall shape of the system. It's also important for reasoning about counterfactuals: what would have happened had the situation been different (as in the sex discrimination example above) even if you have no way actually to make the system different. You can't change the plaintiff's sex, but you can play out the consequences of doing so through the links of causal connections.

One of the themes of this book has been that correlation, as measured by the correlation coefficient between two variables, is a severely limited way to describe relationships. Instead, the book has emphasized the use of model coefficients. These allow you to incorporate additional variables — covariates — into your interpretation of the relationship between two variables. Model coefficients provide more flexibility and nuance in describing relationships than does correlation. They make it possible, for instance, to think about the relationship between variables A and B *while adjusting for* variable C.

Even so, model coefficients describe the global properties of your data. If you want to use them to examine the local, mechanistic connections, there is more work to be done. Presumably, what people mean in saying "correlation is not causation" is that correlation is not on its own compelling evidence for causation.

An example of the difference between local causal connections and global correlations comes from political scientists studying campaign spending. Analysis of data on election results and campaign spending in US Congressional elections shows that increased spending by those running for re-election — incumbents — is associated with lower vote percentages. This finding is counterintuitive. Can it really be that an incumbent's campaign spending causes the incumbent to lose votes? Or is it that incumbents spend money in elections that are closely contested for other reasons? When the incumbent's election is a sure thing, there is no need to spend money on the campaign. So the negative correlation between spending and votes, although genuine, is really the result of external forces shaping the election and not the mechanism by which campaign spending affects the outcome.[36]

19.3 Hypothetical Causal Networks

In order to think about how data can be used as evidence for causal connections, it helps to have a notation for describing local connections. The notation I will use involves simple, schematic diagrams. Each diagram depicts a **hypothetical causal network**: a theory about how the system works. The diagrams consist of **nodes** and **links**. Each node stands for a variable or a component of the system. The nodes are connected by links that show the connections between them. A one-way arrow refers to a causal mechanistic connection. To illustrate, Figure 19.4 shows a hypothetical causal network for campaign spending by an incumbent.

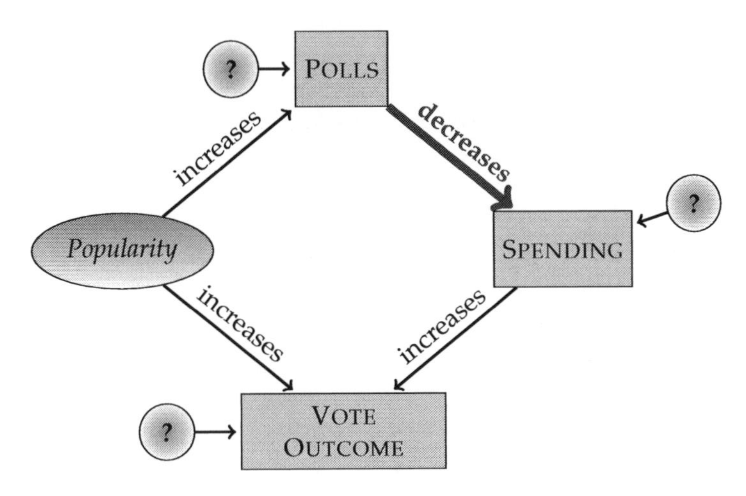

Figure 19.4: A hypothetical causal network describing how campaign spending by an incumbent candidate for political office is related to the vote outcome.

The hypothetical causal network in Figure 19.4 consists of four main components: spending and the vote outcome are the two of primary interest, but the incumbent candidate's popularity and the pre-election poll results are also included. According to the network, an incumbent's popularity influences both the vote outcome and the pre-election polls. The polls indicate how close the election is and this shapes the candidate's spending decisions. The amount spent influences the vote total.

A complete description of the system would describe how the various influences impinging on each node shape the value of the quantity or condition represented by the node. This could be done with a model equation, or less completely by saying whether the connection is positive or negative (as in the above diagram). For now, however, focus on the topology of the network: what's connected to what and which way the connection runs.

Nodes in the diagrams are drawn in two shapes. A square node refers to a variable that can be measured: poll results, spending, vote outcomes. Round nodes are for unmeasured quantities. For example, it seems reasonable to think about a candidate's popularity, but how to measure this outside of the context set by a campaign? So, in the diagram, popularity itself is not directly measured. Instead, there are poll results and the vote outcome itself. Specific but unmeasured variables such as "popularity" are sometimes called **latent variables**.

Often the round, unmeasured nodes will be drawn (?), which stands for the idea the *something* is involved, but no description is being given about what that something is; perhaps it's just random noise.

The links connecting nodes indicate causal influence. Note that every line has a arrow that tells which way causation works. The diagram

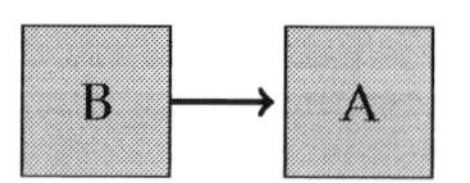

means that B causally influences A. The diagram

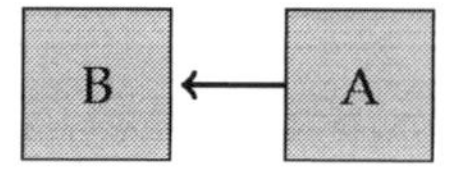

is the opposite: A is causally influencing B.

It's possible to have links running both ways between nodes. For instance, A causes B and B causes A. This is drawn as two different links. Such two way causation produces loops in the diagrams, but it is not necessarily illogical circular reasoning. In economics, for instance, it's conventional to believe that price influences production and that production influences price.

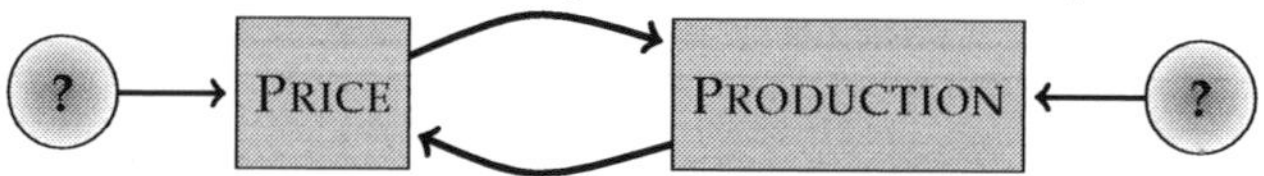

Why two causal links? When some outside event intervenes to change production, price is affected. For example, when a factory is closed due to a fire, production will fall and price will go up. If the outside event changes price — for instance, the government introduces price restrictions — production will change in response. Such outside influences are called **exogenous**. The (?) stands for an unknown exogenous input.

A hypothetical causal network is a model: a representation of the connections between components of the system. Typically, they are incomplete, not attempting to represent all aspects of the system in detail. When an exogenous influence is marked as (?), the modeler is saying, "I don't care to try to represent this in detail." But even so, by marking an influence with (?), the modeler is making an affirmative claim that the influence, whatever it be, is not itself caused by any of the other nodes in the system: it's exogenous. In contrast, nodes with a causal input — one or more links pointing to them — are **endogenous**, meaning that they are determined at least in part by other components of the system.

Often modelers decide not to represent all the links and components that might causally connect two nodes, but still want to show that there is a connection. Such **non-causal links** are drawn as double-headed dashed lines as in Figure 19.5. For instance, in many occupations there is a correlation between age and sex: older workers tend to be male, but the population of younger workers

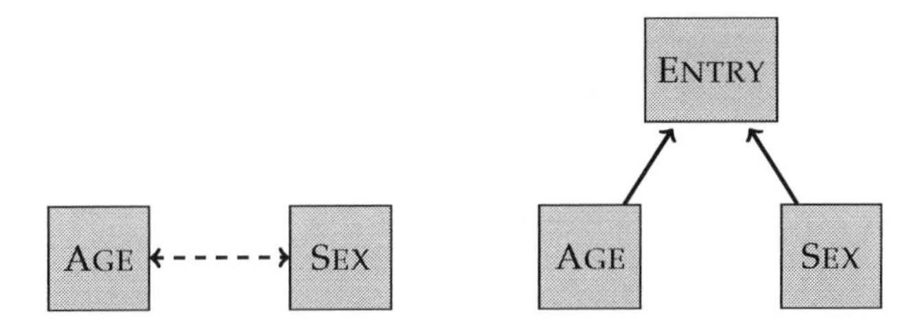

Figure 19.5: A non-causal link and its expansion into a diagram with causal links.

is more balanced. It would be silly to claim a direct causal link from age to sex: a person's sex doesn't change as they age! Similarly, sex doesn't determine age. Instead, women historically were restricted in their professional options. There is an additional variable — entry into the occupation — that is determined by both the person's age and sex. The use of a non-causal link to indicate the connection between age and sex allows the connection to be displayed without including the additional variable.

It's important not to forget the word "hypothetical" in the name of these diagrams. A hypothetical causal network depicts a state of belief. That belief might be strongly informed by evidence, or it might be speculative. Some links in the network might be well understood and broadly accepted. Some links, or the absence of some links, might be controversial and disputed by other people.

19.4 Networks and Covariates

Often, you are interested in only some of the connections in a causal network; you want to use measured data to help you determine if a connection exists and to describe how strong it is. For example, politicians are interested to know how much campaign spending will increase the vote result. Knowing this would let them decide how much money they should try to raise and spend in an election campaign.

It's wrong to expect to be able to study just the variables in which you have a direct interest. As you have seen, the inclusion of covariates in a model can affect the coefficients on the variables of interest.

For instance, even if the direct connection between spending and vote outcome is positive, it can well happen that using data to fit a model vote outcome ~ spending will produce a negative coefficient on spending.

There are three basic techniques that can be used to collect and analyze data in order to draw appropriate conclusions about causal links.

Experiment, that is, intervene in the system to set or influence certain variables and then examine how your intervention relates to the observed outcomes.

Include covariates in order to adjust for other variables.

Exclude covariates in order to prevent those variables from unduly influencing your results.

Experimentation provides the strongest form of evidence. Indeed, many statisticians and scientists claim that experimentation provides the only compelling evidence for a causal link. As the expression goes, "No causation without experimentation."

This may be so, but in order to explain why and when experimentation provides compelling evidence, it's helpful to examine carefully the other two techniques: including and excluding covariates. And, in many circumstances where you need to draw conclusions about causation, experiment may be impossible.

Previous chapters have presented many examples of Simpson's paradox: the coefficient on an explanatory variable changing sign when a covariate is added to a model. Indeed, it is inevitable that the coefficient will change — though not necessarily change in sign — whenever a new covariate is added that is correlated with the explanatory variable.

An important question is this: Which is the right thing to do, include the covariate or not?

To answer this question, it helps to know the right answer! That way, you can compare the answer you get from a modeling approach to the known, correct answer, and you can determine which modeling approaches work best, which rules for including covariates are appropriate. Unfortunately, it's hard to learn such lessons from real-world systems. Typical systems are complex and the actual causal mechanisms are not completely known. As an alternative, however, you can use simulations: made-up systems that let you test your approaches and pick those that are appropriate for the structure of the hypothetical causal network that you choose to work with.

To construct a simulation, you need to add details to the hypothetical causal networks. What's left out of the notation for the networks is a quantitative description of what the link arrows mean: how the variable at the tail of the link influences the variable at the arrowhead. One way to specify this is with a formula. For instance, the formula

$$\mathsf{vote} \leftarrow 0.75\,\mathsf{popularity} + 0.25\,\mathsf{spending} + \mathsf{Normal}(0,5)$$

says that the vote outcome can be calculated by adding together weighted amounts of the level of popularity and spending and the exogenous random component (which is set to be a normal random variable with mean 0 and standard deviation 5).

Notice that I've used $\leftarrow$ instead of $=$ in the formula. The formula isn't just a statement that vote is related to popularity and spending but a description of what causes what. Changing popularity changes vote, but not *vice versa*.

Usually, modelers are interested to deduce the formula from measured data. In a simulation, however, the formula is pre-specified and used to create the simulated data.

Such a simulation can't directly tell you about real campaign spending. It can, however, illuminate the process of modeling by showing how the choice of covariates shapes the implications of a model. Even better, the simulations can guide you to make *correct* choices, since you can compare the results of your modeling of the simulated data to the known relationships that were set up for the simulation.

A simulation can be set up for the campaign spending hypothetical network. Here's one that accords pretty well with most people's ideas about how campaign spending works:

popularity is endogenous, simulated by a random variable with a uniform distribution between 15 and 85 percent, indicating the amount of support the incumbent has.

polls echo popularity, but include a random exogenous component as well: polls $\leftarrow$ popularity + Normal$(0, 3)$.

spending is set based on the poll results. The lower the polls, the more the incumbent candidate spends: spending $\leftarrow 100 -$ polls + Normal$(0, 10)$.

vote is the result of popularity, spending, and an exogenous random input: vote $\leftarrow 0.75$ popularity + 0.25 spending + Normal$(0, 5)$

Remember, these formulas are hypothetical. It's not claimed that they represent the mechanics of actual campaign spending in any detailed way, or that the coefficients reflect the real world (except, perhaps, in sign), or that the size of the random, exogenous influences are authentic. But by using the formulas you can generate simulated data that allows you to test out various approaches to modeling. Then you can compare your modeling results with the formulas to see which approaches are best.

The simulation outlined above was used to generate data on polls, spending, and vote. Then, this simulated data used to fit two different models: one with covariates and one without them. Here are the results from a simulation with $n = 1000$:

	Model	Coef. on spending
1	vote ~ spending	-0.33 ± 0.02
2	vote ~ spending+ polls	0.23 ± 0.03

In the first model, vote ~ spending, the fitted coefficient on spending is negative, meaning the higher spending is associated with lower vote. This negative coefficient correctly summarizes the pattern shown by the data but it is incorrect causally. According to the formulas used in the simulation, higher spending leads to higher vote outcome.

On the other hand, the model vote ~ spending+polls that incorporates polls as a covariate gets it right: the coefficient on spending is positive (as it should be) and is even the right size numerically: about 0.25.

It's tempting to draw this conclusion: always add in covariates. But this general rule is premature and, in fact, wrong. To see why, consider another simulation based in another hypothetical causal network.

Figure 19.6 shows a simple hypothetical causal network that might be used to describe college admissions. This network depicts the hypothesis that whether or not a student gets into college depends on both the intelligence and athletic ability

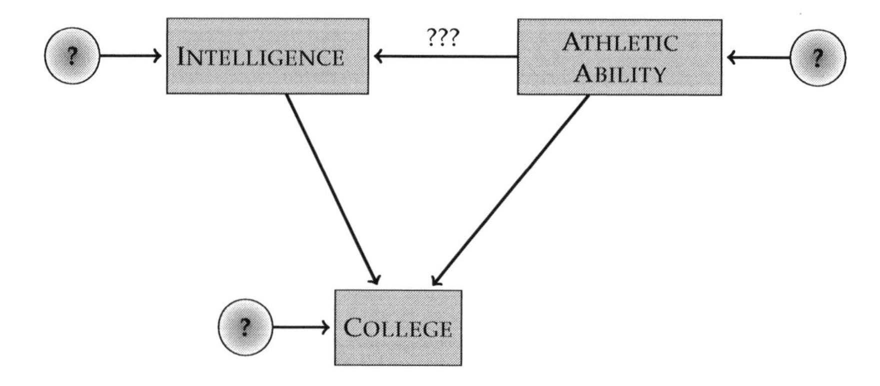

Figure 19.6: A hypothetical causal network about college admissions and athletics.

of the applicant. But, according to the hypothetical causal network, there may also be a link between intelligence and athletic ability.

Suppose someone wants to check whether the link marked ??? in Figure 19.6 actually exists, that is, whether athletic ability affects intelligence. An obvious approach is to build the simple model intelligence ~ athletic ability. Or, should the covariate college be included in the model?

A simulation was set up with no causal connection at all between between athletic ability and intelligence. However, both athletic ability and intelligence combine to determine whether a student gets into college.

Using data simulated from this network (again, with $n = 1000$), models with and without covariates were fit.

	Model	Coef. on spending
1	IQ ~ Athletic	0.03 ± 0.30
2	IQ ~ Athletic+ College	-1.33 ± 0.24

The first model, without college as a covariate, gets it right. The coefficient, 0.03 ± 0.30, shows no connection between athletic and intelligence. The second model, which includes the covariate, is wrong. It falsely shows that there is a negative relationship between athletic ability and intelligence. Or, rather, the second model is false in that it fails to reproduce the causal links that were present in the simulation. The coefficients are actually correct in showing the correlations among athletic, intelligence, and college that are present in the data generated by the simulation. The problem is that even though college doesn't cause intelligence, the two variables are correlated.

The situation is confusing. For the campaign spending system, the right thing to do when studying the direct link between spending and vote outcome was to *include* a covariate. For the college admissions system, the right thing to do when studying the direct link between intelligence and athletic ability was to *exclude* a covariate.

The heart of the difficulty is this: In each of the examples you want to study a single link between variables, for instance these:

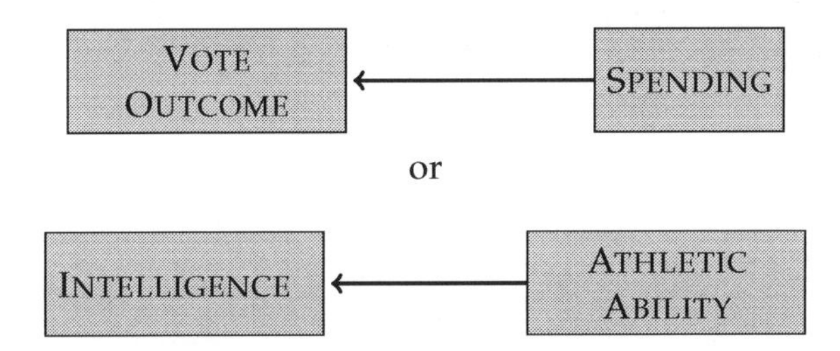

But the nature of networks is that there is typically more than one pathway that connects two nodes. What's needed is a way to focus on the links of interest while avoiding the influence of other connections between the variables.

19.4.1 Pathways

A hypothetical causal network is like a network of roads. The causal links are one-way roads where traffic flows in only one direction. In hypothetical causal networks, as in road networks, there is often more than one way to get from one place to another.

A **pathway** between two nodes is a route between them, a series of links that connects one to another, perhaps passing through some other nodes on the way.

It's helpful to distinguish between two kinds of pathways:

- **Correlating pathways** follow the direction of the causal links.
- **Non-correlating pathways** don't.

To help develop a definition of the two kinds of pathways, consider a simple network with three nodes, A, B, and C, organized with A connected to C, and C connected to B.

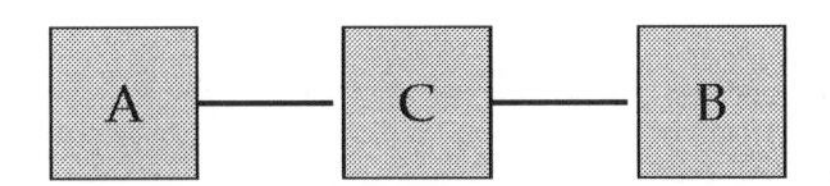

There is a pathway connecting A to B, but in order to know whether it is correlating or non-correlating, you need to know the causal directions of the links. If the links flow as A⇒C⇒B, then the pathway connecting A to B is correlating. Similarly if the pathway is A⇐C⇐B. Less obviously, the pathway connecting A to B in the network A⇐C⇒B is correlating, since from node C you can get to both A and B. But if the flows are A⇒C⇐B, the pathway connecting A and B is non-correlating.

The general rule is that a pathway connecting two variables A and B is correlating if there is some node on the pathway from which you can get to both A and B by following the causal flow. So, in A⇒C⇒B, you can start at A and reach B. Similarly, in A⇐C⇐B, you can start at B and reach A. In A⇐C⇒B you can start at C and reach both A and B. But in A⇒C⇐B, there is no node you can start at from which the flow leads both to A and B.

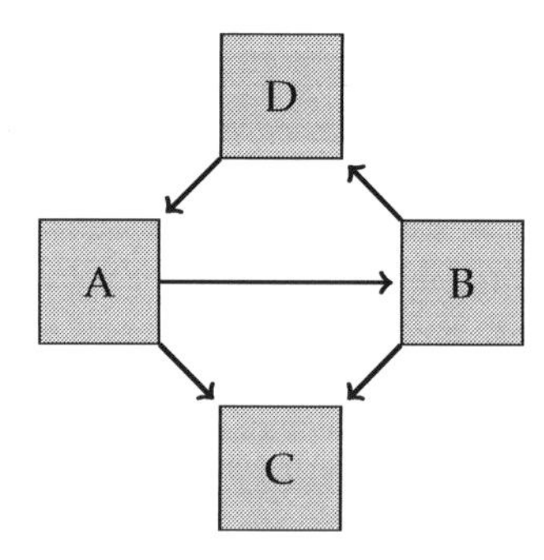

Figure 19.7: A hypothetical causal network with three pathways connecting nodes A and B.

A non-correlating pathway is one where it is not possible to get from one of the nodes on the pathway to the end-points by following the causal links. So, A⇒C⇐B is a non-correlating pathway: Although you can get from B to C by following the causal link, you can't get to A. Similarly, starting at A will get you to C, but you can't get from C to B.

Hypothetical causal networks often involve more than one pathway between variables of interest. To illustrate, consider the network shown in Figure 19.7. There are three different pathways connecting A to B:

- The direct pathway A⇒B. This is a correlating pathway. Starting at A leads to B.
- The pathway through C, that is, A⇒C ⇐B. A non-correlating pathway. There is no node where you can start that leads to both endpoints A and B by following the causal flows.
- The pathway through D, that is, A⇐D ⇐B. This is a correlating pathway — start at B and the flow leads to A.

Of course, correlating pathways can be longer and can involve more intermediate nodes. For instance A⇐C ⇐D ⇒B is a correlating pathway connecting A and B. Starting at D, the flow leads to both A and B.

Pathways can also involve links that display correlation and not causation. When correlations are involved (shown by a double-headed arrow: ⇔), the test of whether a pathway is correlating or non-correlating is still the same: check whether there is a node from which you can get to both end-points by following the flows. For the correlation link, the flows go in both directions. So A ⇔C⇒D ⇒B is correlating; starting at C, the flow leads to both endpoints A and B.

Some hypothetical causal networks include pathways that are causal loops, that is, pathways that lead from a node back to itself following only the direction of causal flow. These are called **recurrent networks**. The analysis of recurrent networks requires more advanced techniques that are beyond the scope of this introduction. Judea Pearl gives a complete theory.[37]

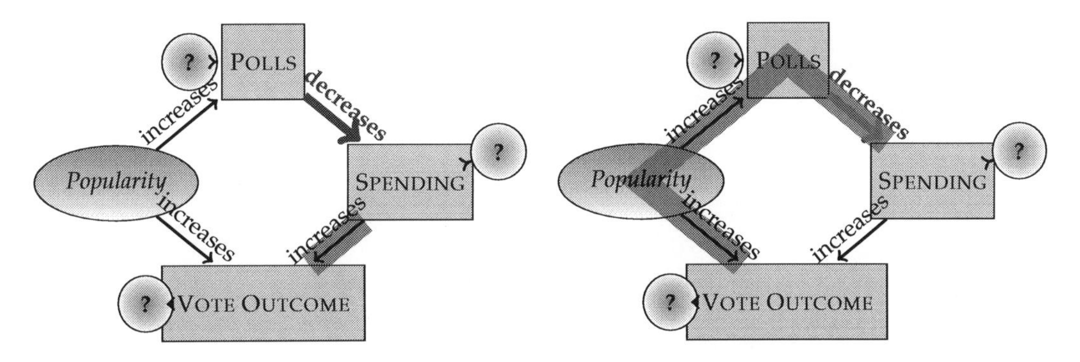

Figure 19.8: Pathways connecting spending to vote outcome. Both of these are correlating pathways.

19.4.2 Pathways and the Choice of Covariates

Understanding the nature of the pathways connecting two variables helps in deciding which covariates to include or exclude from a model. Suppose your goal is to study the direct relationship between two nodes, say between A and B in Figure 19.7. To do this, you need to **block** all the other pathways between those variables. So, to study the direct relationship A⇒B, you would need to block the "backdoor" pathways A⇐D⇐B and A⇒C⇐B.

The basic rules for blocking a pathway are simple:

- For a **correlating pathway**, you must **include** at least one of the interior nodes on the pathway as a covariate.

- For a **non-correlating pathway**, you must **exclude** all the interior nodes on the pathway; do not include any of them as covariates.

Typically, there are some pathways in a hypothetical causal network that are of interest and others that are not. Suppose you are interested in the direct causal effect of A on B in the network shown in Figure 19.7. This suggests the model B ~ A. But which covariates to include in order to block the two backdoor pathways A⇐D ⇐B and A⇒C⇐B? Including D as a covariate will block the correlating pathway A⇐D⇐B, so you should include D in your model. On the other hand, in order to block A⇒C⇐B you need to *exclude* C from your model. So, the correct model is B ~ A+D.

Returning to the campaign spending example, there are two pathways between spending and vote outcome. The pathway of interest is vote⇐spending. The backdoor pathway, pathway vote⇐popularity⇒polls⇒spending is not of direct interest to those concerned with the causal effect of spending on the vote outcome. This backdoor pathway is correlating; starting at node popularity leads to both endpoints of the pathway, spending and vote. To block it, you need to include one of the interior nodes, either polls or popularity. The obvious choice is polls, since this is something that can be measured and used as a covariate in a model. So, an appropriate model for studying the causal connection between vote and spending is the model vote ~ spending + polls.

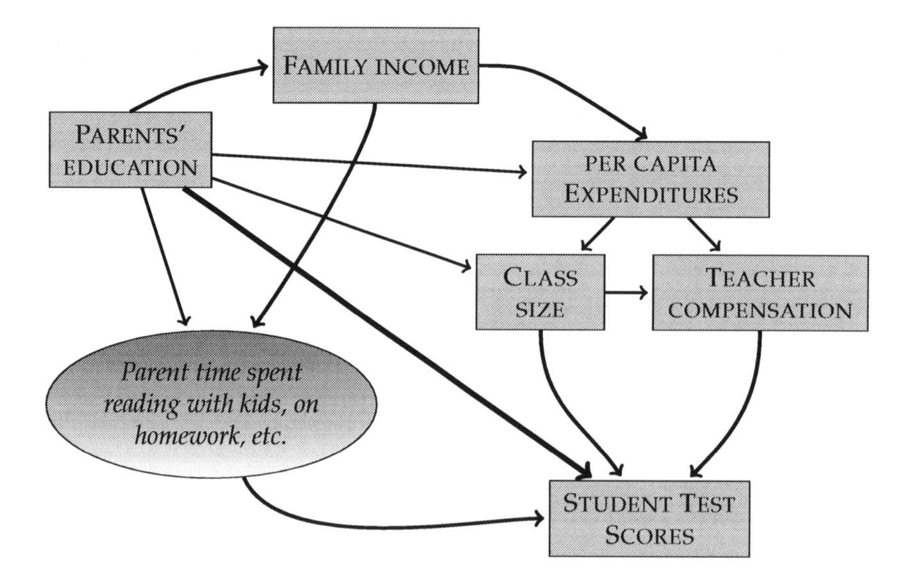

Figure 19.9: A hypothetical causal network relating student test scores to various educational policies and social conditions.

In the college admission network shown in Figure 19.6, there are two pathways connecting athletic ability to intelligence. The direct one is of interest. The backdoor route is not: intelligence⇒college⇐athletic. The backdoor pathway is non-correlating: there is no node from which the flow leads to both intelligence and athletic. In order to block this non-correlating pathway, you must avoid including any interior node, so college cannot be used as a covariate.

Example 19.2: Learning about Learning

Suppose you want to study whether increasing expenditures on education will improve outcomes. You're going to do this by comparing different districts that have different policies. Of course, they likely also have different social conditions. So there are covariates. Which ones should be included in your models?

Figure 19.9 shows what's actually a pretty simple hypothetical causal network. It suggests that there is a link between family income and parents' education, and between parents' education the test scores of their children. It also suggests that school expenditures and policy variables such as class size relate to family income and parents' education. Perhaps this is because wealthier parents tend to live in higher-spending districts, or because they are more active in making sure that their children get personalized attention in small classes.

It hardly seems unreasonable to believe that things work something like the way they are shown in Figure 19.9. So reasonable people might hold the beliefs depicted by the network.

Now imagine that you want to study the causal relationship between per capita expenditures and pupil test scores. In order to be compelling to the people who accept the network in Figure 19.9, you need to block the backdoor pathways between expenditures and test scores. Blocking those backdoor pathways, which are correlating pathways, means including both parents' education and family income as covariates in your models. That, in turn, means that data relating to these covariates need to be collected.

Notice also that both class size and teacher compensation should not be included as covariates. Those nodes lie on correlating pathways that are part of the causal link between expenditures and test scores. They are not part of a backdoor correlating pathway; they are the actual means by which the variable expenditures does it's work. You do not want to block those pathways if you want to see how expenditures connects causally to test scores.

In summary, a sensible model, consistent with the hypothetical causal network in Figure 19.9, is this:

test scores ~ expenditures + family income + parents' education.

19.4.3 Sampling Variables

Sometimes a variable is used to define the sampling frame when collecting data. For example, a study of intelligence and athletic ability might be based on a sampling frame of college students. This seems innocent enough. After all, colleges can be good sources of information about both variables. Of course, your results will only be applicable to college students; you are effectively holding the variable college constant. Note that this is much the same as would have happened if you had included both college and non-college students in your sample and then included college as a covariate, the standard process for adjusting for a covariate when building a model. The implication is that sampling only college students means that the variable college will be implicitly included as a covariate in all your models based on that data. A variable that is used to define your sampling frame is called a **sampling variable**.

In terms of blocking or unblocking pathways, using data based on a sampling variable is equivalent to including that variable in the model. To see this, recall that in the model B ~ A + C, the presence of the variable C means that the relationship between A and B can be interpreted as a *partial relationship*: how B is related to A while holding C constant. Now imagine that the data have been collected using C as a sampling variable, that is, with C constant for all the cases. Fitted to such data, the model B ~ A still shows the relationship between B and A while holding C constant.

In drawing diagrams, the use of a sampling variable to collect data will implicitly change the shape of the diagram. For example, consider the correlating pathway A⇒C⇒D⇒B. Imagine that the data used to study the relationship between A and B were collected from a sampling frame where all the cases had the same value of C. Then the model A ~ B is effectively giving a coefficient on B with C held constant at the level in the sampling frame. That is, A ~ B will give the same B coefficient as A ~ B + C. Since C is an interior variable on the pathway from A to B, and since the pathway is correlating, the inclusion of C as a sampling variable blocks the entire pathway. This effectively disconnects A from B.

Now consider a non-correlating backdoor pathway like A⇒F⇐G⇒B. When F is used as a sampling variable, this pathway is unblocked. Effectively, this translates the pathway into A⇔G⇒B, which is a correlating pathway. (You can get to both endpoints A and B by starting either at G or at A.) So, a sampling variable can unblock a backdoor pathway that you might have wanted to block.

You need to be careful in thinking about how your sampling frame is based on a sampling variable that might be causally connected to other variables of interest. For instance, sampling just college students when studying the link between intelligence and athletic ability in the network shown in Figure 19.6 opens up the non-correlating, backdoor pathway via college. Thus, even if there were no relationship between intelligence and athletic ability, your use of college as a sampling variable could create one.

19.4.4 Disagreements about Networks

The appropriate choice of covariates to include in a model depends on the particular hypothetical causal network that the modeler accepts. The network reflects the modeler's beliefs, and different modelers can have different beliefs. Sometimes this will result in different modelers drawing incompatible conclusions about which covariates to include or exclude. What to do about this?

One possible solution to this problem is for you to adopt the other modeler's network. This doesn't mean that you have to accept that network as true, just that you reach out to that modeler by analyzing your data in the context set by that person's network. Perhaps you will find that, even using the other network, your data don't provide support for the other modeler's claims. Similarly, the other modeler should try to convince you by trying an analysis with your network.

It's particularly important to be sensitive to other reasonable hypothetical causal networks when you are starting to design your study. If there is a reasonable network that suggests you should collect data on some covariate, take that as a good reason to do so even if the details of your own network do not mandate that the covariate be included. You can always leave out the covariate when you analyze data according to your network, but you will have it at hand if it becomes important to convince another modeler that the data are inconsistent with his or her theory.

Sometimes you won't be able to agree. In such situations, being able to contrast the differing networks can point to origins of the dispute and can help you to figure out a way to resolve them.

Example 19.3: Sex Discrimination in Salary

The college where I work is deeply concerned to avoid discrimination against women. To this end, each year it conducts an audit of professors' salaries to see if there is any evidence that the college pays women less than men.

Regrettably, there is such evidence; female faculty earn on average less than male faculty. However, there is a simple explanation for this disparity. Salaries are determined almost entirely by the professor's rank: instructors earn the least, assistant professors next, then associate professors and finally "full" professors. Until recently, there have been relative few female full professors and so relatively few women earning the highest salaries.

Why? It takes typically about 12 years to move from being a starting assistant professor to being a full professor, and another 25 years or so until retirement. It takes five to ten years to go from a college graduate to being a starting assistant professor. Thus, the population of full professors reflects the opportunities for entering graduate students 20 to 45 years ago.

I am writing this in 2009, so 45 years ago was the mid 1960s. This was a very different time from today; women were discouraged even from going to college and certainly from going to graduate school. There was active discrimination against women. The result is that there are comparatively few female full professors. This is changing because discrimination waned in the 1970s and many women are rising up through the ranks.

In analyzing the faculty salary data, you need to make a choice of two simple models. Salary is the response variable and sex is certainly one of the explanatory variables. But should you include rank?

Here's one possible hypothetical causal network:

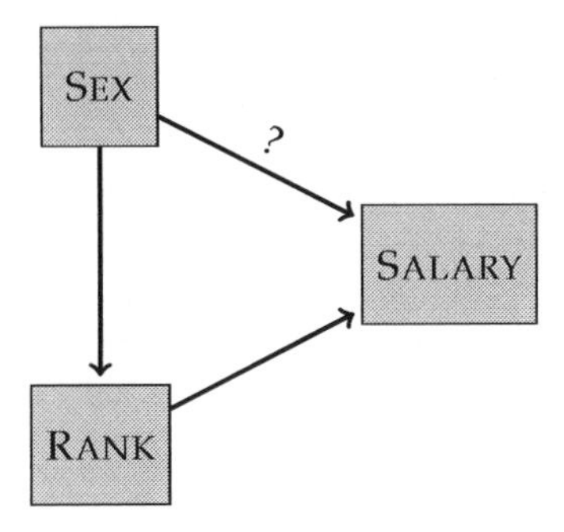

This hypothetical network proposes that there is a direct link between sex and salary — that link is marked with a question mark because the question is whether the data provide support for such a link. At the same time, the network proposes that sex is linked with rank and the network shows this as a causal link.

The consequence of this diagram is that to understand the total effect of sex on salary, you need to include both of the pathways linking the two variables: sex⇒salary as well as the pathway sex⇒rank⇒salary. In order to avoid blocking the second pathway, we should **not** include rank as an explanatory variable. Result: The data do show that being female has a negative impact on salary.

In the above network, the causal direction has been drawn from sex to rank. This might be simply because it's hard to imagine that a person's rank can determine their sex. Besides, in the above comments I already stipulated that there have been historical patterns of discrimination on the basis of sex that determine the population of the different ranks.

On the other hand, discrimination against people in the past doesn't mean that there is discrimination against the people who currently work at the institution. So imagine some other possibilities.

What would happen if the causal direction were reversed? I know this sounds silly, that rank can't cause sex, but suspend your disbelief for a moment. If rank caused sex, there is no causal connection from sex to salary via rank. As such, the pathway from sex to salary via rank ought to be blocked in your models. This is a correlating pathway, so you would block it by including rank as a covariate. Result: The data show that being female does not have a negative impact on salary.

Unfortunately, this means that the result of the modeling depends strongly on your assumptions. If you imagine that sex does not influence rank, then there is no evidence for salary discrimination. But if you presume that sex does influence rank, then the model does provide evidence for salary discrimination. Stalemate.

Fortunately, there is a middle ground. Perhaps rank and sex are not causally related, but are merely correlated. In this situation, too, you would want to block the indirect correlating pathway because it does not tell you about the causal effect of sex on salary. The result would remain that sex does not have a negative impact on salary.

Is your intuition kicking at you? Sex is determined at conception, it can't possibly be caused by anything else in this network. If it can't be caused by anything, it must be the cause of things. Right? Well, not quite. The reason is that all of the people being studied are professors, and this selection can create a correlation.

Here's a simple causal network describing who becomes a professor:

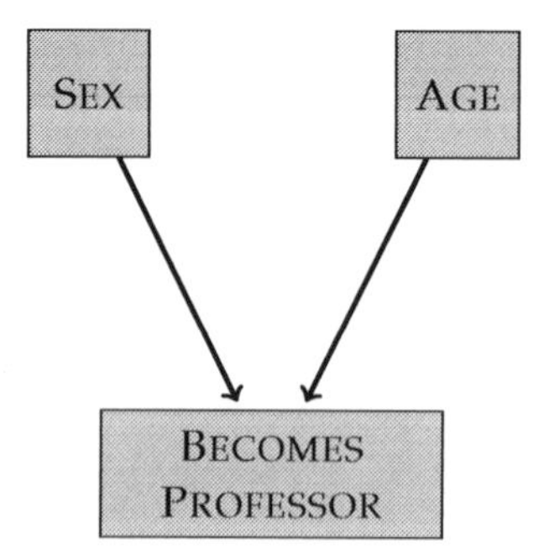

Whether a person became a professor depends on both sex and age. Putting aside the question of whether there is still sex discrimination in graduate studies and hiring, there is no real dispute that there was a relationship in the past. Perhaps this could be modeled as an interaction between sex and age in shaping who becomes a professor. There might be main effects as well.

With this in mind, an appropriate hypothetical causal network might look like this:

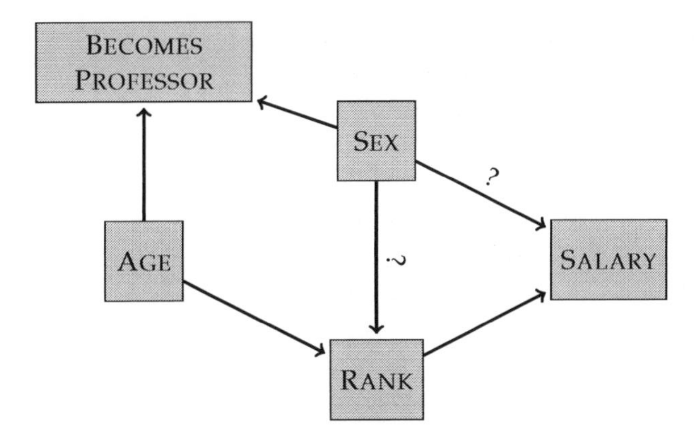

In this network, which people on both sides of the dispute would find reasonable, there is a new backdoor pathway linking sex and salary. This is the pathway sex⇒becomes professor⇐age⇒rank ⇒salary. This pathway is not about our present mechanisms for setting salary, it's about how people became professors in the past. Since the pathway is not of direct interest to studying sex discrimination in today's salaries, it should be blocked when examining the relationship between sex and salary.

The backdoor pathway sex⇒becomes professor⇐age⇒rank ⇒salary is a non-correlating pathway. There is no node on the pathway that flows to the two endpoints. (Of course you can reach salary from sex directly — there is a sex⇒salary pathway — but that is not involved in the backdoor pathway that is under consideration here.)

Here's the catch. Since the data used for the salary audit include only the people working at the college, the variable becomes professor is already included implicitly as a *sampling variable*. This means that the backdoor pathway is unblocked in the model sex $\sim$ salary. That is, the use of becomes professor as a sampling variable creates a correlation between sex and age. The pathway then is equivalent to sex⇔age⇒rank⇒salary. This is a correlating pathway, since you can get from sex to salary by following the flows.

To block the correlating pathway, you need to include as a covariate one of the variables on the pathway. Age offers some interesting possibilities here. Including age as a covariate blocks the backdoor pathway but allows you to avoid including rank. By leaving rank out of the model, it becomes possible for the model to avoid making any assumption about whether there is sex discrimination in setting rank. Thus, it's possible to build a model that will be satisfactory both to those who claim that rank is not the product of sex discrimination and those who claim it is.

Chapter 20

Experiment

The true method of knowledge is experiment. — William Blake (1757-1827)

A theory is something nobody believes, except the person who made it. An experiment is something everybody believes, except the person who made it. — Albert Einstein (1879-1955)

20.1 Experiments

The word "experiment" is used in everyday speech to mean *a course of action tentatively adopted without being sure of the eventual outcome.* [26] You can experiment with a new cookie recipe; experiment by trying a new bus route for getting to work; experiment by planting some new flowers to see how they look.

Experimentation is at the core of the scientific method. Albert Einstein said, "[The] development of Western Science is based on two great achievements — the invention of the formal logical system (in Euclidean geometry) by the Greek philosophers, and the discovery of the possibility to find out causal relationships by systematic experiment (during the Renaissance)." [38, p.1]

Scientific experimentation is, as Einstein said, systematic. In the everyday sense of experimentation, to try out a new cookie recipe, you cook up a batch and try them out. You might think that the experiment becomes scientific when you measure the ingredients carefully, set the oven temperature and timing precisely, and record all your steps in detail. Those are certainly good things to do, but a crucial aspect of systematic experimentation is **comparison**.

Two distinct forms of comparison are important in an experiment. One form of comparison is highlighted by the phrase "**controlled experiment**". A controlled experiment involves a comparison between two interventions: the one you are interested in trying out and another one, called a **control**, that provides a baseline or basis for comparison. In the cookie experiment, for instance, you would compare two different recipes: your usual recipe (the control) and the new recipe (called the **treatment**). Why? So that you can put the results for the new recipe in an appropriate context. Perhaps the reason you wanted to try a new recipe is that you were particularly hungry. If so, it's likely that you will like the new recipe even if were not so good as your usual one. By comparing the two recipes side by side, a controlled experiment helps to reduce the effects of factors such as your state of hunger. (The word "controlled" also suggests carefully maintained experimental conditions: measurements, ingredients, temperature, etc. While such care is important in successful experiments, the phrase "controlled experiment" is really about the comparison of two or more different interventions.)

Another form of comparison in an experiment concerns intrinsic variability. The different interventions you compare in a controlled experiment may be affected by factors other than your intervention. When these factors are unknown, they appear as **random variation**. In order to know if there is a difference between your interventions — between the treatment and the control — you have to do more than just compare treatment and control. You have to compare the observed treatment-vs-control difference to that which would be expected to occur at random. The technical means of carrying out such a comparison is, for instance, the F statistic in ANOVA. When you carry out an experiment, it's important to establish the conditions that will enable such a comparison to be carried out.

A crucial aspect of experimentation involves causation. An experiment does not necessarily eliminate the need to consider the structure of a hypothetical causal network when selecting covariates. However, the experiment actually changes the structure of the network and in so doing can radically simplify the system. As a result, experiments can be much easier to interpret, and much less subject to controversy in interpretation than an observational study.

20.1.1 Experimental Variables and Experimental Units

In an experiment, you, the experimenter, intervene in the system.

- You choose the **experimental units**, the material or people or animals to which the control and treatment will be applied. When the "unit" is a person, a more polite term is **experiment subject**.

- You set the value of one or more variables — the **experimental variables** — for each experimental unit.

The experimenter then observes or measures the value of one or more variables for each experimental unit: a response variable and perhaps some covariates.

The choice of the experimental variables should be based on your understanding of the system under study and the questions that you want to answer. Often, the choice is straightforward and obvious; sometimes it's clever and ingenious. For example, suppose you want to study the possibility that the habit of regularly taking a tablet of aspirin can reduce the probability of death from stroke or heart attack in older people. The experimental variable will be whether or not a person takes aspirin daily. In addition to setting that variable, you will want to measure the response variable — the person's outcome, e.g., the age of their eventual death — and covariates such as the person's age, sex, medical condition at the onset of the study, and so on.

Or, suppose you want to study the claim that keeping a night-light on during sleeping hours increases the chances that the child will eventually need glasses for near-sightedness. The experimental variable is whether or not the child's night light is on, or perhaps how often it is left on. Other variables you will want to measure are the response — whether the child ends up needing glasses — and the relevant covariates: the child's age, sex, whether the parents are nearsighted, and so on.

In setting the experimental variable, the key thing is to create variability, to establish a contrast between two or more conditions. So, in the aspirin experiment, you would want to set things up so that some people regularly take aspirin and others do not. In the nearsightedness experiment, some children should be set to sleep in bedrooms with nightlights on and others in bedrooms with nightlights off.

Sometimes there will be more than one variable that you want to set experimentally. For instance, you might be interested in the effects of both vitamin supplements and exercise on health outcomes. Many people think that an experiment involves setting only one experimental variable and that you should do a separate experiment for each variable. There might be situations where this is sensible, but there can be big advantages in setting multiple experimental variables in the same experiment. (Later sections will discuss how to do this effectively — for instance it can be important to arrange things so that the experimental variables are mutually orthogonal.) One of the advantages of setting experimental variables simultaneously has to do with the amount of data needed to draw informative conclusions. Another advantage comes in the ability to study interactions between experimental variables. For example, it might be that exercise has different effects on health depending on a person's nutritional state.

The choice of experimental *units* depends on many factors: practicality, availability, and relevance. For instance, it's not so easy to do nutritional experiments on humans; it's hard to get them to stick to a diet and it often takes a long time to see the results. For this reason, many nutritional experiments are done on short-living lab animals. Ethical concerns are paramount in experiments involving potentially harmful substances or procedures.

Often, it's sensible to choose experimental units from a population where the effect of your intervention is likely to be particularly pronounced or easy to measure. So, for example, the aspirin experiment might be conducted in people at high risk of stroke or heart attack.

On the other hand, you need to keep in mind the **relevance** of your experiment. Studying people at the very highest risk of stroke or heart attack might not give useful information about people at a normal level of risk.

Some people assume that the experimental units should be chosen to be as homogeneous as possible; so that covariates are held constant. For example, the aspirin experiment might involve only women aged 70. Such homogeneity is not necessarily a good idea, because it can limit the relevance of the experiment. It can also be impractical because there may be a large number of covariates and there might not be enough experimental units available to hold all of them approximately constant.

One argument for imposing homogeneity is that it simplifies the analysis of the data collected in the experiment. This is a weak argument; analysis of data is typically a very small amount of the work and you have the techniques that you need to adjust for covariates. In any event, even when you attempt to hold the covariates constant by your selection of experimental units, you will often find that you have not been completely successful and you'll end up having to adjust for the covariates anyways. It can be helpful, however, to avoid experimental units that are outliers. These outlying units can have a disproportionate influence on the experimental results and increase the effects of randomness.

A good reason to impose homogeneity is to avoid accidental correlations between the covariates and the experimental variables. When such correlations exist, attempts to untangle the influence of the covariates and experimental variables can be ambiguous, as described in Section 17.4. For instance, if all the subjects who take aspirin are women, and all the subjects who don't take aspirin are men, how can you decide whether the result is due to aspirin or to the person's sex?

There are other ways than imposing homogeneity, however, to avoid correlations between the experimental variables and the covariates. Obviously, in the aspirin experiment, it makes sense to mix up the men and women in assigning them to treatment or control. More formally, the general idea is to make the experimental variables orthogonal to covariates — even those covariates that you can't measure or that you haven't thought to measure. Section 20.1.5 shows some ways to do this.

Avoiding a confounding association between the experimental variables and the covariates sometimes has to be done through the experimental protocol. A *placebo*, from the Latin "I will please," is defined as "a harmless pill, medicine, or procedure prescribed more for the psychological benefit to the patient than for any physiological effect." [26] In medical studies, placebos are used to avoid the confounding effects of the patient's attitude toward their condition.

If some subjects get aspirin and others get nothing, they may behave in different ways. Think of this in terms of covariates: Did the subject take a pill (even if it didn't contain aspirin)? Did the subject believe he or she was getting treated? Perhaps the subjects who don't get aspirin are more likely to drink alcohol, upset at having been told not to take aspirin and having heard that a small amount of daily alcohol consumption is protective against heart attacks. Perhaps the patients told to take aspirin tend not to take their vitamin pills: "It's just too much to deal with."

Rather than having your control consist of no treatment at all, use a placebo in order to avoid such confounding. Of course, it's usually important that the subjects be unaware whether they have been assigned to the placebo group or to the treatment group. An experiment in which the subjects are in this state of ignorance is called a **blind experiment**.

There is evidence that for some conditions, particularly those involving pain or phobias, placebos can have a positive effect compared to no treatment. [39, 40] This is the **placebo effect**. Many doctors also give placebos to avoid confrontation with patients who demand treatment. [41] This often takes the form of prescribing antibiotics for viral infections.

It can also be important to avoid accidental correlations created by the researchers between the experimental variables and the covariates. It's common in medical experiments to keep a subject's physicians and nurses ignorant about whether the subject is in the treatment or the control group. This prevents the subject from being treated differently depending on the group they are in. Such experiments are called **double blind**. (Of course, someone has to keep track of whether the subject received the treatment or the placebo. This is typically a statistician who is analyzing the data from the study but is not involved in direct contact with the experimental subjects.)

In many areas, the measured outcome is not entirely clear and can be influenced by the observer's opinions and beliefs. In testing the Salk polio vaccine in 1952, the experimental variable was whether the child subject got the vaccine or not. (See the account in [42].) It was important to make the experiment double-blind and painful measures were taken to make it so; hundreds of thousands of children were given placebo shots of saline solution. Why? Because in many cases the diagnosis of mild polio is ambiguous; it might be just flu. The diagnostician might be inclined to favor a diagnosis of polio if they thought the subject didn't get the vaccine. Or perhaps the opposite, with the physician thinking to himself, "I always thought that this new vaccine was dangerous. Now look what's happened."

Example 20.1: The Virtues of Doing Surgery while Blind In order to capture the benefits of the placebo effect and, perhaps more important, allow studies of the effectiveness of surgery to be done in a blind or double-blind manner, **sham surgery** can be done. For example, in a study of the effectiveness of surgery for osteoarthritis of the knee, some patients got actual arthroscopic surgery while the placebo group had a simulated procedure that included actual skin incisions.[43] Sham brain surgery has even been used in studying a potential treatment for Parkinson's disease.[44] Needless to say, sham surgery challenges the notion that a placebo is harmless. A deep ethical question arises: Is the harm to the individual patient sufficiently offset by the gains in knowledge and the corresponding improvements in surgery for future patients?

Example 20.2: Oops! An accidental correlation! You are a surgeon who wants to test out a new and difficult operation to cure a dangerous condition. Many of the subjects in this condition are in very poor health and have a strong risk of not surviving the surgery, no matter how effective the surgery is at resolving the condition. So you decide, sensibly, to perform the operation on those who are in good enough health to survive and use as controls those who aren't in a position to go through the surgery. You've just created a correlation between the experimental variable (surgery or not) and a covariate (health).

20.1.2 Choosing levels for the experimental variables

In carrying out an experiment, you set the experimental variables for each experimental unit. When designing your experiment, you have to decide what levels to impose on those variables. That's the subject of this section. A later section will deal with how to assign the various levels among the experimental units.

Keep in mind these two principles:

- **Comparison**. The point of doing a systematic experiment is to contrast two or more things. Another way of thinking about this is that in setting your experimental variable, you create variability. In analyzing the data from your experiment you will use this variability in the experimental variables.

- **Relevance**. Do the levels provide a contrast that guide meaningful conclusions. Is the difference strong enough to cause an observable difference in outcomes?

It's often appropriate to set the experimental variable to have a "control" level and a "treatment" level. For instance, placebo vs aspirin, or night-light off vs night-light on. When you have two such levels, a model " response $\sim$ experimental variable" will give one coefficient that will be the difference between the treatment and the control and another coefficient — the intercept — that is the baseline.

Make sure that the control level is worthwhile so that the difference between the treatment and control is informative for the question you want to answer. Try to think what is the interesting aspect of the treatment and arrange a control that highlights that difference. For example, in studying psychotherapy it would be helpful to compare the psychotherapy treatment to something that is superficially the same but lacking the hypothesized special feature of psychotherapy. A comparison of the therapy with a control that involves no contact with the subject whatsoever may really be an experiment about the value of human contact itself rather than the value of psychotherapy.

Sometimes it's appropriate to arrange more than two levels of the experimental variables, for instance more than one type of treatment. This can help tease apart effects from different mechanisms. It also serves the pragmatic purpose of checking that nothing has gone wrong the overall set-up. For instance, an medical experiment might involve a placebo, a new treatment, and an established treatment. In experiments in biology and chemistry, where the measurement involves complicated procedures where something can go wrong (for instance, contamination of the material being studied) it's common to have three levels of the experimental variable: a treatment, a **positive control**, and a **negative control** that gives a clear signal if something is wrong in the procedure or materials.

Often, the experimental variable is naturally quantitative: something that can be adjusted in magnitude. For instance, in an aspirin experiment, the control would be a daily dose of 0mg (delivered *via* placebo pill) and the treatment might be a standard dose of 325 mg per day. Should you also arrange to have treatment levels that are other levels, say 100 mg per day or 500 mg per day? It's tempting to do so. But for the purposes of creating maximal variability in the experimental variable, it's best to set the treatment and control at opposite ends of whatever the relevant range is: say 0 and 325 mg per day.

It can be worthwhile, however, to study the **dose-response** relationship. For example, it appears that a low dose of aspirin (about 80 mg per day) is just as effective as higher dosages in preventing cardiovascular disease.[45] Put yourself in the position of a researcher who wants to test this hypothesis. You start with some background literature — previous studies have established that a dose of 325 mg per day is effective. You want to test a does of one-quarter of that: 80 mg per day. It's helpful if your experimental variable involves at least *three* levels: 0 mg (the control, packaged in a placebo), 80 mg, and 325 mg. Why not just use two levels, say 0 and 80 mg, or perhaps 80 and 325 mg? So that you can demonstrate that your overall experimental set up is comparable to that if the background literature and so that you can, in your own experiment, compare the responses to 80 and 325 mg.

When you analyze the data from such a multi-level experimental variable, make sure to include nonlinear transformation terms in your model. The model " response $\sim$ dose" will not be able to show that 80 mg is just as effective as 325 mg; the linear form of the model is based on the assumption that 80 mg will be only about one-quarter as effective as 325 mg. If your interest is to show that 80 can be just as effective as 325 mg, you must include additional terms, for instance response $\sim$ (dose $>$ 0) + (dose $>$ 80)

20.1.3 Replication

Perhaps it seems obvious that you should include multiple experimental units: **replication**. For example, in studying aspirin, you want to include many subjects, not just a single subject at the control level and a single subject at the treatment level.

But how many? To answer this question, you need to understand the purposes of replication. There are at least two:

- Be able to compare the variability associated with the experimental variables from the background variability that arises from random sampling. This is done, for instance, when looking at statistical significance using an F statistic. (See Section 16.2.)

- Increase the ability of your experiment to detect a difference if it really does exist: the **power** of your experiment.

In setting the number of cases in your experiment, you need first to ensure that you have some non-zero degree of freedom in your residual. If not, you won't be able to compute an F value at all, or, more generally, you won't be able to compare the variability associated with your experimental variables with the variability described as random.

Second, you should arrange things so that the p-value that you would expect, assuming that the experimental variable acts as you anticipate, is small enough to give you a significant result. This relates to the issue of the **power** of your experiment: the probability that you would get a significant result in a world where the experimental variable acts as you anticipate. The relationship between sample size and power is described in Section 17.7.

20.1.4 Experiments vs Observations

Why do an experiment when you can just observe the "natural" relationships? For example, rather than dictating to people how much aspirin they take, why not just measure the amount they take on their own, along with appropriate covariates, and then using modeling to find the relationships?

To illustrate, here is a hypothetical causal network that models the relationship between aspirin intake and stroke:

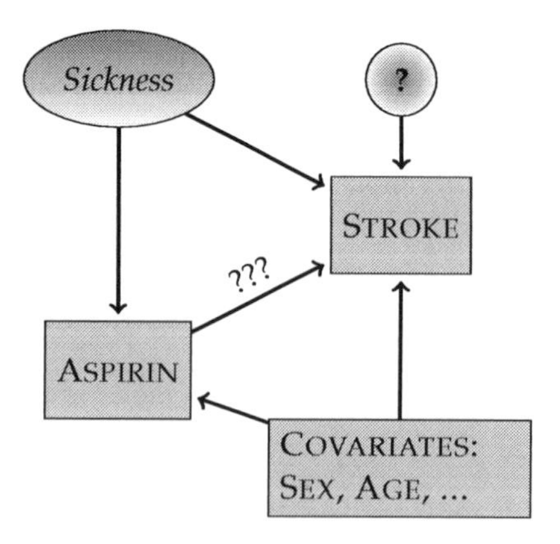

Included in this network is a latent (unmeasured, and perhaps unmeasurable) variable: sickness. The link between sickness and aspirin is intended to account for the possibility that sicker subjects — those who believe themselves to be at a higher risk of stroke — might have heard of the hypothesized relationship between aspirin and risk of stroke and are medicating themselves at a higher rate than those who don't believe themselves to be sick.

There are three pathways between aspirin and stroke. The direct pathway, marked ???, is of principal interest. One backdoor pathway goes through the sickness latent variable: aspirin⇐sickness⇒stroke. Another backdoor pathway goes through the covariates: aspirin⇐covariates⇒stroke.

If you want to study the direct link between aspirin and stoke, you need to block the backdoor pathways. If you don't, you can get misleading results. For instance, the increased risk of stroke indicated by the sickness variable might make it look as if aspirin consumption itself increases the risk of stroke, even if the direct effect aspiring is to decrease the risk. Both the backdoor pathways are correlating pathways. To block them, you need to include sickness and the covariates (sex, age, etc.) in your model, e.g., a model like this:

$$\text{stroke} \sim \text{aspirin} + \text{sickness} + \text{age} + \text{sex}$$

But how do you measure sickness? The word itself is imprecise and general, but it might be possible to find a measurable variable that you think is strongly correlated with sickness, for instance blood pressure or cholesterol levels. Such a variable is called a **proxy** for the latent variable.

Even when there is a proxy, how do you know if you are measuring it in an informative way? And suppose that the hypothetical causal network or the proxy has left something out that's important. Perhaps there is a genetic trait that both increases the risk of stroke and improves the taste of aspirin or reduces negative side effects or increases joint inflammation for which people often take aspirin as a treatment. Any of these might increase aspirin use among those at higher risk. Or perhaps the other way around, decreasing aspirin use among the high risk so that aspirin's effects look better than they really are.

You don't know.

A proper experiment simplifies the causal network. Rather than the possibility of sickness or some other unknown variable setting the level of aspirin use, you know for sure that the level was set by you, the experimenter. This is not a hypothesis; it's a demonstrable fact. With this fact established by the experimenter, the causal network looks like this:

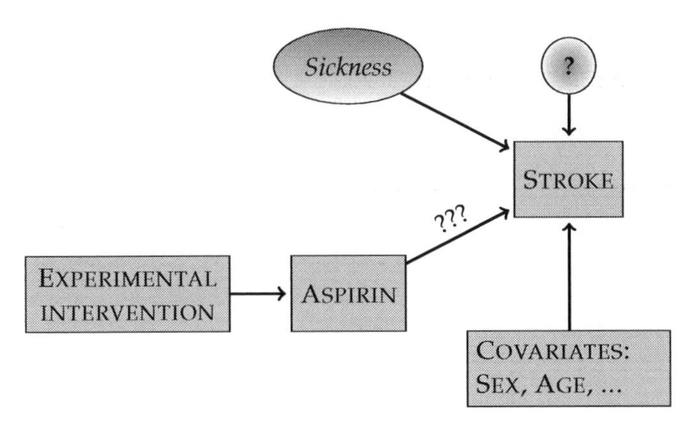

Experimental intervention has broken the link between sickness and aspirin since the experimenter, rather than the subject, sets the level of aspirin. There is

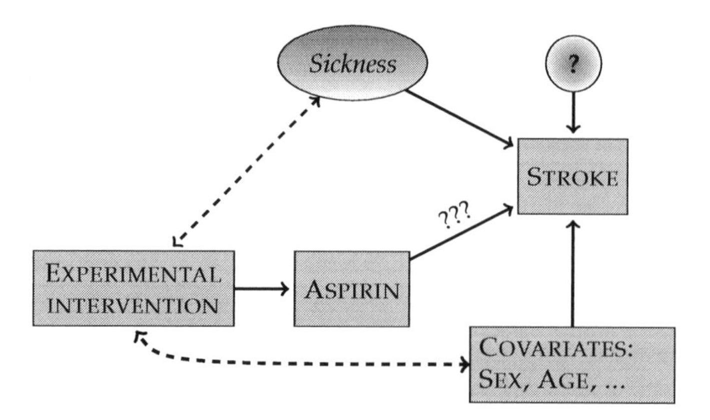

Figure 20.1: If the experimental intervention is influenced by or correlated with other variables in the system, new backdoor pathways will be created.

no longer a backdoor correlating pathway. All the other possibilities for backdoor pathways — e.g., genetics — are also severed. With the experimental intervention, the simple model stroke ~ aspirin is relevant to a wide range of different researchers' hypothetical causal networks.

Details, however, are important. You need to make sure that the way you assign the values of aspirin to the experimental units doesn't accidentally create a correlation with other explanatory variables. Techniques for doing this are described in Section 20.1.5. Another possibility is that your experimental intervention doesn't entirely set the experimental variable. You can tell the subject how much aspirin to take, but they may not comply! You'll need to have techniques for dealing with that situation as well. (See Section 20.2.1.)

20.1.5 Creating Orthogonality

In setting the experimental intervention, it's your intention to eliminate any connections between the experimental variables and other variables in the causal network. Doing this requires some care. It's surprisingly easy to create accidental correlations.

For instance, suppose the person assigning the subjects to either aspirin or placebo thinks it would be best to avoid risk to the sicker people by having them take only the placebo, effectively correlating the treatment with sickness. If such correlations are created, the experimental system will not be so simple: there will still be back doors to be dealt with:

There are two main ways to avoid accidental correlations: random assignment and blocking.

Random Assignment

In **random assignment**, you use a formal random process to determine what level of the experimental variable is to be assigned to each experimental unit.

When there are just two levels, treatment and control, this can be as simple as flipping a coin. (It's better to do something that can be documented, such as having a computer generate a list of the experimental units and the corresponding randomly assigned values.)

Random assignment is worth emphasizing, even though it sounds simple:

Random assignment is of fundamental importance.

Randomization is important because it works for any potential covariate — so long as n is large enough. Even if there is another variable that you do not know or cannot measure, your random assignment will, for large enough n, guarantee that there is no correlation between your experimental variable and the unknown variable. If n is small, however, even random assignment may produce accidental correlations. You can see this in Figure 13.4 on page 227. Referring to that figure, when $n = 25$, the angle between a random vector and another vector — for example, one corresponding to an unmeasured covariate — is very close to $90°$, practically orthogonal, a correlation of near zero. For small n, however, the angle may be quite different from $90°$; an accidental correlation is likely.

Blocking in Experimental Assignment

If you do know the value of a covariate, then it's possible to do better than randomization at producing orthogonality. The technique is called **blocking** . The word "blocking" in this context does not refer to disconnecting a causal pathway; it's a way of assigning values of the experimental variable to the experimental subjects.

To illustrate blocking in assignment, imagine that you are looking at your list of experimental subjects and preparing to assign each of them a value for the experimental variable:

Sex	Age	Aspirin
F	71	aspirin
F	70	placebo
F	69	placebo
F	65	aspirin
F	61	aspirin
F	61	placebo
M	71	placebo
M	69	aspirin
M	67	aspirin
M	65	placebo

Notice that the subjects have been arranged in a particular order and divided into blocks. There are two subjects in each block because there are two experimental conditions: treatment and control. The subjects in each block are similar: same sex, similar age.

Within each block of two, assign one subject to aspirin and the other to placebo. For the purpose of establishing orthogonality of aspirin to sex, it would suffice to make that assignment in any way whatsoever. But since you also want to make aspirin orthogonal to other variables, perhaps unknown, you should do the assignment within each block at random.

The result of doing assignment in this way is that the experimental variable will be **balanced**: the women will be divided equally between placebo and aspirin, as will the men. This is just another way of saying that the indicator vectors for sex will be orthogonal to the indicator vectors for aspirin.

When there is more than one covariate — the table has both age and sex — the procedure for blocking is more involved. Notice that in the above table, the male-female pairs have been arranged to group people who are similar in age. If the experimental subjects had been such that the two people within each pair had identical ages, then the aspirin assignment would be exactly orthogonal to age and to sex. But because the age variable is not exactly balanced, the orthogonality created by blocking will be only approximate. Still, it's as good as it's going to get.

Blocking becomes particularly important when the number of cases n is small. For small n, randomization may not give such good results.

Whenever possible, assign the experimental variable by using blocking against the covariates you think might be relevant, and using randomization within blocks to deal with the potential latent variables and covariates that might be relevant, even if they are not known to you.

Within each block, you are comparing the treatment and control between individuals who are otherwise similar. In an extreme case, you might be able to arrange things so that the individuals are identical. Imagine, for example, that you are studying the effect of background music on student test performance. The obvious experiment is to have some students take the test with music in the background, and others take it with some other background, for instance, quiet.

Better, though, if you can have each student take the test twice: once with the music and once with quiet. Why? Associated with each student is a set of latent variables and covariates. By having the student take the test twice, you can block against these variables. Within each block, randomly assign the student to which background condition to use the first time, with the other background condition being used the second time. This means to pick each student's assignment at random from a set of assignments that's balanced: half the students take the test with music first, the other half take the test with quiet first.

20.2 When Experiments are Impossible

Often you can't do an experiment.

The inability to do an experiment might stem from practical, financial, ethical or legal reasons. Imagine trying to investigate the link between campaign spending and vote outcome by controlling the amount of money candidates have to spend. Where would you get the money to do that? Wouldn't there be opposition from those candidates who were on the losing side of your random assignment of the experimental variables? Or, suppose you are an economist interested in the effect of gasoline taxes on fuel consumption. It's hardly possible to set gasoline taxes experimentally in a suitably controlled way. Or, consider how you would do an experiment where you randomly assign people to smoke or to not smoke in order to understand better the link between smoking, cancer, and other illnesses such as emphysema. This can't be done in an ethical way.

Even when you can do an experiment, circumstances can prevent your setting the experimental variables in the ways that you want. Subjects often don't comply in taking the pills they have been assigned to take. The families randomly admitted to private schools may decline to go. People drop out in the middle of a study, destroying the balance in your experimental variables. Some of your experimental units might be lost due to sloppy procedures or accidents.

When you can't do an experiment, you have to rely on modeling techniques to deal with covariates. This means that your assumptions about how the system works — your hypothetical causal networks — play a role. Rather than the data speaking for themselves, they speak with the aid of your assumptions.

Section 20.2.2 introduces two technical approaches to data analysis that can help to mitigate the impact of your assumptions. This is not to say that the approaches are guaranteed to make your results as valid as they might be if you had been able to do a perfect experiment. Just that they can help. Unfortunately, it can be hard to demonstrate that they actually have helped in any study. Still, even if the approaches cannot recreate the results of an experiment, they can be systematically better than doing your data analysis without using them.

20.2.1 Intent to Treat

Consider an experiment where the assignment of the experimental variables has not gone perfectly. Although you, the experimenter, appropriately randomized and blocked, the experimental subjects did not always comply with your assignments. Unfortunately, there is often reason to suspect that the compliance itself is related to the outcome of the study. For example, sick subjects might be less likely to take their experimentally assigned aspirin.

A hypothetical causal network describing the imperfect compliance aspirin experiment looks like Figure 20.2, with the value of aspirin determined both by the experimental intervention and the subject's state of sickness.

In such a network, the model stroke ~ aspirin describes the total connection between aspirin and stroke. This includes three parts: (1) the direct link aspirin⇒stroke, (2) the backdoor pathway through sickness, and (3) the backdoor pathway through the covariates. By including the covariates as explanatory variables in your model, you can block pathway (3). The experimental intervention has likely reduced the influence of backdoor pathway (2), but because the subjects may not comply perfectly with your assignment, you cannot be sure that your random assignment has eliminated it.

Consider, now, another possible model: stroke ~ experimental intent. Rather than looking at the actual amount of aspirin that the subject took and relating

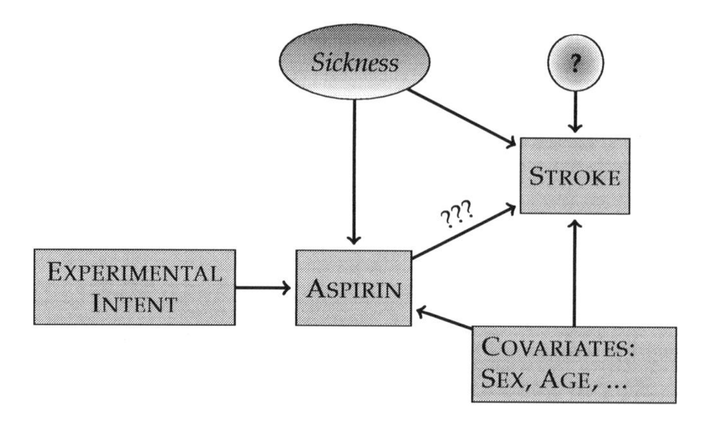

Figure 20.2: In an imperfect experiment, the experimental intent influences the treatment (aspirin, here), but the treatment may be influenced by the covariates and latent variables.

that to stroke, the explanatory variable is the *intent* of the experimenter. This intent, however it might have been ignored by the subjects, remains disconnected from any covariates; the intent is the set of values assigned by the experimenter using randomization and blocking in order to avoid any connection with the covariates.

The hypothetical causal network still has three pathways from experimental intent to stroke. One pathway is experimental intent⇒aspirin⇒stroke. This is a correlating pathway that subsumes the actual pathway of interest: aspirin⇒stroke. As such, it's an important pathway to study, the one that describes the impact of the experimenter's intent on the subjects' outcomes.

The other two pathways connecting intent to stroke are the backdoor pathway via sickness and the backdoor pathway via the covariates. These two pathways cause the trouble that the experiment sought to eliminate but that non-compliance by the experimental subjects has re-introduced. Notice, though, that the backdoor pathways connecting experimental intent to stroke are non-correlating rather than correlating. To see why, examine this backdoor pathway:

experimental intent ⇒aspirin ⇐sickness ⇒stroke

There's no node from which the causal flow reaches both end points. This non-correlating pathway will be blocked so long as sickness is not included in the model. Similarly the back door pathway through the covariates is blocked when the covariates are not included in the model.

As a result, the model stroke ~ experimental intent will describe the causal chain from experimental intent to stroke. This chain reflects the effect of the **intent to treat**. That chain is not just the biochemical or physiological link between aspirin and stoke, it includes all the other influences that would cause a subject to comply or not with the experimenter's intent. To a clinician or to a public health official, the intent to treat can be the proper thing to study. It addresses this question: How much will prescribing patients to take aspirin reduce the amount of stroke, taking into account non-compliance by patients with their

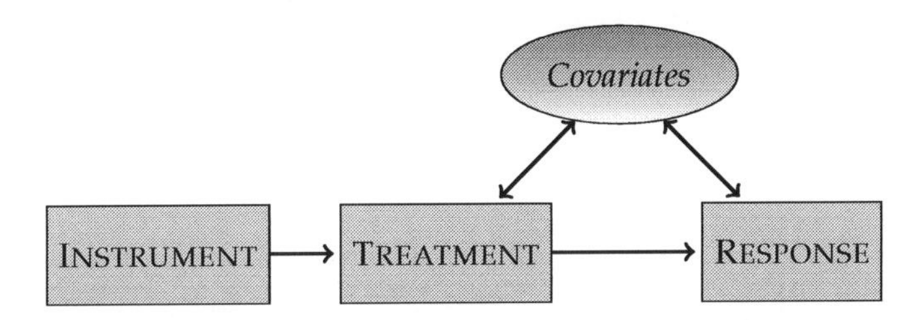

Figure 20.3: An instrument is a variable that is unrelated to the covariates but is connected to the treatment.

prescriptions? This is a somewhat different question than would have been answered by a perfect experiment, which would be the effect of aspiring presuming that the patients comply.

Intent to treat analysis can be counter-intuitive. After all, if you've measured the actual amount of aspirin that a patient took, it seems obvious that this is a more relevant variable than the experimenter's intent. That may be, but aspirin is also connected with other variables (sickness and the covariates) which can have a confounding effect. Even though intent is not perfectly correlated with aspirin, it has the great advantage of being uncorrelated with sickness and the covariates — so long as the experimental variables were properly randomized and blocked.

Of course, pure intent is not enough. In order for there to be an actual connection between intent and the outcome variable, some fraction of the patients has to comply with their assignment to aspirin or placebo.

20.2.2 Destroying Associations

In intent-to-treat analysis, the explanatory variable was changed in order to avoid unwanted correlations between the explanatory variable and covariates and latent variables. The techniques in this section — **matched sampling** and **instrumental variables** — allow you to keep your original explanatory variables but instead modify the data in precise ways in order to destroy or reduce those unwanted correlations.

Figure 20.3 shows the general situation. You're interested in the direct relationship between some treatment and a response, for instance aspirin and stroke. There are some covariates, which might be known or unknown. Also, there can be another variable, an instrumental variable, that is related to the treatment but not directly related to any other variable in the system. An example of an instrumental variable is experimental intent in Figure 20.2, but there can also be instrumental variables relevant to non-experimental systems.

A key question for a modeler is how to construct an appropriate model of the relationship between the treatment and the response. The two possibilities are:

1. response ~ treatment

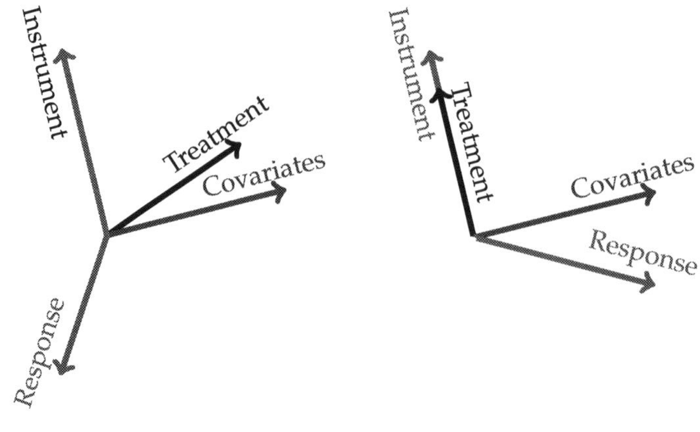

Figure 20.4: In an observational study or an imperfect experiment, the treatment may be correlated with the covariates. Left: Variable-space diagram of an imperfect experiment. Right: A perfect experiment makes the treatment orthogonal to the covariates.

2. response ~ treatment + covariates

The first model is appropriate if the backdoor pathway involving the covariates is a non-correlating pathway; the second model is appropriate if it is a correlating pathway. But it's not so simple; there can be multiple covariates, some of which are involved in correlating pathways and some not. Beyond that, some of the covariates might be unknown or unmeasured — latent variables — in which case the second form of model isn't possible; you just don't have the data.

In terms of model vectors, the situation is shown in Figure 20.4. The treatment and the covariates are aligned because there is a causal connection between them. When there is such alignment, the inclusion of the covariates in a model of the response makes a difference; you would get a different coefficient on treatment depending on whether the covariates were included in the model.

In a proper experimental design, where you can assign the value of treatment, you use randomization and blocking to arrange treatment to be orthogonal to the covariates. Given such orthogonality, it doesn't matter whether or not the covariates are included in the model; the coefficient on treatment will be the same either way. That's the beauty of orthogonality.

In the absence of an experiment, or when the experimental intervention is only partially effective, then treatment will typically not be orthogonal to the covariates. The point of matched sampling and of instrumental variables is to arrange things, by non-experimental means, to make them orthogonal.

Instrumental Variables

An **instrumental variable** is a variable that is correlated with the treatment variable, but uncorrelated with any of the covariates. Its effect on the response vari-

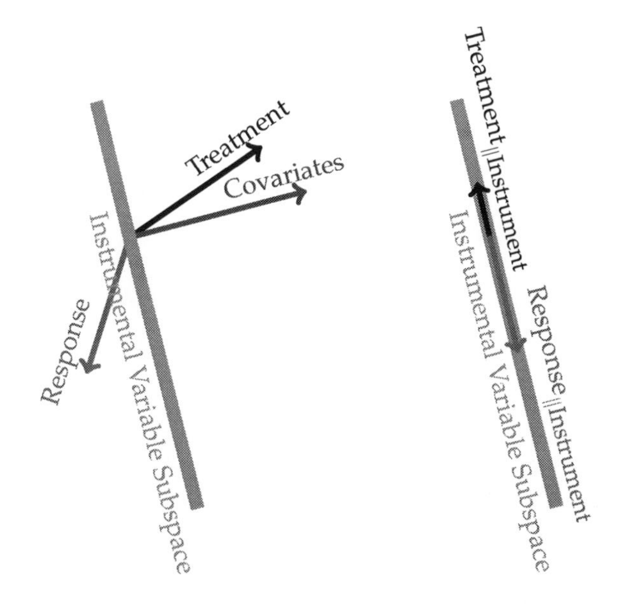

Figure 20.5: An instrumental variable defines a subspace that is orthogonal to the covariates. By projecting other vectors onto this subspace, an estimate can be made of the relationship between the treatment and the response that is not influenced by the covariates.

able, if any, is entirely indirect, via the treatment variable.

Perhaps the simplest kind of instrumental variable to understand is the experimenter's intent. Assuming that the intent has some effect on the treatment — for instance, that some of the subjects comply with their assigned treatment — then it will be correlated with the treatment. But if the intent has been properly constructed by randomization and blocking, it will be uncorrelated with the covariates. And, the intent ordinarily will affect the response only by means of the treatment. (It's possible to imagine otherwise, for example, if being assigned to the aspirin group caused people to become anxious, increased their blood pressure, and caused a stroke. Placebos avoid such statistical problems.)

An instrumental variable is orthogonal to the covariates but not to the treatment, and thus it provides a way to restructure the data in order to eliminate the need to include covariates when estimating the relationship between the treatment variable and the response. The approach is very clever but straightforward: carry out the fitting of the model response $\sim$ treatment in the subspace defined by the instrumental variable. This is illustrated in Figure 20.5 and is often done in two stages. First, the treatment is projected onto the instrument. The resulting vector, $\text{treatment}_{||\text{instrument}}$ is exactly aligned with the instrument and therefore shares the instrument's orthogonality to the covariates. Second, model the response by the $\text{treatment}_{||\text{instrument}}$ variable. In doing this, it doesn't matter whether the covariates are included or not, since they are orthogonal to the projected treatment variable.

If you have an appropriate instrument, the construction of a meaningful model to investigate causation can be straightforward. The challenge is to find an appropriate instrument. For experimenters dealing with imperfections of compliance or drop-out, the experimental intent makes a valid instrument. For others, the issues are more subtle. Often, one looks for an instrument that predates the treatment variable with the idea that the treatment therefore could not have caused the instrument. It's not the technique of instrumental variables that is difficult, but the work behind it, "the gritty work of finding or creating plausible experiments" [46], painstaking "shoe-leather" research[47]. Find a good instrumental variable requires detailed understanding of the system under study, not just the data but the mechanisms shaping the explanatory variables.

Imagine, for instance, a study to relate how the duration of schooling relates to earnings after leaving school. There are many potential covariates here. For example, students who do better in school might be inclined to stay longer in school and might also be better earners. Students from low-income families might face financial pressures to drop out of school in order to work, and they might also lack the family connections that could help them get a higher-earning job.

An instrumental variable needs to be something that will be unrelated to the covariates — skill in school, family income, etc. — but will still be related to the duration a student stays in school. Drawing on detailed knowledge of how schools work, researchers Angrist and Krueger [48] identified one such variable: date of birth. School laws generally require children to start school in the calendar year in which they turn six and require them to stay in school until their 16th birthday. As a result, school dropouts whose birthday is before the December 31 school-entry cut-off tend to spend more months in school than those whose birthday comes on January 1 or soon after. So date of birth is correlated with duration of schooling. But there's no reason to think that date of birth is correlated with any covariate such as skill in school or family income. (Angrist and Krueger found that men born in the first quarter of the calendar year had about one-tenth of a year less schooling than men born in the last quarter of the year. The first-quarter men made about 0.1 percent less than the last-quarter men. Comparing the two differences indicates that a year of schooling leads to an increase in earnings of about 10%.)

Matched Sampling

In matched sampling, you use only a subset of your data set. The subset is selected carefully so that the treatment is roughly orthogonal to covariates. This does not eliminate the actual causal connection between the treatment and the covariates, but when fitting a model to the subset of the data, the coefficient on treatment is rendered less sensitive to whether the covariates are included in the model.

To illustrate, imagine doing an observational study on the link between aspirin and stroke. Perhaps you pick as a sampling frame all the patients at a clinic ten years previously, looking through their medical records to see if their physician has asked them to take aspirin in order to prevent stroke, and recording the patients age, sex, and systolic blood pressure. The response variable is whether or not the patient has had a stroke since then. The data might look like this:

Case #	Aspirin Recommended	Age	Sex	Systolic Blood Pressure	Had a Stroke
1	Yes	65	M	135	Yes
2	Yes	71	M	130	Yes
3	Yes	68	F	125	No
4	Yes	73	M	140	Yes
5	Yes	64	F	120	No
		... and so on ...			
101	No	61	F	110	No
102	No	65	F	115	No
103	No	72	M	130	No
104	No	63	M	125	No
105	No	64	F	120	No
106	No	60	F	110	Yes
		... and so on ...			

Looking at these data, you can see there is a relationship between aspirin and stroke — the patients for whom aspirin was recommended were more likely to have a stroke. On the other hand, there are covariates of age, sex, and blood pressure. You might also notice that the patients who were recommended to take aspirin typically had higher blood pressures. That might have been what guided the physician to make the recommendation. In other words, the covariate blood pressure is aligned with the treatment aspirin. The two groups — aspirin or not — are different in other ways as well: the no-aspirin group is more heavily female and younger.

To construct a matched sample, select out those cases from each group that have a closely matching counterpart in the other group. For instance, out of the small data set given in the table, you might pick these matched samples:

Case #	Aspirin Recommended	Age	Sex	Systolic Blood Pressure	Had a Stroke
5	Yes	64	F	120	No
105	No	64	F	120	No
2	Yes	71	M	130	Yes
103	No	72	M	130	No

The corresponding pairs (cases #5 and #105, and cases #2 and #103) match closely according to the covariates, but each pair is divided between the values of the treatment variable, aspirin. By constructing a subsample in this way, the covariates are made roughly orthogonal to the treatment. For example, in the matched sample, it's not true that the patients who were recommended to take aspirin have a higher typical blood pressure.

In a real study, of course, the matched sample would be constructed to be much longer — ideally enough data in the matched sample so that the power of the study will be large.

There is a rich variety of methods for constructing a matched sample. The process can be computationally intensive, but is very feasible on modern computers. Limits to practicality come mainly from the number of covariates. When there is a large number of covariates, it can be hard to find closely matching cases between the two treatment groups.

A powerful technique involves fitting a **propensity score**, which is the probability that a case will be included in the treatment group (in this example, aspirin recommended) as a function of the covariates. (This probability can be estimated by logistic regression.) Then matching can be done simply using the propensity score itself. This does not necessarily pick matches where all the covariates are close, but is considered to be an effective way to produce approximately orthogonality between the set of covariates and the treatment.

Example 20.3: Returning to Campaign Spending ...

Analysis of campaign spending in the political science literature is much more sophisticated than the simple model `vote` $\sim$ `spending`. Among other things, it involves consideration of several covariates to try to capture the competitiveness of the race. An influential series of papers by Gary Jacobson (e.g., [36, 49]) considered, in addition to campaign spending by the incumbent and vote outcome, measures of the strength of the challenger's party nationally, the challenger's expenditures, whether the challenger held public office, the number of years that the incumbent has been in office, etc. Even including these covariates, Jacobson's least-square models showed a much smaller impact of incumbent spending than challenger spending. Based on the model coefficients, a challenger who spends $100,000 will increase his or her vote total by about 5 percent, but an incumbent spending $100,000 will typically increase the incumbent's share of the vote by less than one percent.

In order to deal with the possible connections between the covariates and `spending`, political scientists have used instrumental variables. (The term found in the literature is sometimes **two-stage least squares** as opposed to instrumental variables.) Green and Krasno [50], in studying elections to the US House of Representatives in 1978, used as an instrument incumbent campaign expenditures during the 1976 election. The idea is that the 1976 expenditures reflects the incument's propensity to spend, but expenditures in 1976 could not have been caused by the situation in 1978. When using this instrument, Green and Krasno found that $100,000 in incumbent expenditures increased his or her vote total by about 3.7 percent — much higher than the least squares model constructed without the instrument. (Jacobson's reports also included use of instrumental variables, but his instruments were too collinear and insufficiently correlated with `spending` to produce stable results.)

Steven Levitt (whose later book *Freakonomics* [51] became a best-seller) took a different perspective on the problem of covariates: matched sampling. (Or, to use the term in the paper, **panel data**.) He looked at the elections to the US House between 1972 and 1990 in which the same two candidates — incumbent and challenger — faced one another two or more times. This was only about 15% of the total number of House elections during the period, but it is a subset where the covariates of candidate quality or personality are held constant for each pair. Levitt found that the effect of spending on vote outcome was small for both challengers and incumbents: an additional $100,000 of spending is associated with a 0.3 percent vote increase for the challenger and 0.1 percent for the incumbent — neither spending effect was statistically significant in an F test.

So which is it? The different modeling approaches give different results. Jacobson's original work indicates that challenger spending is effective but not incumbent spending; Green and Krasno's instrumental variable approach suggests that spending is effective for both candidates; Levitt's matched sampling methodology signals that spending is ineffective for both challenger and incumbent. The dispute focuses attention on the validity of the instruments or the ability of the matching to compensate for changing covariates. Is 1976 spending a good instrument for 1978 elections? Is the identity of the candidates really a good matching indicator for all the covariates or lurking variables? Such questions can be challenging to answer reliably. If only it were possible to do an experiment!

20.3 Conclusion

It's understandable that people are tempted to use correlation as a substitute for causation. It's easy to calculate a correlation: Plug in the data and go! Unfortunately, correlations can easily mislead. As in Simpson's paradox, the results you get depend on what explanatory variables you choose to include in your models.

To draw meaningful conclusions about causation requires work. At a minimum, you have to be explicit about your beliefs about the network of connections and examine which potential explanatory variables lie on correlating pathways and which on non-correlating pathways. If you do this properly, your modeling results can be correctly interpreted in the context of your beliefs. But if your beliefs are wrong, or if they are not shared by the people who are reading your work, then the results may not be useful. There's room for disagreement when in comes to causation because reasonable people can disagree about the validity of assumptions.

Other analysis techniques — for instance matched sampling or instrumental variables — can ease the dependence of your results on your assumptions and beliefs. But still there is room for interpretation. Is the instrument really uncorrelated with the covariates? How complete is the set of covariates included in your matched sampling?

Ultimately, for a claim about causation to be accepted by skeptics, you need to be able to demonstrate that your assumptions are true. This can be particularly challenging when the skeptics have to take statistical methodology on faith. You'll find that many decision makers have little or no understanding of methodology, and some critics are willfully ignorant of statistics. On the other hand, it's appropriate for the modeler to show some humility and to work honestly to identify those assumptions that are not themselves adequately justified by the data, problems with the sampling process, and the potential misinterpretation of p-values. A good experiment can be invaluable, since the experimenter herself determines the structure of the system and since the statistical analysis can be straightforward.

It's sometimes said that science is a process of investigation, not a result. The same is true for statistical models. Each insight that is demonstrably supported by data becomes the basis for refinement in the way you collect new data, the variables and covariates you measure in future work, the incisive experimental interventions that require labor, resources, and expertise. However well your data have been collected and your models have been structured and built, the conclusions drawn from statistical models are still inductive, not deductive. As such, they are not certain. Even so, they can be useful.

Further Readings & Bibliography

The reader interested in learning more about statistics will do well to start with some history. David Salsburg's *The Lady Tasting Tea* [52] gives an entertaining and readable survey of 20th century statistics and statisticians. Stephen Stigler's *Statistics on the Table* [53] reaches back a century further in history and has a more conceptual focus. Ian Hacking's *The Emergence of Probability* goes back, as the title says, to the emergence of the concept of probability in the seventeenth century.

Chapter 1. Overview. For essays about the application of statistics in context and in practice, the various editions of *Statistics: A Guide to the Unknown* are hard to beat.[54, 55] Each *Guides* essays is oriented around a particular social, political, medical, or scientific issue. In contrast, most statistics textbooks ([42, 56, 57, 58] are among the good choices) are arranged around topics in statistical methodology. The cited books cover a complete gamut of statistical topics — sampling, descriptive statistics, statistical tests — but emphasize the statistics of single variables or the relationship between pairs of variables. They build on high-school algebra and do not require college-level calculus. At a somewhat more advanced mathematical level, the well-titled *Statistical Sleuth* introduces multivariate regression. At a still higher level, for those who are comfortable with college-level linear algebra, an excellent multivariate modeling text is [59]. For classroom learning activities in introductory statistics, [60, 61] provide many ideas.

An excellent, non-statistical introduction to constructing models, with a strong emphasis on models as "purposeful representations," is [62].

Many materials about the R package [5] are available at the project web site: `www.r-project.org`. Good references for R (and the language it is based on, S) are [63, 64, 65]. Some introductions to statistics using R are [66, 67, 68].

Chapters 2 & 3. Data and Description. Introductory statistics books (e.g. ([42, 56, 57, 58]) provide a good introduction to methods and issues in collecting data and describing variables numerically and graphically. For advanced modeling methods with units of analysis at different levels, see [69].

The series of books by Tufte [70, 71] has been rightfully influential in showing how to present data graphically (and beautifully). For technical displays of statistical data, see Cleveland's *Visualizing Data*.[72]

Important criticisms of the emphasis on conventional descriptions of center and spread are provided by the best-selling books by Gould [73] and Taleb [74].

The elementary but classic *How to Lie with Statistics* [75] has many examples of the use of quantitative information in a way that misleads; it's not so much about statistics as it is about lying.[76]

The somewhat obscure statistic "unalikeability" is described in [77] and [78].

Chapters 4 - 8. Modeling and Correlation. The modeling notation is described in an influential advanced book by Chambers and Hastie.[65] Box, Hunter, & Hunter give good reasons to take fitted models with a grain of salt. [2, pp. 397-407]

Stigler provides an account of the invention of correlation.[79] For a variety of different ways to interpret correlation, see [80].

Gawande writes for a general reader about why simple model formulas can be more useful for prediction and diagnosis than human judgment.[81]

Chapters 9 - 13. Geometry. The idea of using geometry to inform statistical reasoning started perhaps with Ronald Fisher at the start of the 20th century.[82] It was unfortunately seen as obscure and statistics pedagogy was rooted in algebra. Fisher's daughter, Joan Fisher Box, writes about the reaction of the famous statistician William Gosset:

> .... Gosset found the geometric representation always highly mysterious; but he accepted the mystification, and jokingly would remark, "I take it that whatever it is follows at once 'obviously' from a consideration of n-dimensional geometry," or demand of Fisher whether a certain equation came out of n-dimensional space "or what is much the same thing, your head." [83]

Starting in the 1980s, David Saville and Graham Wood have been writing about the uses of geometry in teaching statistics. [84, 85, 86]. Other articles appear from time to time, e.g., [87, 88]. The classic book *Statistics for Experimenters* [2] contains a thread using geometry to complement algebraic explanations. See also [89].

The idea of a random walk has been important in science. Robert Brown's observation of the movement of pollen grains in 1827 is reported in [20]. One of the famous trio of papers from Albert Einstein's 1905 *annus mirabilis* is about Brownian motion, [22]; Perrin [21] won a Nobel prize for his work confirming Einstein's theory experimentally. The accessible book by Berg [90] reveals many aspects of random walks in biology. There are also best-selling accounts, e.g., [91, 92, 93].

Chapters 14-18. Statistical Inference. Descriptions of the classic tests — t-tests and so on — are the bread and butter of statistics textbooks. A discussion of the search for significance and how this can render p-values uninterpretable is the subject of an editorial by Freedman.[17] Other papers on multiple comparisons and inconsistency are [27, 94]; important critiques of statistical inference are [95, 96]. Good and Hardin's book, *Common Errors in Statistics (and How to Avoid Them)*, is a useful reference.[97]

The work of Bradley Efron introduced randomization methods such as bootstrapping to many social and natural scientists and statistical workers. [98, 99, 100] An early attempt to introduce randomization into teaching statistics at an introductory level is [101]. Hardly any introductory statistics textbook even mentions resampling or bootstrapping, though thankfully there are signs this is starting to change.

Saville and Wood make a case for the angle between vectors as a primary statistic. [102, 103] Figure 16.7 is inspired by Fig. 10.A.2.b in [2] and similar figures in Appendix D of [86].

Chapters 19 & 20. Causation and Experiment. The work of Judea Pearl on causation is the inspiration for the material on hypothetical causal networks. A book-length treatment is [37]; the last chapter should perhaps be read first as an introduction. Many of Pearl's papers are available on the web. The "gentle introduction" in [104] is a good place to start.

Box, Hunter, and Hunter [2] is a well regarded survey of experimental design, artfully combining theory and practice. Cobb [105] provides a carefully graded path through experimental design and analysis accessible; it's sophisticated but accessible to a non-mathematical reader.

For background on instrumental variables, see [46, 106]. For matched sampling: [107, 108, 109, 110]. The propensity-score method has been tested in situations where there were parallel studies — one experimental and one observational: informed skepicism is called for. [111, 112, 113] The best-selling book *Freakonomics* [51] entertainingly reports many clever attempts to extract information about causation from observational data.

References

[1] George Cobb. The introductory statistics course: a Ptolemaic curriculum? *Technology Innovations in Statistics Education*, 1(1), 2007.

[2] George E.B. Box, J. Stuart Hunter, and William G. Hunter. *Statistics for Experimenters: Design, Innovation, and Discovery*. Wiley, 2nd edition, 2005.

[3] Robert Hughes. *The Fatal Shore: the epic of Australia's founding*. Vintage, 1988.

[4] John M Chambers. *Software for data analysis: Programming with R*. Springer, 2008.

[5] R Development Core Team. *R: A language and environment for statistical computing*. R Foundation for Statistical Computing, Vienna, Austria, 2005. ISBN 3-900051-07-0.

[6] Daniel T Kaplan. Computing and introductory statistics. *Technology Innovations in Statistics Education*, 1(1):Article 5, 2007. repositories.cdlib.org/uclastats/cts/tise/iss1/art5.

[7] David Bellhouse. Ann landers survey on parenthood. web pamphlet: www.stats.uwo.ca/faculty/bellhouse/stat353annlanders.pdf, 2008. accessed Jan. 2008.

[8] James A. Hanley. "Transmuting" women into men: Galton's family data on human stature. *American Statistician*, 58(3):237–243, 2004.

[9] John M Chambers and Trevor J Hastie. *Statistical Models*. Wadsworth, 1992.

[10] Deborah Lynn Guber. Getting what you pay for: the debate over equity in public school expenditures. *Journal of Statistics Education*, 7(2), 1999.

[11] DWJ Thompson. A large discontinuity in the mid-twentieth century in observed global-mean surface temperature. *Nature*, 453:646–649, 29 May 2008.

[12] Francis Galton. Co-relations and their measurement, chiefly from anthropometric data. *Proceedings of the Royal Society of London*, pages 135–145, 1888.

[13] College Board. Research report no. 2008-5: Validity of the sat for predicting first-tyear college grade point average. Technical report, College Board, 2008.

[14] A.A. Bailey and P.L. Hurd. Finger length ratio (2d:4d) correlates with physical aggression in men but not in women. *Biol. Psychol.*, 68(3):215–222, 2005.

[15] Rachel Carson. *Silent Spring*. Houghton, 1962.

[16] John Tierney. Fateful voice of a generation still drowns out real science. New York Times, June 5 2007.

[17] David A Freedman. Editorial: Oasis or mirage? *Chance*, 21(1):59–61, 2008.

[18] http://www.infoplease.com/ipa/a0883976.html, 2008.

[19] E W Montroll and M F Schlesinger. Maximum entropy formalism, fractals, scaling phenomena, and $1/f$ noise: a tale of tails. *Journal of Statistical Physics*, 32(209), 1983.

[20] Robert Brown. *Miscellaneous Botanical Works*, volume 1. London, Royal Society, 1866. quoted in MP Crosland (1971) *The Science of Matter*.

[21] M. Jean Perrin. *Brownian Movement and Molecular Reality*. Taylor and Francis, London, 1910.

[22] Albert Einstein. Investigations on the theory of the brownian movement. In A.D. Cowper, editor, *Collected Papers*, volume 2. Dover Publications, 1956.

[23] James Surowiecki. Running numbers. *The New Yorker*, page 29, Jan. 21 2008.

[24] Roger A. Pielke Jr. Who decides? forecasts and responsibilities in the 1997 red river flood. *Applied Behavioral Science Review*, 7(2):83–101, 1999.

[25] Ronald A. Fisher. On a distribution yielding the error functions of several well known statistics. *Prod. Int. Cong. Math., Toronto*, 2:805–813, 1924.

[26] New oxford american dictionary.

[27] David J. Saville. Multiple comparison procedures: the practical solution. *The American Statistician*, 44(2):174–180, 1990.

[28] Office of Federal Contract Compliance Programs. *Federal Contract Compliance Manual*, chapter Chapter III: Onsite Review Procedures. US Department of Labor, 1993, revised 2002.

[29] Richard Shavelson Stephen Klein, Roger Benjamin. The collegiate learning assessment: Facts and fantasies. Technical report, Council for Aid to Education, http://www.cae.org/content/pro_collegiate_reports_publications.htm, 2007.

[30] Paul Basken. Test touted as 2 studies question its value. *The Chronical of Higher Education*, 54(39):A1, June 6 2008.

[31] S. Koutros et al. Meat and meat mutagens and risk of prostate cancer in the agricultural health study. *Cancer epidemiology biomarkers and prevention*, 17:80–87, 2008.

[32] Rod Minchin. personal communication. 2008.

[33] `http://www.prostateline.com/prostatelinehcp`. Web site, 2009.

[34] Charles Dana Wilber. *The Great Valleys and Prairies of Nebraska and the Northwest.* Daily Republican Print, Omaha, Neb., 3rd edition, 1881.

[35] Leonard J. Schmidt, Brooke Warner, and Peter A. Levine. *Panic: Origins, Insight, and Treatment.* North Atlantic Books, 2002.

[36] Gary C. Jacobson. The effects of campaign spending in congressional elections. *The American Political Science Review*, 72(2):469–491, 1978.

[37] Judea Pearl. *Causation: Models, Reasoning, and Inference.* Princeton Univ. Press, 2000.

[38] Daniel J. Boorstin. *Cleopatra's Nose: Essays on the Unexpected.* Vintage, 1995.

[39] A Hrobjartsson and PC Gotzcsche. Is the placebo powerless? update of a systematic review with 52 new randomized trials comparing placebo with no treatment. *Journal of Internal Medicine*, 256(2):91–100, 2004.

[40] BE Wampold, T Minami, SC Tierney, TW Baskin, and KS Bhati. The placebo is powerful: estimating placebo effects in medicine and psychotherapy from randomized clinical trials. *Journal of Clinical Psychology*, 61(7):835–854, 2005.

[41] A Hrobjartsson and M Norup. The use of placebo interventions in medical practice — a national questionnaire survey of danish clinicians. *Evaluation and the Health Professions*, 26(2):153–165, 2003.

[42] David Freedman, Roger Purves, and Robert Pisani. *Statistics.* WW Norton, 4th edition, 2007.

[43] J Bruce Moseley et al. A controlled trial of arthroscopic surgery for osteoarthritis of the knee. *New England Journal of Medicine*, 347:81–88, 2002.

[44] TB Freeman. Use of placebo surgery in controlled trials of a cellular-based therapy for parkinson's disease. *New England Journal of Medicine*, 341(13):988–992, 1999.

[45] CL Campbell, S Smyth, G Montalescot, and S R Steinhubl. Aspirin dose for the preventage of cardiovascular disease: a systematic review. *Journal of the American Medical Association*, 18:2018–2024, 2007.

[46] Joshua D. Angrist and Alan B. Krueger. Instrumental variables and the search for identification: from supply and demand to natural experiments. *Journal of Economic Perspectives*, 15(4):69–85, 2001.

[47] David A. Freedman. Statistical models and shoe leather. In Peter Marsden, editor, *Sociological Methodology 1991.* American Sociological Association, 1991.

[48] Joshua D. Angrist and Alan B. Krueger. Does compulsory school attendance affect schooling and earnings? *Quarterly Journal of Econometrics*, 106(4):979–1014, 1991.

[49] Gary C. Jacobson. The effects of campaign spending in house elections: new evidence for old arguments. *American Journal of Political Science*, 34(2):334–362, 1990.

[50] Donald P Green and Jonathan S Krasno. Salvation for the spendthrift incumbent: Reestimating the effects of campaign spending in house elections. *American Journal of Political Science*, 32(4):884–907, 1988.

[51] Steven D Levitt and Stephen J Dubner. *Freakonomics.* William Morrow, 2005.

[52] David Salsburg. *The Lady Tasting Tea: How statistics revolutionized science in the twentieth century.* W.H. Freeman, New York, 2001.

[53] Stephen M Stigler. *Statistics on the Table: The History of Statistical Concepts and Methods*. Harvard Univ. Press, 1999.

[54] Roxy Peck, George Casella, George Cobb, Roger Hoerl, Deborah Nolan, Robert Starbuck, and Hal Stern. *Statistics: A Guide to the Unknown*. Duxbury Press, 4th edition, 2005.

[55] Judith M. Tanur, Frederick Mosteller, William H. Kruskal, Erich L. Lehmann, Richard F. Link, Richard S. Pieters, and Gerald R. Rising. *Statistics: A Guide to the Unknown*. Duxbury Press, 3rd edition, 1989.

[56] David S Moore, George P McCabe, and Bruce Craig. *Introduction to the Practice of Statistics*. W.H. Freeman, 6th edition, 2007.

[57] Anne E Watkins, Richard L Scheaffer, and George W Cobb. *Statistics in Action: Understanding a World of Data*. Key College Publishing, 2004.

[58] David E Bock, Paul F Velleman, and Richard D De Veaux. *Stats: Modeling the World*. Addison Wesley, 2nd edition, 2006.

[59] David A. Feedman. *Statistical Models: Theory and Practice*. Cambridge University Press, 2005.

[60] Allan J Rossman and Beth Chance. *Workshop Statistics: Discovery with Data*. Wiley, 2nd edition, 2008.

[61] Andrew Gelman and Deborah Nolan. *Teaching Statistics: A Bag of Tricks*. Oxford Univ. Press, 2002.

[62] Anthony M. Starfield, Karl A. Smith, and Andrew L. Blelock. *How to model it*. McGraw-Hill, 1990.

[63] William N Venables and Brian D Ripley. *Modern Applied Statistics with S*. Springer, 4th edition, 2002.

[64] Phil Spector. *Data Manipulation with R*. Springer, 2008.

[65] John M Chambers and Trevor J Hastie. *Statistical Models in S*. Chapman & Hall, 1992.

[66] Peter Dalgaard. *Introductory Statistics with R*. Springer, 2nd edition, 2008.

[67] John Verzani. *Using R for Introductory Statistics*. Chapman & Hall, 2005.

[68] Michael J Crawley. *Statistics: An Introduction using R*. Wiley, 2005.

[69] Andrew Gelman and Jennifer Hill. *Data analysis using regression and multilevel/hierarchical models*. Cambridge Univ. Press, 2007.

[70] Edward R Tufte. *The Visual Display of Quantitative Information*. Graphics Press, 2nd edition, 2001.

[71] Edward R Tufte. *Beautiful Evidence*. Graphics Press, 2006.

[72] William S Cleveland. *Visualizing Data*. Hobart Press, 1993.

[73] Stephen J. Gould. *The Mismeasure of Man*. W.W. Norton, 1996.

[74] Nassim Taleb. *The Black Swan*. Random House, 2007.

[75] Darell Huff. *How to Lie with Statistics*. Norton, 1954. illustrated by Irving Geis.

[76] J. Michael Steele. Darell Huff and fifty years of how to lie with statistics. *Statistical Science*, 20(5):205–209, 2005.

[77] Gary D. Kader and Mike Perry. Variability for categorical variables. *Journal of Statistics Education*, 15(2), 2007. http://www.amstat.org/publications/jse/v15n2/kader.html.

[78] Alan Agresti. *Categorical Data Analysis.* John Wiley and Sons, 1990.

[79] Stephen M Stigler. Francis galton's account of the invention of correlation. *Statistical Science*, 4(2):73–79, 1989.

[80] Joseph L Rodger and W Alan Nicewander. Thirteen ways to look at the correlation coefficient. *The American Statistician*, 42(1):59–66, 1988.

[81] Atul Gawande. *Complications: A Surgeon's Notes on an Imperfect Science.* Metropolitan/Henry Holt, 2002.

[82] David G. Herr. On the history of the use of geometry in the general linear model. *The American Statistician*, 34(1):43–47, 1980. see 2682995.pdf.

[83] Joan Fisher Box. Guinness, Gosset, Fisher, and small samples. *Statistical Science*, 2(1):45–52, 1987.

[84] David J. Saville and Graham R. Wood. A method for teaching statistics using n-dimensional geometry. *The American Statistician*, 40(3):205–214, 1986.

[85] David J. Saville and Graham R. Wood. *Statistical Methods: The Geometric Approach.* Springer, 1997.

[86] David J. Saville and Graham R. Wood. *Statistical Methods: A Geometric Primer.* Springer, 1996.

[87] Johan Bring. A geometric approach to compare variables in a regression model. *The American Statistician*, 50(1):57–62, 1996.

[88] David Bock and Paul F. Velleman. Why variances add — and why it matters. In Gail F. Burrill and Portia C. Elliot, editors, *Thinking and reasoning with data and chance.* National Council of Teachers of Mathematics, 2006.

[89] Michael J. Wichura. *The Coordinate-free Approach to Linear Models.* Cambridge University Press, 2006.

[90] Howard C Berg. *Random Walks in Biology.* Princeton Univ. Press, 1993.

[91] Burton G Malkiel. *A Random Walk Down Wall Street: The Time-Tested Strategy for Successful Investment.* W.W. Norton, 2007.

[92] Leonard Mlodinow. *The Drunkard's Walks: How Randomness Rules Our Lives.* Pantheon, 2008.

[93] Nassim Nicholas Taleb. *Fooled by Randomness: The Hidden Role of Chance in Life and in the Markets.* Random House, 2008.

[94] David J. Saville. Basic statistics and the inconsistency of multiple comparison procedures. *Canadian Journal of Experimental Psychology*, 57(3):167–175, 2003.

[95] Charmont Wang. *Sense and Nonsense of Statistical Inference.* CRC, 1992.

[96] Richard A. Berk. *Regression analysis: A constructive critique.* Sage publications, 2004.

[97] Phillip I. Good and James W. Hardin. *Common Errors in Statistics (and How to Avoid Them).* Wiley Interscience, 2003.

[98] Bradley Efron. Bootstrap methods: Another look at the jackknife. *The Annals of Statistics*, 7:1–26, 1979.

[99] Bradley Efron. Computers and the theory of statistics: Thinking the unthinkable. *SIAM Review*, 21:460–480, 1979.

[100] Bradley Efron and Robert J. Tibshirani. *An Introduction to the Bootstrap.* Chapman and Hall, 1993.

[101] Julian L Simon. *Resampling: The New Statistics*. Resampling Stats, 1974-1992.

[102] Graham R Wood and David J Saville. The ubiquitous angle. *Journal Royal Statistical Society A*, 168(1):95–107, 2005.

[103] Dave Saville and Graham Wood. The geometry of the p-value. In *Collaborations, Designs, and Explorations: A festschrift for Peter Johnstone*, pages 99–111. 2006. ISBN 0-478-20909-6.

[104] Judea Pearl. Causal inference in statistics: A gentle introduction. In *Computing Science and Statistics, Proceedings of Interface '01*, v. 33, 2001.

[105] George W Cobb. *Introduction to Design and Analysis of Experiments*. Springer, 1998.

[106] Sander Greenland. An introduction to instrumental variables for epidemiologists. *International Journal of Epidemiology*, 29:722–729, 2000.

[107] D.B. Rubin. Using multivariate matched sampling and regression adjustment to control bias in observational studies. *Journal of the American Statistical Association*, 74:318–328, 1979.

[108] P.R Rosenbaum and D.B. Rubin. Constructing a control group using multivariate matched sampling methods that incorporate the propensity score. *The American Statistician*, 39:33–38, 1985.

[109] P.R. Rosenbaum. *Observational Studies*. Springer-Verlag, 2nd ed. edition, 2002.

[110] B.B. Hansen. Full matching in an observational study of coaching for the sat. *Journal of the American Statistical Association*, 99(467):609:618, 2004.

[111] K Arceneaux, A S Gerber, and D P Green. Comparing experimental and matching methods using a large-scale voter mobilization experiment. *Political Analysis*, 14:37–62, 2006.

[112] D A Lawlor, G DaveySmith, D Kundu, and et al. Those confounded vitamins: what can we learn from the differences between observational vs randomised trial evidence. *The Lancet*, 363:1724–1727, 2004.

[113] R Kunz and A D Oxman. The unpredictability paradox: review of empirical comparisons of randomised and non-randomized clinical trials. *British Medical Journal*, 317:1185–1190, 1998.

Datasets

College Grade Database

There are three tables in this data set: `grades.csv`, `courses.csv`, and `grade-to-number.csv`.

`grades.csv` has one row for each course taken by a 2005 graduate in his or her college career. To preserve confidentiality, the data are a 50% random sampling of the complete listing. The variables are:

- sid — A unique ID for each student.
- sessionNo — A unique ID for each course session. If a course has multiple sections, that is, different groups of students who meet at different times, each section will have a unique session number.

- grade — The student's letter grade in that session: A, A-, B+, and so on through D-. Other codes: I=incomplete, NC=failed/no credit, S=passed (for courses taken on a pass/fail basis), AU=audit (taken not for credit). The numerical equivalents for grade-point calculations are given in the file `grade-to-number.csv`, described below.

DATA FILE `courses.csv`

`courses.csv` has one row for each course session listed in the grades table. Since there are many students in each session, the number of courses is much smaller than the number of cases in the grades table. For each course session, the variables are:

- sessionID — The unique session ID corresponding to the entries in the grades table.
- iid — A unique ID for each instructor.
- dept — The name of the department in which the course was offered. To preserve confidential information, a single-letter code is used for each department.
- sem — The semester in which the course session took place.
- enroll — The number of students enrolled in the course.
- level — A number indicating the "level" of the course. 100-level courses are the most elementary, 200 is more advanced, then 300, then 400. Courses marked with 600 are research courses, internships, and so on.

To preserve confidentiality, the courses table contains a random sample of the courses offered, and includes only courses with an enrollment of 10 or greater. (For this reason, there are very few 600-level courses listed.)

`grade-to-number.csv` contains the translation from a letter grade to a number.

The coding and subsampling done was done for reasons of confidentiality. The version of the data set retained for internal study at the institution retains all the data and proper identifying labels.

Galton's Height Data

Context In the 1880's, Francis Galton was developing ways to quantify the heritability of traits. As part of this work, he collected data on the heights of adult children and their parents.

Variables Each child is one case.

- height the child's height (as an adult) in inches.
- sex the child's sex: F or M
- mother the mother's height in inches
- father the father's height in inches
- nkids the number of adult children in the family, or, at least, the number whose heights Galton recorded.
- family a numerical code to identify the members of one family.

Entries were deleted for those children whose heights were not recorded numerically by Galton, who sometimes used entries such as "tall", "short", "idiotic", "deformed" and so on.

Sources The data were transcribed by J.A. Hanley [8], who has published them at `http://www.medicine.mcgill.ca/epidemiology/hanley/galton/`.

Kids' Foot Measurements

Context These data were collected by a statistician, Mary C. Meyer, in a fourth grade classroom in Ann Arbor, MI, in October 1997. They are a convenience sample — the kids who were in the fourth grade.

Quoted from the source: "From a very young age, shoes for boys tend to be wider than shoes for girls. Is this because boys have wider feet, or because it is assumed that girls, even in elementary school, are willing to sacrifice comfort for fashion? To assess the former, a statistician measures kids' feet."

Variables Each case is one child.

- name — the child's first name.
- birthmonth — the month of birth
- birthyear — the year of birth
- length — Length of longer foot (cm)
- width — Width of longer foot (cm), measured at widest part of foot
- sex — boy or girl
- biggerfoot — Which foot was longer (right or left)
- domhand — Right- or left-handedness

Sources Mary C. Meyer (2006) "Wider Shoes for Wider Feet?" *Journal of Statistics Education* 14(1), `www.amstat.org/publications/jse/v14n1/datasets.meyer.html`

Marriage Licenses

Context Marriage records from the Mobile County, Alabama, probate court. Records were picked quasi-randomly by Alan Eisinger.

Variables Each case is one person getting married. Both the bride and groom are included in the data set.

- BookpageID: The book and page in the county register on which the marriage is recorded. Used as a unique identifier of the marriage.
- Appdate: The date on which the application was filed, in m/d/y format.
- Ceremonydate: The date on which the marriage ceremony took place, in m/d/y format.
- Delay: The number of days between the application and the ceremony.
- Officialtitle: The listed title of the official who conducted the marriage.
- Person: Which of the bride or groom is represented (byperson only)
- Dob: The date of birth of the person, in m/d/y format.
- Age: The age of the person, in years and fractions of year.
- Race: The race of the person, as listed on the application.
- Prevcount: The number of previous marriages of the person, as listed on the application.
- Prevconc: The way the last marriage ended, as listed on the application.
- Hs: The number of years High School education, as listed on the application.
- College: The number of years College education, as listed on the application. Where no number was listed, this field was left blank, unless less than 12 years High School was reported, in which case it was entered as 0.
- Dayofbirth: The day of birth, as a number from 1 to 365 counting from January 1.

- Sign: The astrological sign, calculated by using dayofbirth. May not correctly sort people directly on the borders between signs. This variable is not part of the original record.

Sources The records were collected through `http://www.mobilecounty.org/probatecourt/recordssearch.htm`

SATs by State

Context These data were assembled for a statistics education journal article on the link between SAT scores and measures of educational expenditure.[10]

Variables
- `State` — Name of state (in quotation marks)
- `expend` — Expenditure per pupil in average daily attendance in public elementary and secondary schools, 1994-95 (in thousands of dollars)
- `ratio` — Average pupil/teacher ratio in public elementary and secondary schools, Fall 1994
- `salary` — Estimated average annual salary of teachers in public elementary and secondary schools, 1994-95 (in thousands of dollars)
- `frac` — Percentage of all eligible students taking the SAT, 1994-95
- `verbal` — Average verbal SAT score, 1994-95
- `math` — Average math SAT score, 1994-95
- `sat` — Average total score on the SAT, 1994-95

Sources From Deborah Lynn Guber (1999) "Getting What You Pay For: The Debate Over Equity in Public School Expenditures" *Journal of Statistics Education* 7(2) available from `http://www.amstat.org/publications/jse/secure/v7n2/datasets.guber.cfm`

Smoking and Death

Context Data on age, smoking, and mortality from a one-in-six survey of the electoral roll in Whickham, a mixed urban and rural district near Newcastle upon Tyne, in the UK. The survey was conducted in 1972-1974 to study heart disease and thyroid disease. A follow-up on those in the survey was conducted twenty years later. This dataset contains a subset of the survey sample: women who were classified as current smokers or as never having smoked.

Variables Each case is one woman. Three variables are measured:
- `Smoker` — Whether the woman identified herself as a smoker during the original survey.
- `Outcome` — Whether the woman was alive or not at the time of the follow up survey. (Needless to say, only living women were included in the original survey.)
- `Age` — The woman's age at the time of the first survey.

Sources DR Appleton, JM French, MPJ Vanderpump, *American Statistician* 50:4, 1996, pp.340-341. I have synthesized the data set from the summary description tables given in that paper.

Swimming World Records

Context World record times in the 100m freestyle swimming race.

Variables

- time the record time, in seconds
- sex whether this is the women's or the men's record.
- year the year in which the record was set.

Sources Record information is widely available on the Internet. A history is available on http://en.wikipedia.org/wiki/World_record_progression_100m_freestyle

Ten-Mile Race

DATA FILE ten-mile-race.csv

Context The Cherry Blossom 10 Mile Run is a road race held in Washington, D.C. in April each year. (The name comes from the famous cherry trees that are in bloom in April in Washington.) The results of this race are published. The file contains the results from the 2005 race.

Variables Each case is one runner who completed the race course.

- sex —
- age —
- time — The official time from starting gun to the finish line.
- net — The recorded time from when the runner crossed the starting line to when the runner crossed the finish line. This is generally less than the official time because of the large number of runners in the race: it takes time to reach the starting line after the gun has gone off.
- state — The state of residence of the runner. This can indicate how far the runner travelled to participate in the race. Many runners come from areas close to DC in Maryland (MD) and Virginia (VA).

Another dataset, "running-longitudinal.csv," contains the results the races from 1999 to 2008 for those runners who ran in multiple years. In that dataset, each case is one runner in one year, so there are multiple cases for each runner; 41,248 cases altogether with 14,434 different runners. The variable id identifies each runner uniquely across all years. The variable nruns tells how many different time each runner participated.

Sources Data are from http://www.cherryblossom.org/

Trucking Jobs

DATA FILE truckingjobs.csv

Context A dataset from a mid-western US trucking company on annual earnings of its employees in 2007. Datasets like this are used in audits by the Federal Government to look for signs of discrimination.

Variables Each case is one employee.

- sex: M or F
- earnings: Annual earnings, in dollars. Hourly wages have been converted to a full-time basis.
- age in years
- title the job title
- hiredyears how long the employee has been working for the company.

Sources For reasons of confidentiality, the source of these data cannot be revealed.

Utility Bills

Context Gas and electricity bills on the author's house in Saint Paul, Minnesota 55116.

Variables

- month — The number of the month on which the meter was read
- day — The day of the month on which the meter was read
- year — The year.
- temp — The average temperature during the billing period.
- kwh — The number of kilowatt hours of electricity used.
- ccf — Cubic feet of natural gas used
- thermsPerDay — The utilities calculation of therms of energy in the natural gas per day of the billing period. A therm is a measure of energy equivalent to 10,000 British Thermal Units (BTU). One BTU raises the temperature of one pound of water by one degree fahrenheit.
- dur — How many days in the billing period.
- totalbill — The total utility bill: gas + electricity
- gasbill — The amount of the gas bill during that period
- electricbill — The amount of the electric bill during that period

Note that the meter might not always have been read accurately. Records for 5/30/200 and 3/25/2000 are outliers easily explained by having too low a first reading followed by a correct reading, giving an apparently heavy gas usage during the second period.

A similar, but smaller version of this may have happened with 1/28/2000 and 2/26/2000.

Notes indicate when changes were made to the house, for example replacing the furnace and water heater, both of which are fueled by natural gas. Cooking is fueled by both electricity (oven, microwave, appliances) and gas (stove-top, outdoor grill).

Sources Monthly utility bills from XCEL Energy received by the author.

Used Car Prices

Two different files, used-hondas.csv and used-fords.csv

DATA FILE
used-hondas.csv

Context Prices of used Honda Accord LXs as advertised on cars.com in October 2007. Prices of used Ford Tauruses as advertised on cars.com in February 2009.

Variables

- Price — The price, in US dollars, asked for the car.
- Year — The car's model year.
- Mileage — The number of miles the car has been driven.
- Location — The city where the car is located.
- Color — The car's body color.
- Age — The age of the car: 2007 – Year for the Hondas or 2009 – Year for the Fords.

Sources These data were compiled by Macalester College students for a class project. Aleksander Azarnov collected the Honda data; Elise delMas, Emiliano Urbina, and Candace Groth collected the Ford data.

Wages from the Current Population Survey

DATA FILE
cps.csv

Context The Current Population Survey (CPS) is used to supplement census information between census years. These data consist of a random sample of 534 persons from the CPS, with information on wages and other characteristics of the workers, including sex, number of years of education, years of work experience, occupational status, region of residence and union membership. (Quoted from `http://lib.stat.cmu.edu/datasets/CPS_85_Wages`.)

Questions:

- Are wagesrelated to these other characteristics?
- Is there a gender gap in wages?

Variables Each case is one person.

- educ Number of years of education.
- south Indicator variable for living in a southern region: S=lives in south, NS=doesn't
- sex M or F
- exper Number of years of work experience. This was inferred from age and education
- union Indicator variable for union membership.
- wage Wage (dollars per hour).
- age Age (years).
- race Race: White (W), Nonwhite (NW).
- sector Sector of the economy: clerical, const, management, manufacturing, professional, sales, service, other
- married Marital Status: Married or Single

Sources Data are from `http://lib.stat.cmu.edu/datasets/CPS_85_Wages` which sites as the original source, Berndt, ER. *The Practice of Econometrics* 1991. Addison-Wesley. The data file `cps.csv` is recoded from the original, which had entirely numerical codes.

Data suggested by Prof. Naomi Altman.

Dataset Index

R Operator Index

Items in `typewriter font` are R operators or arguments.
A * identifies functions defined in the `ISM.Rdata` workspace, not standard R.

Concept Index

Made in the USA
Lexington, KY
18 September 2011